THE BEHAVIOUR AND INFLUENCE OF FLUIDS IN SUBDUCTION ZONES

THE BEHAVIOUR AND INFLUENCE OF FLUIDS IN SUBDUCTION ZONES

PROCEEDINGS OF
A ROYAL SOCIETY DISCUSSION MEETING
HELD ON 8 AND 9 NOVEMBER 1990

ORGANIZED AND EDITED BY J. TARNEY,
K. T. PICKERING, R. J. KNIPE AND
J. F. DEWEY

LONDON
THE ROYAL SOCIETY
1991

Printed in Great Britain for the Royal Society
by the
University Press, Cambridge

ISBN 0 85403 437 4

First published in *Philosophical Transactions of the Royal Society of London*,
series A, volume 335 (no. 1638), pages 225–418

Copyright

British Library Cataloguing in Publication Data

The behaviour and influence of fluids in
subduction zones.
I. Tarney, J.
551.9

ISBN 0-85403-437-4

Published by the Royal Society
6 Carlton House Terrace, London SW1Y 5AG

PREFACE

Unique among the terrestrial planets, the Earth has a substantial hydrosphere, which has had an extraordinary influence on the way the planet has evolved. Whereas volcanic activity is the main route by which fluids escape from depth to reach the atmosphere/hydrosphere, the dominant route by which fluids may enter the deeper levels of the Earth is via subduction zones. How deep can these fluids penetrate? More importantly, what influence do fluids have in controlling geological processes at depth? These processes include the formation of continental crust: both the primary magmatic additions to the crust and the lateral additions in the form of accreted sedimentary material. This Discussion Meeting was called to assess the role of fluids in the subduction environment, first at shallow levels where material is entering the subduction system, but then following processes down to progressively deeper levels to the zone of melting and beyond. Below, we summarize and focus on some of the more important issues arising from the discussions. The collection of papers in the volume serve as both a review or synthesis of what recent research has achieved, and highlight a number of research directions which require more attention in the future analysis of fluids in subduction zones.

1. INTRODUCTION

Subduction zones can be likened to a gigantic press where low density geological materials saturated with seawater are progressively squeezed between the over-riding and under-riding lithospheric plates and the fluids expelled. The abyssal sediment blanket on the ocean floor may be 0.5–1 km thick and contain as much as 50 % fluid, the underlying basaltic ocean crust may have been altered to depths of perhaps 4 km during ridge hydrothermal activity and have been replaced by clays, zeolites, chlorites and amphiboles with water contents ranging from 15 to 2 %, and the underlying harzburgite peridotite may similarly have been variably converted to serpentine with 12 % water. With subduction rates of up to 10 cm a^{-1} it is possible that expulsion rates of 40 km^3 Ma^{-1} of seawater per unit kilometre of trench may be exceeded. Obviously a majority of the available fluid is expelled in the early stages and at shallow levels, which is why discharge effects are so manifest in accretionary prisms. Conversely, because some hydrous minerals are now known to be stable to great depths, fluid expulsion can continue over a very considerable depth range, though on a diminishing scale.

Of course a subduction zone differs from a simple press analogue in that fluid is not just being squeezed out mechanically, but is also being driven out thermally as low-temperature hydrous minerals transform at deeper levels to less-hydrous or anhydrous minerals. This balance between mechanical and thermal dehydration changes with time as continued subduction of cold oceanic lithosphere depresses the isotherms to deeper levels in the mantle, at least until some sort of thermal equilibrium is reached. Some of the fluid can be mopped up by new hydrous minerals (hornblende and phlogopite) growing in the mantle wedge, or at the boundary with the over-riding plate (phengitic micas and glaucophane).

2. DEFORMATION AND FLUID EXPULSION IN ACCRETIONARY PRISMS

Seismic reflection techniques are capable of imaging structural features, lithological changes and the presence of bottom simulating reflectors (BSRs) caused by the presence of gas hydrate phase changes. The variation in seismic velocity can be used as a measure of porosity changes (Westbrook). Where borehole data are available the seismic reflection techniques are most potent for constraining structural models to explain the architecture of accretionary prisms and their physical development.

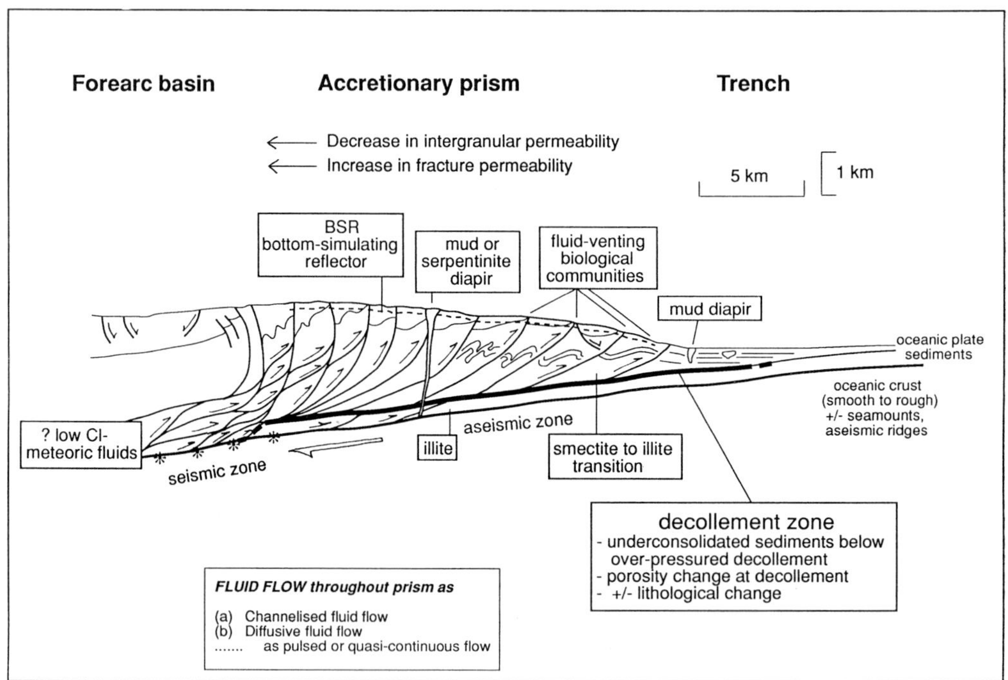

Figure 1. Cartoon illustrating some of the more important features of accretionary sedimentary wedges formed in the upper parts of subduction zones.

Fluid expulsion is an important feature of accretionary prisms. The nature of fluid pathways is complex (figure 1) and depends, among other parameters, upon permeability, porosity and hydraulic gradient. Permeability is the main control on fluid flow within accretionary prisms, yet it is one of the most difficult parameters to accurately determine. The rate of fluid flow is determined by the hydraulic gradient together with the permeability. Within accretionary prisms, permeability values appear to vary by 6–7 orders of magnitude, because of the lithological variations, whereas the hydraulic gradient changes by less than an order of magnitude (Moore *et al.*). Permeability is a dynamic parameter since it changes with the evolution of the prism, being determined by factors such as (*a*) earthquakes, (*b*) compaction, (*c*) deformation pathways, (*d*) diagenesis, (*e*) diapirism, and (*f*) metamorphism. Given the paramount importance in obtaining reliable *in situ* measurements of permeability, there is a need to glean much more data from modern accretionary prisms, from core recovery through drilling, and using geophysical techniques.

The characterization of fluid pathways is an essential ingredient in understanding the hydrology and, therefore, the rheology of accretionary prisms. A variety of sediment and crustal deformation associated with fluid activity has been observed in modern forearc slopes and trenches (Taira & Pickering). Fault zones are efficient conduits for focusing fluid flow and for fluid expulsion, something that is well documented for the Barbados and Oregon, Cascadia (Carson *et al.*) and the Nankai, Japan and Izu-Bonin accretionary prisms (Taira & Pickering).

Microstructural analyses on material recovered from the ocean drilling programs (DSDP, IPOD and ODP) provides constraints on (*a*) porosity/permeability evolution of different fluid migration pathways; (*b*) deformation mechanisms and the history of deformation, and (*c*) the inter-connectivity of migration pathways (Knipe *et al.*).

The nature and origin of fluids in accretionary prisms can be inferred using geochemical techniques, especially by the study of stable isotopes (Kastner *et al.*). Pore fluids with lower than seawater Cl^- concentrations are characteristic of the accretionary prisms studied to date. The sources of such low salinity fluids may be from (*a*) the release of water during the breakdown of hydrous minerals, (*b*) mechanical compaction associated with a reduction in porosity; (*c*) diagenetic and metamorphic reactions which release water, for example the illitization of smectites; (*d*) the ingress of continental meteoric waters from adjacent continental margins, as in the Peru–Chile margin, and (*e*) deeper thermogenic fluid sources.

The main diagenetic reactions that modify fluid compositions are (Kastner *et al.*) (*a*) carbonate recrystallization and precipitation; (*b*) bacterial and thermal degradation of organic matter; (*c*) formation and dissociation of gas hydrates; (*d*) hydration and transformation of hydrous minerals, especially of clay minerals and opal-A and (*e*) alteration of volcanic ash and the upper oceanic crust, principally as zeolitization and clay mineral formation. Using direct and indirect measurements of fluid in the toe of the Barbados, Central Oregon, northern Cascadia and Nankai accretionary prisms, Le Pichon *et al.* suggest large-scale non-steady-state fluid flow, with fluid convection driven by the reduced salinity of less saline fluids of deep origin, i.e. thermogenic fluids and/or the long-distance transport of freshwater from adjacent sedimentary basins, with recharge mechanisms that may include seismic pumping.

The abundance of quartz veins in uplifted accretionary prism complexes attests to massive transport of silica, probably over considerable distances, and therefore to an abundance of hydrous fluids as a transport medium. Many of these quartz veins are highly deformed, and as fluids greatly facilitate recrystallization reactions, it is not surprising that accretionary prisms include some of the most intensely and pervasively deformed rocks making up the continents. Beyond this, at increasing temperatures and pressures, we enter the realm of metamorphism.

3. Thermal structure of subduction zones

The whole thermal structure of a subduction zone is critically dependent on a number of factors: (a) the angle of subduction, (b) the amount of sediment on the downgoing plate, (c) the age of the subducting lithosphere (young and warm, or old and cold) and the maturity of the subduction system, (d) whether the overriding plate is composed of thin, hot and rheologically weak oceanic lithosphere (island arc) or thick, cool and refractory continental lithosphere (continental arc), and (e) whether the region is one of relative extension (where everything, including the forearc is being subducted) or relative compression (where much of the sediment is being scraped off). The important point to stress is that the thermal structure which eventually develops in and around a subduction zone is metastable and strongly dependent on continued subduction. Thus any change in the angle of subduction, thermal character of the subducting plate (e.g. ridge subduction), thermal character of the over-riding plate (e.g. extension permitting uprise of hot asthenosphere), or even cessation of subduction or subduction flip, would perturb this metastable arrangement and lead to massive thermal breakdown of hydrous minerals in and bordering the mantle wedge. The release of fluids could trigger intensive melting or, at higher levels in the subduction zone, perhaps temporarily reduce the density of metamorphic rocks sufficiently to promote uplift (e.g. of blueschists).

The contributions by Davies & Bickle, and by Peacock, illustrate the important advances that have been made in understanding and constraining the thermal structure of subduction zones under many different conditions, and in particular the conditions under which blueschists may form, and where melting can occur. It is becoming clear that the conditions under which the slab may melt are very restricted indeed, and may only be attained during the initial stages of subduction. Dehydration processes in the slab result in a hydrous fluid flux from the slab into the mantle wedge which permits growth of hydrous phases such as hornblende and phlogopite. Induced convection in the mantle wedge is now recognized as an important process, dragging hornblende-bearing material downwards across the hornblende-out partial melting reaction. This releases hydrous melts which can further react with mantle at higher levels to form hydrous minerals, but this mantle is then continually dragged down to repeat the process, allowing considerable scope for geochemical enrichments. Presumably when the volume of magma becomes large enough to gain sufficient buoyancy to overcome constraining forces, the magma may escape to be erupted or emplaced. The extent to which induced convection can occur in older, cooler, more rigid sub-continental lithosphere is still a matter of debate. The models predict average degrees of wedge melting between 2 and 8%, and that higher degrees of melting should occur where the asthenosphere is involved. This is encouraging because magmas in intraoceanic environments, where asthenospheric involvement is most likely, are basaltic, whereas those at continental margins tend to be dominantly andesitic or dacitic.

4. Geochemical constraints from subduction zone magmas

The compositions of subduction zone magmas are unusually varied. Island arc tholeiites and high-Mg andesites (boninites) tend to be erupted at the start of subduction, and these are quite different compositionally from the calc-alkaline andesites which dominate more mature arc systems, and from other types of high-Mg andesites (bajaites) which erupt following ridge-subduction, or from post-orogenic shoshonitic lavas. It takes considerable ingenuity by petrologists to derive all these magma types from the subduction zone environment, but their compositional diversity illustrates the complex fractionation processes which must occur. One compositional characteristic shared by all subduction magmas which is not easy to explain is their low ratio of high field strength (HFS) to large ion lithophile (LIL) elements, particularly manifest in their low contents of Nb and Ta. This issue was addressed by a number of contributors, and in discussion. The crucial point is to identify the minerals which are causing these element fractionations and where these minerals are stable in the subduction system.

The paper by Ayers & Watson is very valuable here in that it summarizes the relevant experimental work on the solubility of minor phases such as apatite, monazite, rutile and zircon in hydrous fluids under the P–T–X conditions appropriate to subduction zones. The solubilities are low under normal conditions, which would enable HFS elements such as P, Th, Ti (Nb, Ta) and Zr to be retained in the slab during the dehydration phase and permit the required LIL/HFS element fractionation to occur. However, rutile is more soluble at higher P–T, which would decouple the behaviour of Ti (Nb, Ta) from the other HFS elements, a feature which is generally not observed. So, whereas the required element fractionations can take place in subduction zones, many of the details remain to be elucidated.

The potential of subduction zones as major sites of chemical fractionation throughout Earth history are considered in the final three papers by Saunders *et al.*, by Hawkesworth *et al.* and by McDonough. Together, they use a considerable global database on the trace element and isotopic characteristics of subduction and other igneous rocks in an attempt to identify where the fractionations are taking place, and also whether the residues returned to the mantle from past subduction processes can help account for the development of the various distinctive mantle chemical components (MORB, HIMU, EM1, EM2, etc.), and whether components from these reservoirs (as well as a possible 'prevalent mantle' PREMA reservoir, and subducted sediment) are involved in the production of magmas in subduction zones. The issues are very complex, and not easy to resolve. The balance of evidence does suggest, in modern subduction zones, that the slab may provide fluids and LIL elements through dehydration processes, but is unlikely to melt significantly, except following initiation of subduction. Moreover, if the mantle wedge is the source of subduction zone magmas, then induced convection is necessary to replenish the magma wedge source regions with chemical components (figure 2). To this extent there is

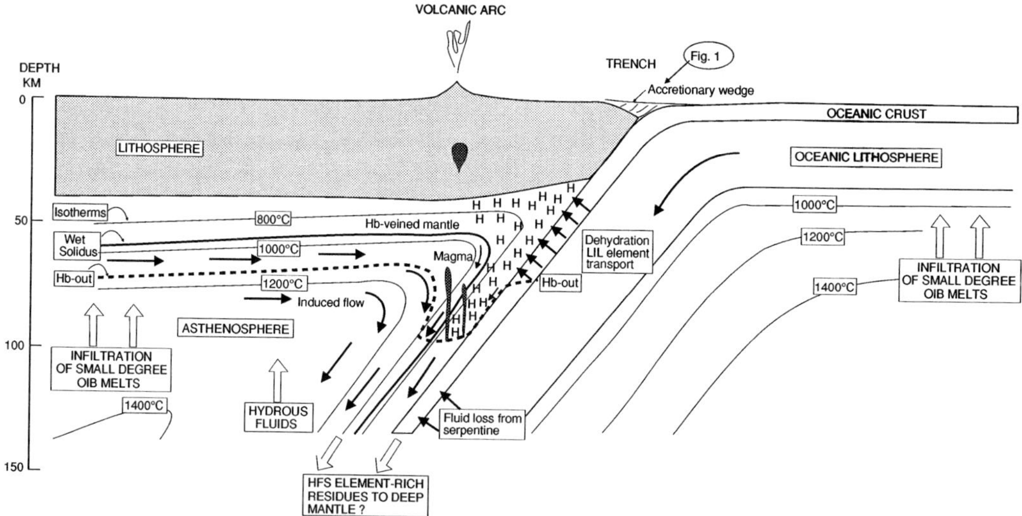

Figure 2. Cartoon illustrating the critical aspects of the deeper parts of subduction zones. Slab dehydration transports LIL elements and silica into mantle wedge, leading to hornblende (H) growth in harzburgitic wedge as shown. Induced convection continually drags wedge downward leading to breakdown of hornblende below the Hb-out (dashed) curve at *ca.* 100 km depth. Released fluids promote melting to left of and below Wet Solidus curve (full line), but melts may solidify to give hornblende-rich veins above, and the process repeated. Mantle wedge is refertilized through 'enriched' mantle being dragged along into the melting zone as a result of induced convection, the enrichment having been brought about through continual infiltration by small degree melts of OIB composition. Hydrous fluids released at deeper levels may promote phlogopite growth below the 'Hb-out' zone, which would by a similar mechanism lead to generation of more potassic magma compositions distal from the trench.

concordance between the thermal modelling, the experimental data on the solubilities of minor phases, and the deductions made from the geochemical and isotopic data on arc volcanic products. The residues resulting from magma extraction from the mantle wedge at subduction zones, which must carry a positive budget of HFS elements such as Nb and Ta, are returned deep into the mantle.

Induced convection provides an alternative means of replenishment of wedge source regions in subduction zone magmagenesis and so overcomes the problem of deriving large volumes of magma repeatedly from a rather refractory wedge and requiring that replenishment must come from melting of the slab. There are a number of interesting consequences from this which are explored in individual contributions.

Finally, this was planned as a Discussion Meeting, and so it turned out to be, with every one of the scheduled discussion periods, in addition to those following every individual presentation, being filled to the brim, and to the last bell. We should like to take this opportunity to thank all those who participated so constructively in the discussions, the chairmen of the respective sessions, and the contributors themselves for their talks and their responses, that together made the Meeting a success. We also thank our many reviewers who responded so promptly and helped us adhere to the tight six-month publication schedule.

March 1991

John Tarney
Kevin Pickering
Rob Knipe
John Dewey

CONTENTS

xiv   CONTENTS

Geophysical evidence for the role of fluids in accretionary wedge tectonics

By G. K. Westbrook

*School of Earth Sciences, University of Birmingham, Edgbaston,
Birmingham B15 2TT, U.K.*

The seismic-reflection technique, by imaging lithological and structural boundaries, can largely define the framework upon which models of the fluid-flow régimes of accretionary wedges are hung. The distribution of fluid loss from the sediments that form the accretionary wedge and that lie beneath it, can be estimated from variation in seismic velocity as a measure of change in porosity, in conjunction with the interpretation of the structural evolution derived from seismic-reflection sections. Seismic techniques have detected regions of pronounced undercompaction believed to be associated with overpressured pore fluid, and detailed modelling has defined zones of localized fluid overpressuring, such as the decollements beneath wedges. The measurement of heat flow, directly, or indirectly from the methane-hydrate seismic reflector, can be used to detect the outflow of fluids, and map its variation in relation to structure and lithology. Geophysical techniques will achieve their full potential in constraining models of the behaviour of accretionary wedges, when calibrated from borehole measurements.

1. Seismic techniques

Seismic techniques have provided the principal means of determining the structure and physical properties of accretionary wedges. It is doubtful whether most of the current concepts of processes in accretionary wedges would have been developed if it had not been for the seismic reflection sections across convergent plate margins. Certainly, seismic reflection sections show to a greater or lesser degree, the internal architecture of an accretionary wedge and can provide information on changes in physical properties that are related to its fluid-flow régime. Although seismic reflection images are sometimes very limited in what they show of the internal structure of accretionary wedges, there is no comparable technique to define the configuration of the structures and beds from which may be inferred the distribution of permeability and the mechanisms of deformation that generate the excess pore fluid pressure that drives fluid flow. To define the hydrogeological system of an accretionary wedge, measurements of the hydrogeological properties of samples and *in situ* tests of fluid pressure and permeability from boreholes with measurements of fluid outflow at the surface, are needed to complement the structural information obtained from seismic reflection. Even without this important calibration, however, seismic reflection can be used as a guide to major elements of the fluid-flow régime (figure 1). The position of the decollement horizon in the sedimentary section can strongly influence where porewater is expelled from the sediment by compaction. If

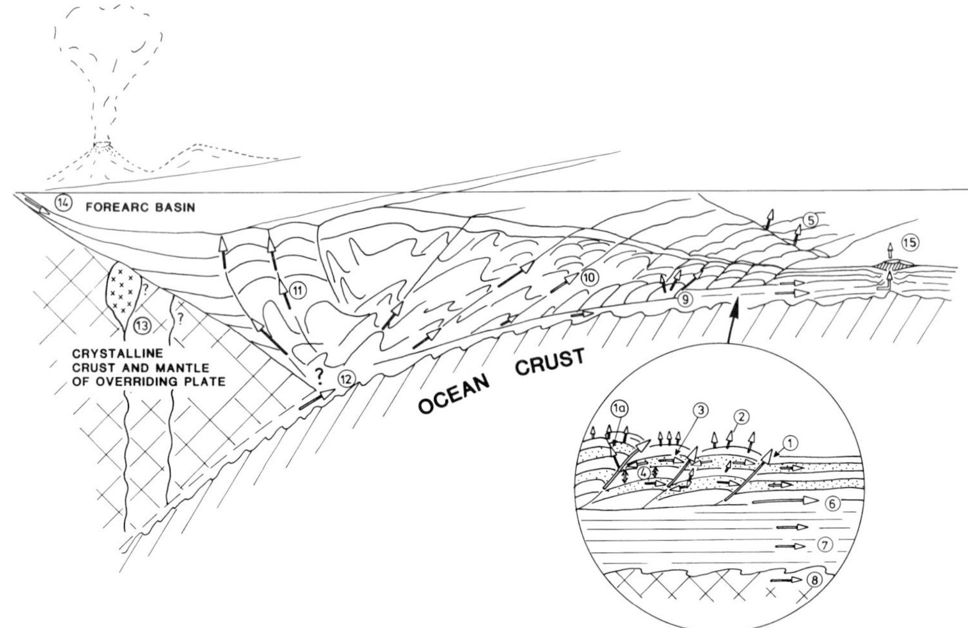

Figure 1. Illustration of modes of fluid outflow from an accretionary wedge (after Westbrook *et al.* 1987). (1) Fluid escape along thrust fault associated with accretion at the front of the wedge, and (1a) along faults antithetic to these. (2) Diffuse flow out of the sea bed in response to compaction. Fractures produced by local tensile stress at crests of anticlines may aid outflow. (3) Lateral flow along more permeable units in accreted sequence, feeding flow to faults. (4) Flow from less permeable units into more permeable units. (5) Outflow from vertical strike–slip faults intersecting the accretionary wedge. (6) Horizontal flow along the decollement beneath the wedge. (7) Outward flow in sequence beneath wedge in response to the load of the wedge. (8) Flow in permeable upper part of oceanic igneous crust. (9) Release of water into the wedge as a consequence of the deformation of subcreted (underplated) packets of sediment in duplexes. (10) Flow along major out-of-sequence landward dipping, thrusts that enable the wedge to maintain its critical taper. Local dilation and vein deposits from outflowing water may be the cause of acoustic impedance contrasts that gives these faults seismic visibility. (11) Flow along backthrusts into the forearc basin. (12) Upflow from deeper parts of the subduction zone of water released by dehydration reactions. This will be dependent to some degree on subduction rate. (13) Water from deeper part of subduction zone penetrating fractures in overriding lithosphere. Fractures are probably radial to the island arc. Hydration of the ultramafic rock may cause formation of serpentinite diapirs. (14) Meteoric groundwater from island arc or cordilleran continental margin seeping into horizons in the forearc basin, and then out into basin. (15) Outflow through mud volcano on ocean floor in front of wedge.

the decollement is deep and most of the section becomes accreted to the toe of the wedge then much of the fluid will be released near the toe (Davis *et al.* 1990). A shallow decollement above a thick section of sediment introduces the possibility of a larger proportion of the fluids in the sedimentary section being underthrust farther beneath the wedge (Westbrook *et al.* 1988), to be released into the wedge above, to flow back oceanward into the undeformed section along the higher permeability units in the underthrust sequence or the decollement (Westbrook & Smith 1983; Moore *et al.* 1988), or to be carried down into the mantle. Off Costa Rica, the landward convergence of reflectors in the underthrust sequence of hemipelagic and pelagic sediments beneath the outermost 4 km of the wedge indicates rapid compaction and water loss (Shipley & Moore 1986; Shipley *et al.* 1990). If the original thickness of the layer and the relationship between seismic velocity and porosity are known, the amount of compaction and fluid loss can be calculated, which for the Costa Rica example is $0.0025 \text{ m}^3 \text{ m}^{-2} \text{ a}^{-1}$ over the surface area of the underthrust section. Expressed differently, the hemipelagic and pelagic layers lose 44 % and 58 % of their original content of water by the time they have moved a horizontal distance of 4 km beneath the toe of the wedge at a rate of 90 km Ma^{-1}. At the toe of the northern part of the Barbados accretionary wedge, the compaction undergone by the underthrust section is slight, and difficult to measure accurately on the seismic sections (Bangs *et al.* 1990). Measurements of the porosity of samples obtained from drilling there show that only 4 % of the original volume of pore fluid is lost in the first 4 km of underthrust section (Moore *et al.* 1988). This approach to estimating compaction depends strongly on knowing the original thickness of the layer or assuming that it was the same as that in the undeformed section in front of the wedge. This assumption may often not be justified, particularly if the underthrust section includes part of the turbidite fill of the trench.

2. Seismic velocity

Variation in seismic velocity has been used widely to estimate changes in porosity caused by compaction of sediments in accretionary wedges. These changes in porosity may be used to derive, with decreasing certainty, fluid outflow and fluid pressure. The techniques used to measure seismic velocity, fall mostly into two categories: (*a*) velocity analysis of precritical reflections from multichannel seismic lines run across the strike of the structures in a wedge, (*b*) analysis of the pre- and post-critical reflections and refracted phases from lines run parallel to the strike. Velocity analyses from multichannel seismic surveys have been used to estimate porosity by Bray & Karig (1985), Minshull & White (1989), Davis *et al.* (1990), and Bangs *et al.* (1990), but only in the work reported by Bangs *et al.* for the Barbados Ridge accretionary wedge, was the aperture of the seismic experiment (5 km) designed with velocity derivation as a specific objective and a method of analysis employed to minimize the statistical and deterministic errors in the data (every common midpoint gather was used for velocity analysis and velocities were corrected for dip). The analyses of data from wide-angle reflection and refraction experiments shot along strike usually provide more precisely determined functions of velocity with depth than those obtained from precritical reflections, but these velocity/depth functions can be quite inaccurate. This is because the techniques commonly used, expanding spread seismic profiles (ESPs) and unreversed lines shot to sonobuoys, have no inherent control on the effect of dip on velocity (ESPs are better than simple unreversed lines), and so depend on remaining perfectly parallel to strike unless seismic reflection lines show the dips of major seismic boundaries. The advantages of these techniques are that they can yield accurate velocities to greater depths in the wedge, the inversion of the travel time data to velocity/depth is simpler because the problems of lateral variation of the velocity field are avoided, and, perhaps most importantly, the refracted phases can yield the velocity structure in regions where there are no distinct reflectors. The disadvantages are that the techniques cannot be used to map the variation in velocity continuously across the wedge, and cannot be used to examine porosity variation within individual structures. If refracted phases alone are used, reversals of velocity with depth can cause problems for inversion schemes. Commonly, such as in the Barbados Ridge accretionary complex, the

[3]

11-2

analyses of data from ESPS and sonobuoys, generally show higher velocities than those from the multichannel reflection data. The wider offset ESPS and sonobuoys, have a much greater horizontal component in their ray paths, and where there are closely spaced, layers of high and lower velocity, the average horizontal velocity is biased towards the higher velocities (in fact inversion of the refraction data usually only determines the horizontal velocity). The most generally useful type of velocity data for investigating accretionary wedges is that which comes from a multichannel reflection survey, provided that it has sufficient aperture, and provided that there are sufficient seismic reflectors. The velocity analyses, however, need to be much more extensive than those usually conducted to produce a seismic reflection section. They should be on data which has been corrected for dip moveout, and ray tracing and velocity determination coupled to migration may be required, especially if large lateral velocity contrasts occur within an accretionary wedge.

Velocity data from accretionary wedges, even though they vary in type and quality, have already provided important information on the response of sediments to being accreted and to being underthrust. Porosity has been derived from velocity through empirically based relationships such as that of Hamilton (1978) for silt, mudstone and turbidite lithologies. This is not an accurate method of determining absolute values of porosity, but it is adequate for showing changes in porosity in formations of predominantly the same lithology. Data from wedges that have accreted turbidites with a high sand fraction, such as off the Makran (Fowler *et al.* 1985; Minshull & White 1989) and off Vancouver (Davis *et al.* 1990), indicate that water is lost rapidly from the accreted section near the toe of the wedge, and that the sedimentary section is overcompacted relative to undeformed section in the trench. Off Barbados, in the north where the accreted section is predominantly pelagic clays, the porosity/depth profile in the first 20 km of the wedge is one of undercompaction, even though all the accreted sediments have undergone some compaction relative to their pre-accreted state (Bangs *et al.* 1990; Taylor & Leonard 1990). The reason for this effect is that sediments buried more deeply by tectonic thickening have not responded fully in reducing their porosities to those normally appropriate for their depths, because low permeability has inhibited the release of porewater. Farther south, where the accreted section is predominantly composed of distal turbidites it shows no significant change in the porosity/depth profile in the first 20 km, even though it increases in thickness by 1.1 km (a proportional increase of 1.6). To understand the history of compaction and dewatering of the sediment, the mechanics of thickening of the wedge must be known. This is illustrated with an example from the Barbados Ridge accretionary complex (figure 2). Line 480 crosses the wedge at 14° 22′ N (Bangs *et al.* 1990; Ladd *et al.* 1990), where a 1.9 km thick section of distal turbidites in the Orinoco submarine fan is being accreted above a 1.4 km thick section of predominantly pelagic and hemipelagic sediments. The section has been thickened primarily by thrusting along faults that dip at 30° through most of the section, but which curve toward horizontal in the upper part of the section where the least consolidated part of the section is overridden. Porosity in the section was derived from seismic velocity (Bangs *et al.* 1990). The pre-accretion porosity/depth profile, derived from the section in front of the wedge (after a small correction for the westward stratigraphic thickening of the section before accretion), was then subtracted from each thrust slice, following the internal stratigraphy, to map the change in porosity. The proportional changes in volume were calculated from $((\phi_0-\phi_1)/(1-\phi_1))$, where ϕ_0 is the initial porosity and ϕ_1 is the final porosity. The

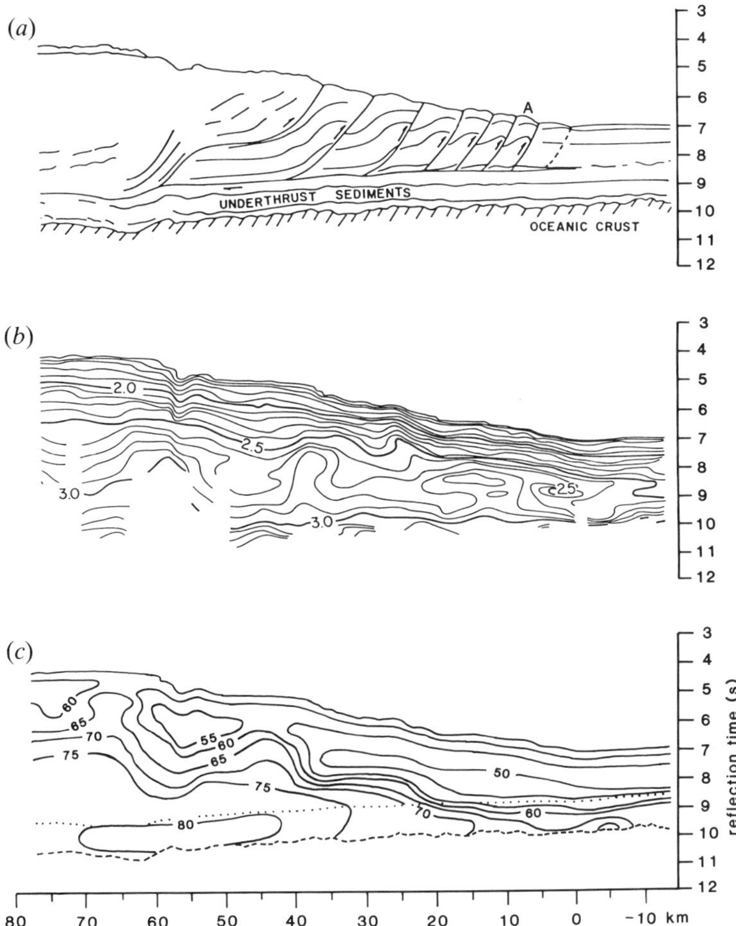

Figure 2. Illustration of the relationship between structure and fluid pressure inferred from seismic velocity, at the front of the Barbados Ridge accretionary complex, seismic line 480, 14° 22′ N. (*a*) Line drawing of major reflectors and interpretation of structure. (*b*) Contours of P-wave, seismic velocity, at 0.1 km s⁻¹ intervals. (*c*) Contours of λ (ratio of fluid pressure to lithostatic pressure) expressed as percentages. Fluid pressures are derived through the assumption that velocity reflects the effective stress in a reference sedimentary section in which fluid is at hydostatic pressure throughout. Dotted line shows the position of the decollement. Dashed line shows the position of the oceanic igneous basement.

resulting section of volume loss (illustrated for one thrust slice in figure 3) shows no loss from the axial regions of the uplifted anticlines, and maximum volume loss in the footwall sections of the thrusts especially in the upper, and initially least consolidated parts where it can be as much as 30%. Simple models of accretion used to model fluid flow (Screaton *et al.* 1990) and heat flow (Langseth *et al.* 1990), or both (Le Pichon *et al.* 1990) are often ones in which the accreted sediment flows into a fixed wedge taper, slowing as its streamlines diverge to follow the sides of the wedge. In these models the sediment is usually assumed to have an exponential porosity/depth function and as it thickens in the wedge, each unit of sediment increases its depth beneath the wedge surface and becomes more compacted, releasing fluid. The maximum rate of fluid loss occurs about 1 km beneath the surface, because the rate of change depth with increasing horizontal distance into the wedge increases with

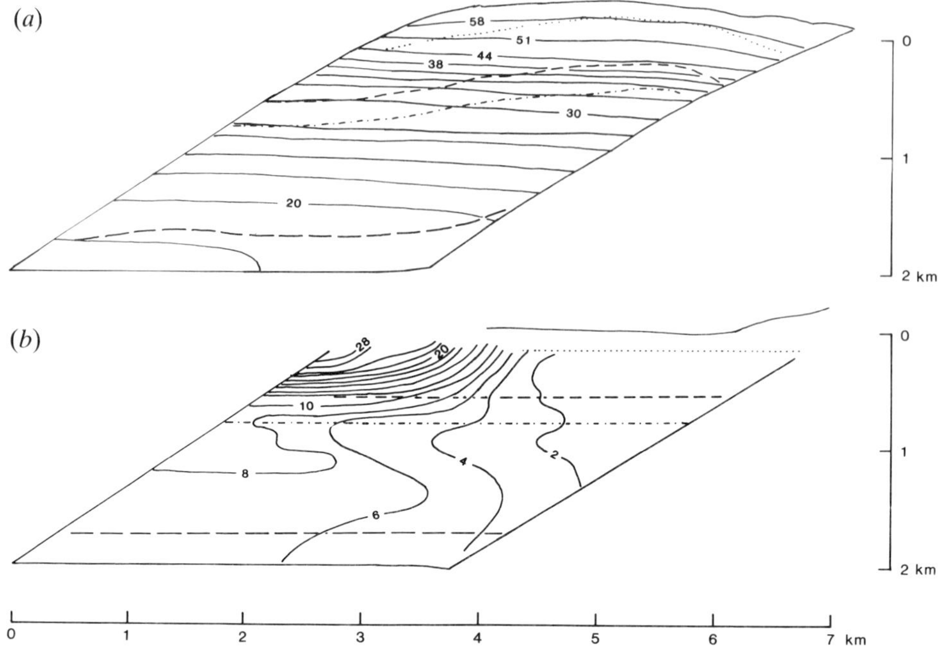

Figure 3. Porosity variation in and fluid loss from a thrust slice at the toe of the Barbados Ridge accretionary wedge imaged on seismic line 480, marked A in figure 2. (*a*) Shape of thrust slice with contours of porosity expressed as percentages, derived from seismic velocity. Major seismic reflectors marked by dashed and dotted lines. (*b*) Thrust slice restored to shape before deformation, with contours of percentage volume decrease from original volume.

depth, whereas porosity decreases with depth. This is not the same distribution of fluid loss as that derived from the porosity data from seismic velocity (figure 3). The important difference of the model is that it has no mechanism for strongly loading the near-surface section. The data and the model show about the same overall loss of fluid, but the difference in distribution is significant. Fluids derived from deeper in the section will have a much greater effect on heat flow and their pore-fluid chemistry could be significantly different, depending on thermally dependent reactions and variation in lithology with depth.

Pore fluid pressures may be estimated from the differences between actual porosity and porosity in a normal sedimentary section at the same depth, given two important assumptions. These assumptions are that sediments are steadily compacting (i.e. water is leaving the pore spaces under a load which has always been increasing) and that the reference (expected) porosity represents that developed at hydrostatic pressure. Porosity and effective stress (difference between lithostatic pressure and hydrostatic pressure) can be related through the compressibility of the sediment, as Shi & Wang (1988) have done in their modelling of pore fluid pressures in accretionary wedges, but compressibility in clays changes with degree of consolidation and effective stress and does not take the same value during unloading that it has during loading (Taylor & Leonard 1990). To avoid this complication, porosity was related to effective stress through the reference porosity/depth curve (Bangs *et al.* 1990), assuming that the water was at hydrostatic pressure throughout, that its density was 1050 kg m^{-3} and the sediment grain density was 2700 kg m^{-3}. As the porosity/velocity relationship is defined (Hamilton 1978), the effective stress can be directly related to velocity, so contours of effective stress follow those of velocity. Using velocity to indicate effective stress, and calculating lithostatic stress by integrating the density of the overlying sediments over their thickness, the ratio of fluid pressure to lithostatic pressure (λ) can be calculated. This was carried out on seismic data from the Barbados Ridge Complex, and shows that values of λ just reach 0.8 in the underthrust section of line 480, 45 km from the toe of the wedge (figure 2). Values beneath the first 10 km of the wedge lie in the range 0.55–0.70. These values are much less than predicted by models such those of Davis *et al.* (1983) which give $\lambda = 0.92$, but in these models λ at the decollement was assumed to be the same as λ in the wedge, for lack of information on what the variation in λ might be. In the frontal part of the wedge fluid pressure appears to be hydrostatic with λ varying between 0.62 near the seabed, and 0.49, just above the decollement. Lower fluid pressures in the wedge strengthen it and increase the shear stresses required at the base to give the wedge its taper, consequently lessening the degree of overpressuring required at the base. Even so, the values of λ are still too low to satisfy the model. A narrow zone of high porefluid pressures would, however, satisfy the tectonic constraints, but would not show up clearly in the type of velocity data used here. In the section of Line 465, where the sediments are predominantly clays, values of λ in excess of 0.85 are predicted in the region of the decollement 12–25 km from the toe of the wedge (figure 4).

Values of fluid pressure are likely to be underestimated by this method and so should be viewed with caution, because the assumption that the reference velocity/depth function derived from velocity data from the western Atlantic and the Barbados Ridge region (Bangs *et al.* 1990; Houtz 1981) represents a section in which pressures are hydrostatic may not be true. Any overpressuring in the regions from which the reference data were derived, will mean that the effective stresses derived from the reference function on the basis that they are hydrostatic will be overestimates. Also, uplift and compaction induced by tectonic stress will produce rocks that are overcompacted in relation to fluid pressures within them, according to the simple method of derivation of effective stress used here, and so erroneously low fluid pressures will be predicted. (The seismic velocity of a sample of rock from 1 km depth will not return to the value that it had shortly after deposition at the seabed when brought to the seabed.) Tectonic stress will increase the mean stress over that predicted from thickness of overburden. The Mohr–Coulomb model of the wedge of Davis *et al.* (1983) requires it to be at failure throughout, so the ratio the maximum to minimum principal effective stresses can be calculated from $(1 + \sin \phi)/(1 - \sin \phi)$ where ϕ is the angle of friction (Dahlen 1984). This ratio would typically be about 3. If λ is 0.6 then the effective stress is 40% of the total mean stress, and so the ratio of maximum to minimum principal stress would be 1.5. In the absence of tectonic stresses it is common to assume that all stresses are equal to the lithostatic load. In the presence of tectonic stresses, the lithostatic load will be an underestimate of the mean stress and this alone will lead to an underestimate of fluid pressure. Taking the limiting case of the lithostatic load being the minimum principal stress, if a value of $\lambda = 0.6$ is calculated from this rather than the mean stress, then the true value of $\lambda = 0.67$.

The zones of high fluid pressure correspond (as they must do given their manner of derivation) to regions of undercompaction. The greatest undercompaction occurs in the region of the decollement and just beneath it on this and the other two seismic lines north and south of it (Bangs *et al.* 1990). This is not surprising, because the

[7]

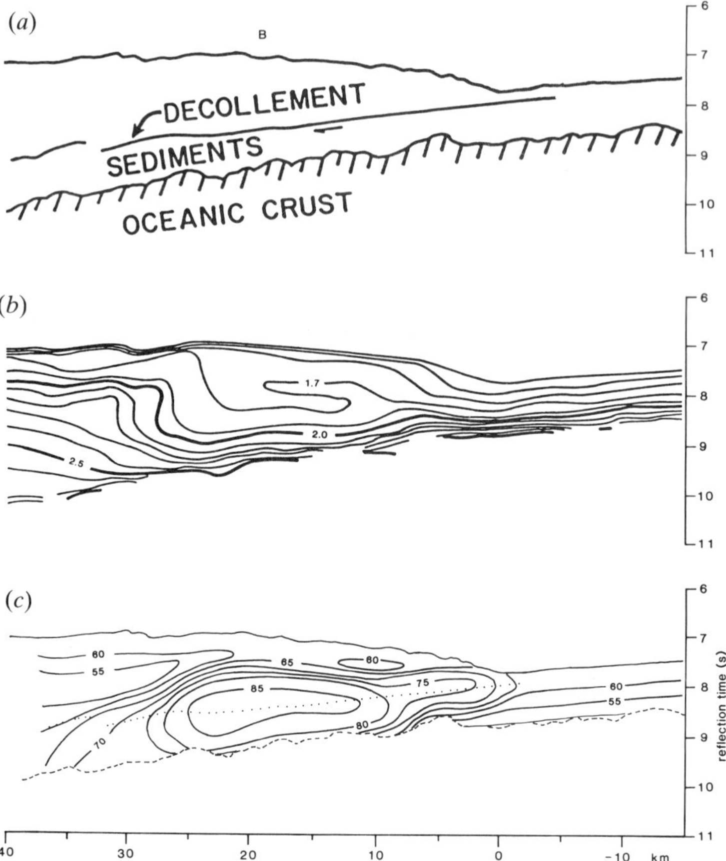

Figure 4. Relationship between fluid pressure inferred from seismic velocity and structure at the front of the Barbados Ridge Complex, seismic line 465, 16° 12′ N. The legend is as for figure 2. The sediments at 16° 12′ N are predominantly clays, whereas the upper 1.9 km of sediments at 14° 22′ N are the distal part of a submarine fan, incorporating silts and possibly sands.

sediments are among the least permeable in the section, and undergo the greatest rate of increase in load. Moore *et al.* (1990) report a similar situation in the accretionary wedge of the Nankai Trough off Japan, from preliminary results of seismic experiments. There, the underthrust pelagic section beneath the decollement shows little or no velocity increase with depth, and there is a downward velocity decrease across the decollement, contrasting with a downward velocity increase across the same stratigraphic level in the sediments beneath the trench.

3. Amplitudes of seismic reflectors

The amplitudes of reflections can be used to derive the changes in acoustic impedance (product of velocity and density) across boundaries. If these are boundaries of zones such as faults or decollements that act as fluid conduits, then the difference in fluid pressure may be inferred from the change in acoustic impedance. Cloos (1984) suggested that the seismic visibility of landward dipping reflectors in many accretionary wedges was a consequence of their activity as fluid conduits. Aoki *et al.* (1985) inverted a seismic-reflection section across the Nankai Trough, using well log data from Deep Sea Drilling Project (DSDP) site 583F as a control, to obtain an acoustic impedance section. This showed a reduction in acoustic impedance beneath the decollement at the base of the toe of the accretionary wedge and zones of low impedance along thrusts cutting through the wedge. These zones of low impedance could be produced by a relatively high porosity associated with fluid overpressuring.

The decollement reflector beneath the Barbados Ridge accretionary wedge at $16°$ $12'$ N (figure 4) shows a progressive change in amplitude and character as it extends westward beneath the wedge. At 20 km landward from the toe of the wedge, modelling of the seismic reflection section (Bangs & Westbrook 1991) indicates that the reflector is a 20 m thick layer of lower acoustic impedance than the rocks above and below (figure 5). At Ocean Drilling Project (ODP) site 671 where the decollement zone was penetrated it was found to be a zone of sheared clays 40 m thick (Moore *et al.* 1988). The porosity of the clay in the decollement zone was similar to (in fact, slightly less than) the porosities of clays immediately above and below it (Taylor & Leonard 1990). The proposed explanation for the seismic reflector is that the inner 20 m of the decollement zone has raised pore fluid pressure with consequently reduced effective stress and greater porosity, relative to the sediments above and below. The necessary reduction in acoustic impedance is given by a reduction of seismic velocity from 1.95 to 1.75 km s^{-1} with a concomitant reduction in density from 2023 to 1796 kg m^{-3}, an increase in porosity from 0.41 to 0.55. This implies a volume increase, under the effects of increased fluid pressure, of 30% of original volume.

The range of published values for the compressibility of clay, 10^{-6}–10^{-8} Pa^{-1} (Freeze & Cherry 1979) and the value used by Shi & Wang (1988) in modelling fluid pressures, indicate that a probable value for the clay in the decollement zone would be 10^{-7} Pa^{-1}. If so, an increase in fluid pressure of 2.6 MPa is required to inflate the decollement zone. An effective stress of 3.1 MPa, and lithostatic load of 23.1 MPa in the sediment directly overlying the decollement were estimated in the same manner as for Line 480 giving a value for λ of 0.87. Consequently, in the decollement zone the increased fluid pressure reduces effective stress to 0.5 MPa and raises λ to 0.98. Consolidation tests on core samples from the Barbados accretionary wedge (Taylor & Leonard 1990) show that if the initial effective stress were 3.1 MPa, a reduction to about 0.05 MPa would be needed to give the required change in porosity, implying fluid pressure very close to lithostatic, with $\lambda = 0.998$. At high fluid pressure (low effective mean stress), the effect of deviatoric stresses in producing dilation can be significant, and so the values of fluid pressure derived from simple elastic relaxation as above, are likely to be in error. Even so, the data indicate the value of λ for the decollement zone is above 0.95.

There are lateral variations over several hundred metres in the amplitude of the decollement reflector which have been shown by Bangs & Westbrook (1991) to be caused by changes in the acoustic impedance rather than changes in the thickness of the decollement layer. If this is correct, and the amplitude variation is not produced by some three-dimensional effect not accounted for in the two-dimensional model, then lateral changes in fluid pressure could be the cause of these changes in amplitude. The lateral variations in velocity of 1.75–1.88 km s^{-1} requires a volume change of 20% of original volume, which would imply a change in pressure of 1.8 MPa if compressibility is 10^{-7} Pa. Using the Leonard & Taylor data, a reduction

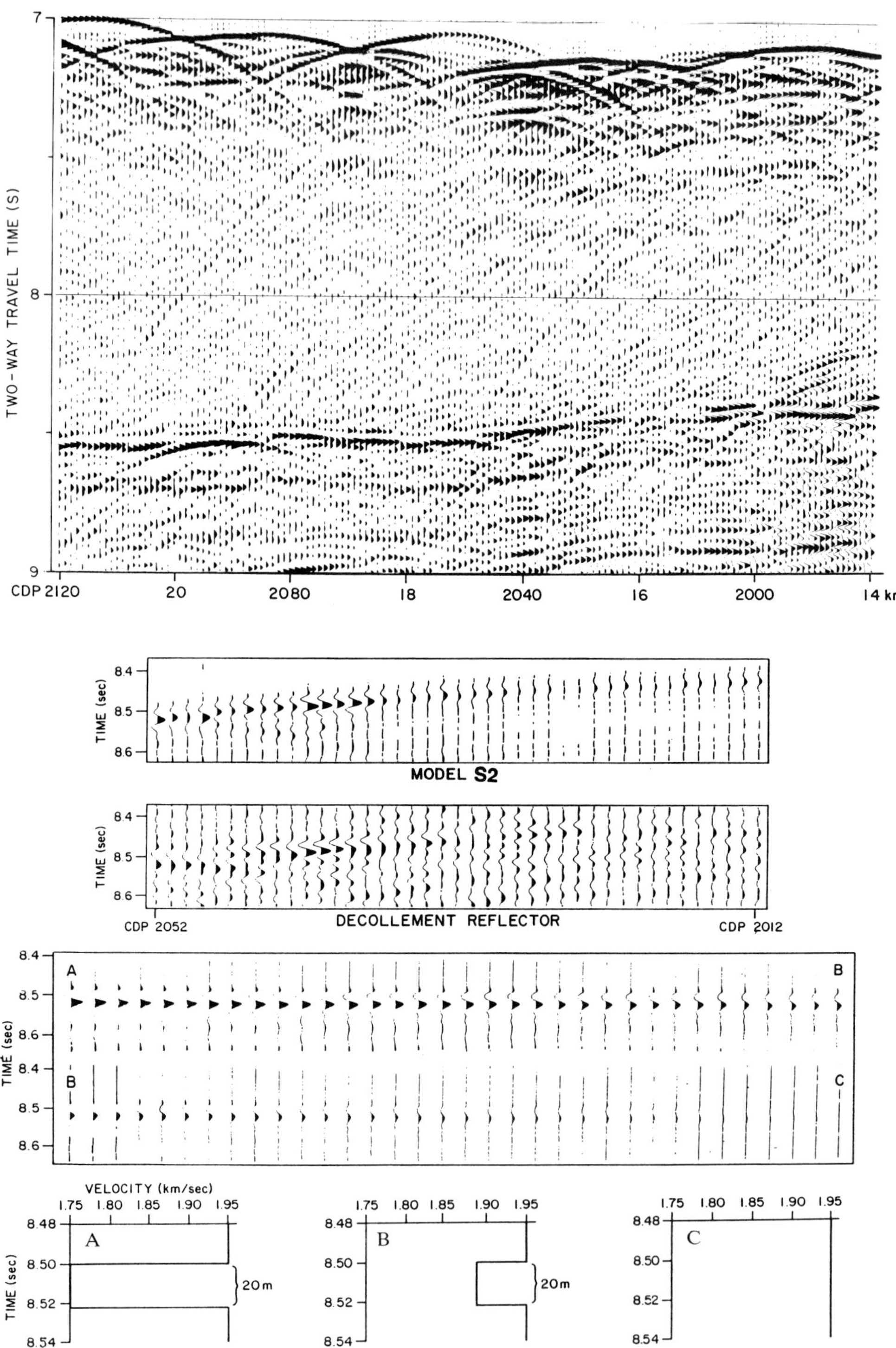

Figure 5. For description see opposite.

in pressure of about 2 MPa is indicated. This pressure change occurs over about 500 m, giving a horizontal pressure gradient of 4000 Pa m^{-1}. Leonard & Taylor measured a permeability of 6.48×10^{-21} m^2 (for viscosity of water of 10^{-3} Pa s) for a claystone close to but not from the decollement. If this value were representative of the decollement, the inferred pressure gradient would drive fluid at a Darcian velocity of 2.6×10^{-14} m s^{-1}, or 5.2×10^{-14} m s^{-1} average linear velocity (the rock has 0.5 porosity). The rate of subduction is 6.3×10^{-10} m s^{-1}, so with this low permeability the decollement could never transmit fluid to the toe of the wedge, but that has been shown to happen by the geochemistry of pore water in the ODP Leg 110 area. Screaton *et al.* (1990) on similar grounds assign a permeability of 10^{-14} m^{-2} to the decollement layer in their models. This higher permeability would yield any average flow rate of 8×10^{-8} m s^{-1} (2.5 m a^{-1}). Fisher & Hounslow (1990) estimate flow rates of 6.2 m a^{-1} along the decollement from variations in fluid temperature along the decollement measured in ODP Leg 110. If the value of 6.48×10^{-21} m^2 is representative of the intrinsic permeability of clay in the decollement layer then the bulk permeability of the layer must be controlled by the fracture permeability of the many shear surfaces developed parallel to the layer.

4. Heat flow

The advection of heat by water flowing out of the seabed produces an upwardly convex geothermal gradient than can be detected by heat flow probe measurements if the rate of flow (Darcian velocity) exceeds about 3×10^{-9} m s^{-1} (0.1 m a^{-1}) (Langseth & Herman 1981). The rates of diffuse outflow of water from wedges produced by compaction, predicted by a variety of models, do not exceed 3×10^{-10} m s^{-1} (0.01 m a^{-1}) (Minshull & White 1989; Langseth *et al.* 1990; Screaton *et al.* 1990; Le Pichon *et al.* 1990). Consequently, measurements of heat flow at the seabed do not offer a means of determining the 'background' rate of dewatering of accretionary wedges. The focused outflow of water through faults and other features can be detected by heat-flow measurements. If flow rates exceed 0.1 m a^{-1} they can be detected from the curvature of the thermal gradient, but more commonly it is the effect upon surface heat flow of introducing heat into the subsurface structure that is seen. Warm water flowing up a thrust fault will increase heat flow in the overlying thrust slice with the increase being greatest at the tip of the thrust slice. This gives the opposite heat-flow pattern to that produced by thrusting alone upon the conductive thermal régime, which is to reduce heat flow at the tip of the thrust slice because it is moving over the cool near-surface sediments. These effects have been modelled by Shi *et al.* (1988) off Oregon, and by Henry *et al.* (1989) for local heat-flow variations around seeps in the Nankai Trough.

An example of the relationship of the variation of heat flow in relation to thrusting comes from the southern Barbados Ridge complex at 12° 18′ N (figure 6). Heat-flow data from Foucher *et al.* (1990) and unpublished results from cruise 2907 of the

Figure 5. Part of seismic section 465 from the Barbados Ridge accretionary complex (see B on figure 4), with detail of the decollement reflector which can be modelled as a layer of low acoustic impedance 20 m thick (Bangs & Westbrook 1991). Variations in the amplitude of the reflector are best modelled by varying the acoustic impedance of the layer rather than its thickness. This variation is illustrated through models A, B, C. The model section, S2, is compared with the real section in the two central panels.

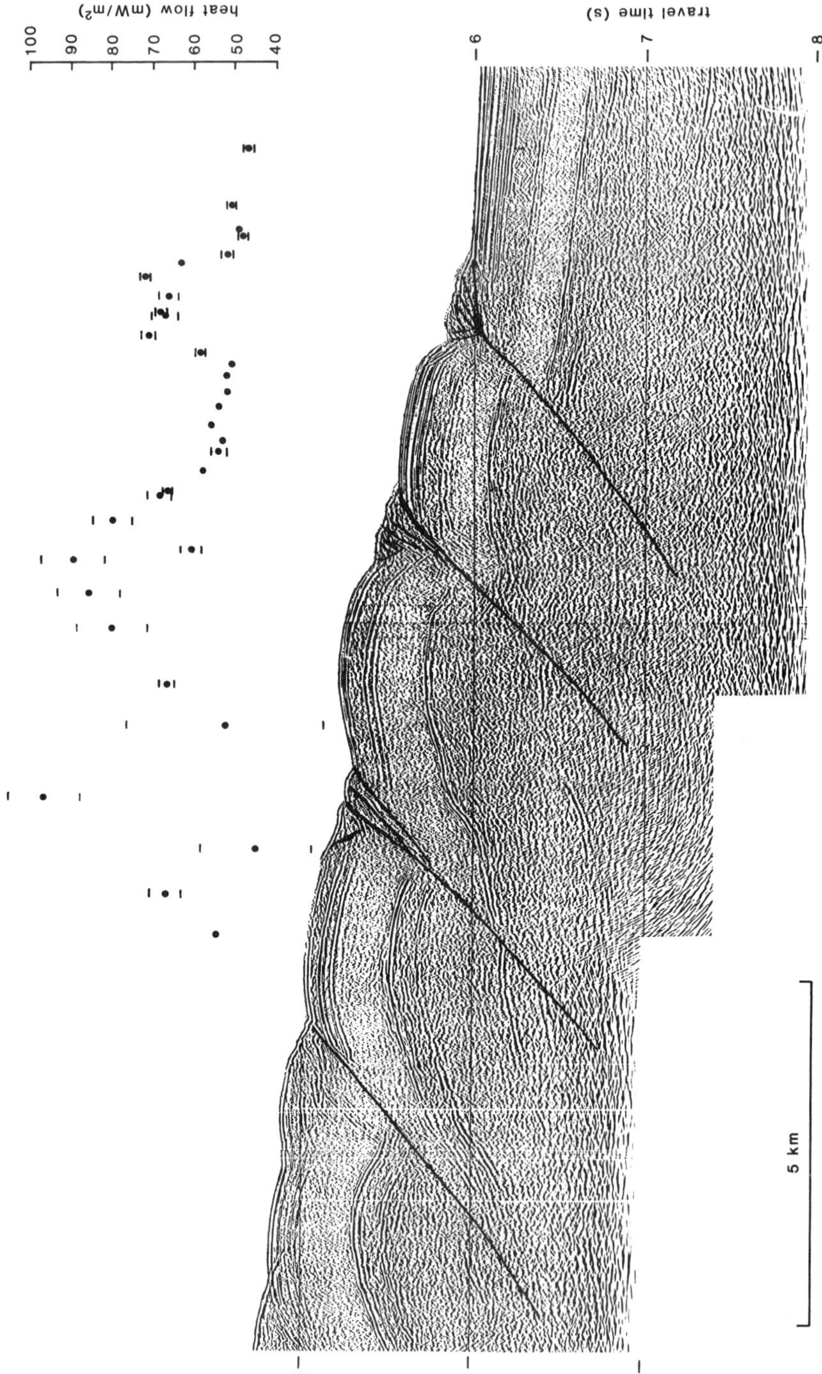

Figure 6. Heat flow across the toe of the Barbados Ridge accretionary complex at 12° 18′ N measured along a high resolution seismic profile (Westbrook *et al.* 1984). Heat flow from the thrust-outcrop wedge at the toe of the wedge is raised by 40% relative to the ocean floor before it. This is attributed to warm fluids flowing up the thrust. Raised heat flow at the tip of the second thrust slice is more broadly distributed as a consequence of the longer period of warming. Error bars indicate standard deviations of heat flow measurements greater than 1 mW m^{-2}.

Robert D. Conrad show raised values of heat flow in the regions of the outcrops of the thrusts. For the frontal thrust all the high values come from the thrust-outcrop wedge. They are about 40 % higher than those from the section in front of the accretionary wedge. The measured thermal gradients are not noticeably nonlinear. The higher heat flow at this thrust-outcrop wedge has been explained by warm fluids migrating up the thrust fault and through several small thrusts within the thrust-outcrop wedge (Foucher *et al.* 1990). Heat flow from the first anticline returns to nearly the same value as from the ocean floor, and rises again over the second thrust-outcrop wedge. Heat flow does not, however, decrease so quickly over the second anticline. This rise in heat flow over the forward part of the anticline is probably caused by the conductive diffusion of heat from warm water flowing along the second thrust, or it may be produced directly by upward flow of the warm water. In either case, time is required for the effect to be seen at the surface, which is why it is shown in the second anticline rather than the first, which it predates in age of formation by about 100 000 years (the characteristic time for thermal diffusion from a source at 1 km depth is about 75 000 years). The average increase in heat flow from the second thrust-outcrop wedge and the second anticline is 20 mW m^{-2}. Over the total width of the two features of 4.5 km the total heat flow is 90 W (per metre of strikelength). The rate of flow of water along the thrust required to produce this is 11 m^3 a^{-1}, assuming that heat flow is steady state and that the water comes from the decollement at a depth of 2 km with an initial temperature of 60 °C and cools to the temperature of seawater by the time it flows out from the seabed. This value, which is an underestimate of the flow, is between one third and one sixth of the volume of fluid in the pore space of sediment accreted to or underthrust beneath the accretionary wedge each year (Foucher *et al.* 1990). Measurement of temperatures by Fisher & Hounslow (1990) across thrusts intersected by drill holes in the northern part of the Barbados Ridge Complex, shows that fluids travel along the faults in pulses of limited duration (tens of years).

Heat-flow data for accretionary wedges are quite sparse. Measurements of heat flow in the seabed made with a thermistor probe are time consuming. If the seabed is hard it may be impossible to get the probe to penetrate it. The three regions where most direct measurements have been made are Barbados (Langseth *et al.* 1988; Langseth *et al.* 1990; Foucher *et al.* 1990), Nankai (Yamano *et al.* 1984; Kinoshita & Kasumi 1988), and Vancouver (Davis *et al.* 1990). Smaller numbers of measurements exist for southern Chile (Cande *et al.* 1987) and Washington-Oregon (Shi *et al.* 1988). Heat flow can be estimated from the depths of bottom simulating seismic reflectors (BSRs) which are common in accretionary wedges (Shipley *et al.* 1979). The reflector is produced at the base of a solid methane hydrate which is stable in the upper few hundred metres, where the temperature is low in relation to pressure, because of the depth of water. The rise of temperature with increasing depth makes the hydrate unstable. The BSR occurs along the phase boundary between solid hydrate and methane and water, and so is a measure of temperature at the depth of the BSR. The temperature dependence may be used to derive the thermal gradient and hence heat flow. Yamano *et al.* (1982) applied the approach to the Nankai Trough and it has been applied to a limited extent elsewhere (Cande *et al.* 1987; Minshull & White 1989; Davis *et al.* 1990; Ferguson *et al.* 1991). Minshull & White (1989) used the shallowing of the BSR in the regions of thrust faults in the accretionary wedge of the Makran to infer the flow of warm pore fluids along the faults. Off Vancouver, Davis *et al.* (1990) found the temperature estimated for the BSR, by using the phase diagram for methane hydrate and water, to be 7 °C less than that estimated from downward extrapolation of the thermal gradient at the seabed. This difference can be explained if there is significant advection of heat by upward fluid flow. Davis *et al.* estimated a rate of 8×10^{-10} m s^{-1} for this flow, which is very large, and inconsistent with outflow estimated from compaction models by an order of magnitude (E. Davis, personal communication). Uncertainties over the mode of formation of hydrate in nature, including pore-water chemistry, have probably reduced the extent to which people have been prepared to apply this approach with confidence. For example, if the methane in the Vancouver margin is thermogenic, which is quite possible given the high geothermal gradient, then propane could be present. Only 1 % propane is required to raise the temperature for hydrate stability by 7 °C (Sloan 1990). The presence of a BSR may also be taken as an indicator of fluid flow, if fluid flow is required to transport sufficient methane into the near-surface sediments to form a hydrate, rather than those sediments being sufficiently rich in organic carbon to generate the hydrate *in situ*. The degree to which flow is required to generate a BSR is an unresolved issue. Also, the formation of a hydrate will, to a varying extent, reduce the permeability of the sediment within which it is formed by filling pore spaces.

5. Summary

Seismic data have been of great value in providing images of the subsurface structure of accretionary wedges to identify possible fluid pathways, the state of compaction of sediments, the structural mechanisms by which compaction may be produced, and limiting estimates of fluid pressures. In the future, alongside continued progress in imaging, will be a greater and more detailed employment of velocity, amplitude and attenuation information to derive the influence of fluid on the behaviour of sediments in accretionary wedges from its effect on their physical properties. Although difficult to use in the submarine settings of accretionary wedges, this future work should make use of S-waves to better constrain changes in fluid content and to detect the alignment of the structural fabric which guides fluid flow.

Heat-flow data can be used directly or indirectly to derive rates of fluid flow. They are limited by the difficulty of obtaining measurements rapidly, but they can provide powerful constraints upon models, especially when three-dimensional information is obtained through boreholes. Gas hydrate BSRs, potentially provide a powerful approach for examining heat-flow variation over wide areas, if uncertainties over their mode of formation can be adequately resolved.

References

Aoki, Y., Kinoshita, H. & Kagami, H. 1985 Evidence of a low-velocity layer beneath the accretionary prism of the Nankai Trough: inferences from a synthetic sonic log. *Initial Rep. DSDP* **87**, 727–735.

Bangs, N. L. B., Westbrook, G. K., Ladd, J. W. & Buhl, P. 1990 Seismic velocities from the Barbados Ridge Complex: indicators of high pore-fluid pressures in an accretionary complex. *J. geophys. Res.* **95**, 8767–8782.

Bangs, N. L. B. & Westbrook, G. K. 1991 Seismic modelling of the decollement zone at the base of the Barbados Ridge accretionary complex. *J. geophys. Res.* (In the press.)

Bray, C. J. & Karig, D. E. 1985 Porosity of sediments in accretionary prisms and some implications for dewatering processes. *J. geophys. Res.* **90**, 768–778.

Cande, S. C., Leslie, R. B., Parra, J. C. & Hobart, M. 1987 Interaction between the Chile Ridge and the Chile Trench: geophysical and geothermal evidence. *J. geophys. Res.* **92**, 495–520.

Cloos, M. 1984 Landward-dipping reflectors in accretionary wedges: active dewatering conducts? *Geology* **12**, 519–522.

Dahlen, F. A. 1984 Noncohesive critical Coulomb wedges: an exact solution. *J. geophys. Res.* **89**, 10125–10133.

Davis, D. M., Suppe, J. & Dahlen, F. A. 1983 Mechanics of fold-and-thrust belts and accretionary wedges. *J. geophys. Res.* **88**, 1153–1172.

Davis, E. E., Hyndman, R. D. & Villinger, H. 1990 Rates of fluid expulsion across the northern Cascadia accretionary prism: constraints from new heat flow and multichannel seismic reflection data. *J. geophys. Res.* **95**, 8869–8889.

Ferguson, I. J., Westbrook, G. K., Langseth, M. G. & Thomas, G. P. 1991 Heat flow from the Barbados Ridge accretionary complex: comparison with conductive models. *J. geophys. Res.* (In the press.)

Fisher, A. T. & Hounslow, M. W. 1990 Transient fluid flow through the toe of the Barbados accretionary complex: constraints from Ocean Drilling Program Leg 110 heat flow studies and simple models. *J. geophys. Res.* **95**, 8845–8858.

Foucher, J. P., Le Pichon, X., Lallemant, S., Hobart, M. A., Henry, P., Benedetti, M., Westbrook, G. K. & Langseth, M. G. 1990 Heat flow tectonics, and fluid circulation at the toe of the Barbados Ridge accretionary prism. *J. geophys. Res.* **95**, 8859–8867.

Fowler, S. R., White, R. S., & Louden, K. E. 1985 Sediment dewatering in the Makran accretionary prism. *Earth planet. Sci. Lett.* **75**, 427–438.

Freeze, R. A. & Cherry, J. A. 1979 *Groundwater*. Englewood Cliffs, New Jersey: Prentice Hall.

Hamilton, L. 1978 Sound velocity-density in seafloor sediments and rocks. *J. acoust. Soc. Am.* **63**, 366–377.

Henry, P., Lallemant, S. J., Le Pichon, X. & Lallemand, S. F. 1989 Fluid venting along Japanese trenches: tectonic context and thermal modelling. *Tectonophysics* **160**, 277–291.

Houtz, R. E. 1981 Comparison of velocity-depth characteristics in western North Atlantic and Norwegian Sea sediments. *J. acoust. Soc. Am.* **68**, 1409–1414.

Kinoshita, M. & Kagumi, Y. 1988 Heat flow measurements in the Nankai Trough area. In *Preliminary Report of the Hakuko Maru Cruise KH 86-5* (ed. A. Taira), University of Tokyo Ocean Research Institute, 190–206.

Ladd, J. W., Westbrook, G. K., Buhl, P. & Bangs, N. 1990 Wide-aperture seismic profiles across the Barbados Ridge Complex. In *Proc. Ocean Drilling Program Initial Rep. b* **110**, 3–6.

Langseth, M. G. & Herman, B. M. 1981 Heat transfer in the oceanic crust of the Brazil Basin. *J. geophys. Res.* **86**, 10805–10819.

Langseth, M. G., Westbrook, G. K. & Hobart, M. A. 1988 Geophysical survey of a mud volcano seaward of the Barbados Ridge accretionary complex. *J. geophys. Res.* **93**, 1049–1061.

Langseth, M. G., Westbrook, G. K. & Hobart, M. A. 1990 Contrasting geothermal regimes of the Barbados Ridge accretionary complex. *J. geophys. Res.* **95**, 8829–8843.

Le Pichon, X., Henry, P. & Lallemant, S. 1990 Water-flow in the Barbados accretionary complex. *J. geophys. Res.* **95**, 8945–8967.

Moore, G. F., Shipley, T. H., Stoffa, P. L., Karig, D. E., Taira, A., Kuramoto, S., Tokugama, H. & Suyehiro, K. 1990 Structure of the Nankai Trough accretionary zone from multichannel seismic reflection data. *J. geophys. Res.* **95**, 8753–8765.

Moore, J. C., Mascle, A., Taylor, E. & Shipboard Party. 1988 Tectonics and hydrogeology of the northern Barbados Ridge: results from Ocean Drilling Program Leg 110. *Geol. Soc. Am. Bull.* **100**, 1578–1593.

Minshull, T. A. & White, R. S. 1989 Sediment compaction and fluid migration in the Makran accretionary prism. *J. geophys. Res.* **94**, 7387–7402.

Screaton, E. J., Wuthrich, D. R. & Dreiss, S. J. 1990 Permeabilities, fluid pressures, and flow rates in the Barbados Ridge Complex. *J. geophys. Res.* **95**, 8997–9007.

Shi, Y. & Wang, C. Y. 1988 Generation of high pore pressures in accretionary prisms: inferences from the Barbados subduction complex. *J. geophys. Res.* **93**, 8839–8910.

Shi, Y., Wang, C. Y., Langseth, M. G., Hobart, M. A. & von Huene, R. 1988 Heat flow and thermal structures of the Washington–Oregon accretionary prism – a study of the lower slope. *Geophys. Res. Lett.* **15**, 1113–1116.

Shipley, T. H., Houston, M. H., Buffler, R. T., Shaub, F. J., McMillen, K. J., Ladd, J. W. & Worzel, J. L. 1979 Seismic evidence for widespread possible gas hydrate horizons on continental slopes and rises. *Am. Ass. Petrol. Geol. Bull.* **63**, 2204–2213.

Shipley, T. H. & Moore, G. F. 1986 Sediment accretion, subduction and dewatering at the base of the trench slope off Costa Rica: a seismic reflection view of the decollement. *J. geophys. Res.* **91**, 2019–2028.

Shipley, T. H., Stoffa, P. L. & Dean, D. F. 1990 Underthrust sediments, fluid migration paths, and mud volcanoes associated with the accretionary wedge of Costa Rica: Middle America Trench. *J. geophys. Res.* **95**, 8743–8752.

Sloan, E. D. 1990 *Clathrate hydrates of natural gases.* New York: Marcel Dekker.

Taylor, E. & Leonard, J. 1990 Sediment consolidation and permeability at the Barbados forearc. In *Proc. ODP Sci. Results* (ed. J. C. Moore *et al.*) 110: College Station, TX (Ocean Drilling Program), pp. 289–308.

Westbrook, G. K. & Smith, M. J. 1983 Long decollements and mud volcanoes: evidence from the Barbados Ridge Complex for the role of high pore-fluid pressure in the development of an accretionary complex. *Geology* **11**, 279–285.

Westbrook, G. K., Ladd, J. W., Buhl, P., Bangs, N. & Tiley, G. 1988 Cross section of an accretionary wedge: Barbados Ridge Complex. *Geology* **76**, 631–635.

Westbrook, G. K., Mascle, A. & Biju-Duval, B. 1984 Geophysics and structure of the Lesser Antilles forearc. *Initial Rep. DSDP* A**78**, 24–38.

Westbrook, G. K., Boulegue, J., Bowers, T. S., Cathles, L. M., Davis, E. E., Kastner, M., Langseth, M., Leinen, M. & Ohta, S. 1987 Fluid Circulation in the Crust and the Global Geochemical Budget. In *Rep. 2nd Conf. Scientific Ocean Drilling (COSODII)*. European Science Foundation, Strasbourg, pp. 67–86.

Yamano, M. S., Honda, S. & Uyeda, S. 1984 Nankai Trough: a hot trench. *Mar. Geophys. Res.* **6**, 187–203.

Yamano, M. S., Uyeda, S., Aoki, Y. & Shipley, T. H. 1982 Estimates of heat flow derived from gas hydrates. *Geology* **10**, 339–343.

Fluids in convergent margins: what do we know about their composition, origin, role in diagenesis and importance for oceanic chemical fluxes?

By M. Kastner[1], H. Elderfield[2] and J. B. Martin[1]

[1] Scripps Institution of Oceanography, University of California, San Diego, La Jolla, California 92093-0212, U.S.A.

[2] Department of Earth Sciences, University of Cambridge, Cambridge CB2 3EQ, U.K.

The nature and origin of fluids in convergent margins can be inferred from geochemical and isotopic studies of the venting and pore fluids, and is attempted here for the Barbados Ridge, Nankai Trough and the convergent margin off Peru. Venting and pore fluids with lower than seawater Cl^- concentrations characterize all these margins. Fluids have two types of source: internal and external. The three most important internal sources are: (1) porosity reduction; (2) diagenetic and metamorphic dehydration; and (3) the breakdown of hydrous minerals. Gas hydrate formation and dissociation, authigenesis of hydrous minerals and the alteration of volcanic ash and/or the upper oceanic crust lead to a redistribution of the internal fluids and gases in vertical and lateral directions. The maximum amount of expelled water calculated can be *ca.* $7 \text{ m}^3 \text{ a}^{-1} \text{ m}^{-1}$, which is much less than the tens to more than $100 \text{ m}^3 \text{ a}^{-1} \text{ m}^{-1}$ of fluid expulsion which has been observed. The difference between these figures must be attributed to external fluid sources, mainly by transport of meteoric water enhanced by mixing with seawater.

The most important diagenetic reactions which modify the fluid compositions, and concurrently the physical and even the thermal properties of the solids through which they flow are: (1) carbonate recrystallization, and more importantly precipitation; (2) bacterial and thermal degradation of organic matter; (3) formation and dissociation of gas hydrates; (4) dehydration and transformation of hydrous minerals, especially of clay minerals and opal-A; and (5) alteration, principally zeolitization and clay mineral formation, of volcanic ash and the upper oceanic crust.

1. Introduction

Fluids play a central role in the deformational, thermal and geochemical evolutions of active convergent plate margins (von Huene 1984). Here, considerable fluid–solid diagenetic and metamorphic reactions take place and significant fluid volumes are expelled which may play an important role in global geochemical budgets. The most important internal fluid sources are: (1) pore fluids of the sediment and oceanic crust, present at all depths, and expelled by porosity reduction processes (Carson 1977; Bray & Karig 1985; Fowler *et al.* 1985; Bangs *et al.* 1990; Davis *et al.* 1990); (2) fluids derived from diagenetic and metamorphic dehydration; and (3) fluids from breakdown reactions of hydrous minerals. Additional sources are decarbonation of calcareous phases, and biogenic or thermogenic decomposition of organic matter

(Powers 1967; Perry & Hower 1972; Claypool & Kaplan 1974; Kastner *et al.* 1990; Peacock 1990; Vrolijk *et al.* 1990; Taira *et al.* 1991).

The most important external fluid sources probably are meteoric water transport and seawater mixing through seismic pumping or density inversion induced convection (Kastner *et al.* 1990; Le Pichon *et al.* 1990*a*, *b*). Widespread venting, supporting extensive benthic biological communities (Kulm *et al.* 1986; Boulegue *et al.* 1987), mud volcanoes and diapirs, carbonate chimneys (Westbrook & Smith 1983; Ritger *et al.* 1987; Brown & Westbrook 1988; Langseth *et al.* 1988; Carson *et al.* 1990) and heat flow anomalies (Fisher & Hounslow 1990; Foucher *et al.* 1990) are the important direct manifestations of localized fluid expulsion in accretionary complexes and seaward of the deformation front. In addition, indirect information about both localized and dispersed fluid expulsion is provided by means of seismic studies, e.g. of arcward thinning of the underthrusted sediments (Bray & Karig 1985; Fowler *et al.* 1985; Bangs *et al.* 1990), mineralized veins, carbonate crusts and cements (Kemp 1990; Lindsley-Griffin *et al.* 1990; Kulm *et al.* 1990; Sample 1990; Vrolijk 1987), and the presence of thermogenic methane at shallow depths, such as in the Barbados Ridge decollement (Vrolijk *et al.* 1990).

One of the striking observations central to the problem of the role of fluids in convergent plate margins is the marked discrepancy between the fluid flow rates of tens to over 100 m a^{-1}, obtained at some of the sites of localized fluid discharge or deduced from heat flow measurements, and the calculated maximum rates of fluid discharge, of the order of millimetres per year, assuming steady-state compactive dewatering (Carson *et al.* 1990; Foucher *et al.* 1990; Le Pichon *et al.* 1990*a*, *b*). Even the addition of all other internal fluid sources, particularly from hydrous minerals dehydration and/or breakdown reactions, does not significantly diminish this discrepancy. Hence the need to invoke external sources for the observed 'excess' fluid seems inescapable. In the Cascadia accretionary complex, geochemical evidence suggests a shallow source (less than 100–750 mBSF) for this 'excess' fluid (Carson *et al.* 1990). Discerning fluid sources and budgets, even in such complex systems where original geochemical and isotopic characteristics are modified by fluid–solid exchange reactions and pervasive fluid mixing, hence decoupling of signatures, should be possible with a proper mix of geochemical and isotopic tracers.

2. Background of Barbados, Nankai and Peru convergent margin sites

The summaries of the main geophysical, sedimentological and of some geochemical features, characterizing each of these margins, provided in tables 1 and 2, clearly indicate their large diversity. Notwithstanding, surface manifestations of fluid venting exist at each. In the Nankai Trough, however, they seem to be confined to the eastern, and not to the western region that was drilled. The drill sites at these margins are marked in figure 1. At the Barbados Ridge, Ocean Drilling Project (ODP) Site 671 is located *ca.* 4 km arcward of the toe of the accretionary complex. The decollement was penetrated at *ca.* 500 mBSF and comprises a zone *ca.* 40 m thick (Moore *et al.* 1988). At the western Nankai Trough, ODP Site 808 was penetrated through the accretionary prism, frontal thrust and decollement to basaltic basement. The decollement was encountered at *ca.* 960 mBSF and it comprises a zone *ca.* 20 m thick (Taira *et al.* 1991). In the Peru margin Sites 685 and 688, the sites closest to the trench were chosen because their pore fluid isotopic composition suggests the presence of two distinct fluid régimes in this margin. Site 685 (*ca.* 9° S) is the only

Table 1. *Comparison of data for the Barbados, Nankai and Peru subduction zones*

	Barbados	Nankai	Peru
age of subducting plate/Ma	*ca.* 90	*ca.* 15–17	*ca.* 30 north of 10° S *ca.* 40 south of 10° S
convergence rate/ (mm a^{-1})	slow, *ca.* 20	slow, *ca.* 25	fast, *ca.* 90
geothermal gradient/ (°C km^{-1})	28 at site 674, at 92–434 mBSF[a] 36 at Site 671, at 36–168 mBSF	*ca.* 110 at Site 808[b]	47–50 at four sites[c] (683, 685, 682, 688)
sediment type	fine-grained hemipelagic clays and muds	ashy silt-sand turbidites hemipelagic muds	organic-rich diatomaceous muds and silts
organic C content (%)	<0.2	0.1–0.6	3–4
fate of incoming sediments	much accreted, some subducted or subcreted	much accreted, some subducted or subcreted	much subducted or subcreted, some accreted
manifestations of venting			
(*a*) at sites drilled	mineralized veins and faults	none[b]	mineralized veins; fluid escape structures[d]
(*b*) in and seaward of the accre- tionary complex	thermal regime, mud volcanoes and diapirs[e]	benthic communities, thermal régime, carbonate cements and chimneys and mud volcanoes[b]	dredged clams[f]
gas hydrates	no seismic evidence for BSR, but indirect geochemical evidence for possible presence of gas hydrates in region drilled; further south present at 600–800 mBSF[g]	present; BSR is very shallow, at *ca.* 200 mBSF[b]	ubiquitous, BSR at *ca.* 470 mBSF near site 682 to *ca.* 610 mBSF near site 688[h]

[a] Fisher & Hounslow 1990. [b] Taira *et al.* 1991. [c] Yamano & Uyeda 1990. [d] Kemp 1990; Lindsley-Griffin *et al.* 1990. [e] Langseth *et al.* 1988; Foucher *et al.* 1990; Le Pichon *et al.* 1990 *b*. [f] Kulm *et al.* 1986. [g] Bangs *et al.* 1990; Gieskes *et al.* 1990. [h] Kvenvolden & Kastner 1990.

accretionary complex site drilled, and Site 688 (*ca.* 11° S) exhibits non-steady state geochemical profiles, resembling those obtained at the Barbados Ridge Site 671 (Suess *et al.* 1988; Elderfield *et al.* 1990; Kastner *et al.* 1990).

3. The chloride problem

Despite the large disparity in the overall geologic characteristics of these convergent margins, the majority of the interstitial and discharged fluids analysed to date are characterized by lower than seawater chloride concentrations. Lower than seawater Cl^{-} pore fluids are ubiquitous in convergent margin sediments; seawater Cl^{-} dilutions of 10 to over 50% have been reported. The lowest Cl^{-} values, shown in figures 2 and 3, are given in table 3. Another conspicuous characteristic of most Cl^{-} depth profiles, illustrated in figure 2, is their distinctly non-steady-state nature. Many of the Cl^{-} minima are situated at unconformities, faults, particularly

Table 2. *Synthesis of characteristics of pore fluid* Cl^-, $\delta^{18}O$ *and* $^{87}Sr/^{86}Sr$ *depth profiles for the Barbados, Nankai and Peru accretionary complexes*

	Barbados Ridge site 671	Nankai Trough site 808	Peru Margin site 688	site 685
Cl^- concentrations	decrease with depth; non-steady-state profile	increase in the trench-fill turbidite section; then decrease toward decollement; below, increase again toward oceanic basement	decrease with depth; non-steady-state profile	decrease with depth
$\delta^{18}O$ values	decrease with depth; non-steady-state profile	not available	decrease with depth; non-steady-state profile	decrease with depth; non-steady-state profile
$^{87}Sr/^{86}Sr$ ratios	non-steady-state profile	decrease with depth to 'proto-decollement', then increase at decollement	decrease with depth	more radiogenic than seawater; non-steady-state depth profile
	Cl^- and $^{87}Sr/^{86}Sr$ depth profiles may be coupled Cl^- and $\delta^{18}O$ depth profiles are decoupled	Cl^- and $^{87}Sr/^{86}Sr$ depth profiles are decoupled	Cl^- and $^{87}Sr/^{86}Sr$ depth profiles are decoupled the profiles of $^{87}Sr/^{86}Sr$ and $\delta^{18}O$ appear to be coupled at depths >300 m	Cl^- and $^{87}Sr/^{86}Sr$ depth profiles are decoupled the profiles of $^{87}Sr/^{86}Sr$ and $\delta^{18}O$ appear to be coupled, inversely correlated, at depths >80 m

in the decollement, and are associated with methane and other geochemical and isotopic anomalies as summarized in table 4 (Gieskes *et al.* 1990; Kvenvolden & Kastner 1990; Kastner *et al.* 1990; Le Pichon *et al.* 1990*a*; Taira *et al.* 1991). Some are also associated with thermal anomalies (Fisher & Hounslow 1990; Foucher *et al.* 1990). The occurrence of anomalies at high permeability horizons, at lithological and structural discontinuities, and the coincidence of the geochemical and thermal minima and maxima, indicate lateral fluid flow. Diffusion–advection calculations imply that the important Cl^- minima horizons at Sites 671, 672, 674, 683, and 688 (figure 2) cannot be older than several to tens of thousands of years. For fluid flow along the decollement horizon at Site 671, Barbados Ridge, a thermal model by Fisher & Hounslow (1990) suggests fluid flow velocity of 10^{-7} m s^{-1}. Because this velocity is 100 times faster than predicted by steady-state models, they concluded that fluid flow is transient. The question of transient against steady-state fluid flow in convergent margins directly hinges upon the major problem of the large disparity between observed rates of fluid discharge and those calculated, assuming steady-state flow. Preliminary modelling of the Cl^- concentrations depth profile at Site 808, western Nankai Trough (figure 2*b*) supports transient fluid flow also along this decollement zone. Simple diffusion–advection calculations suggest that a pulse of low

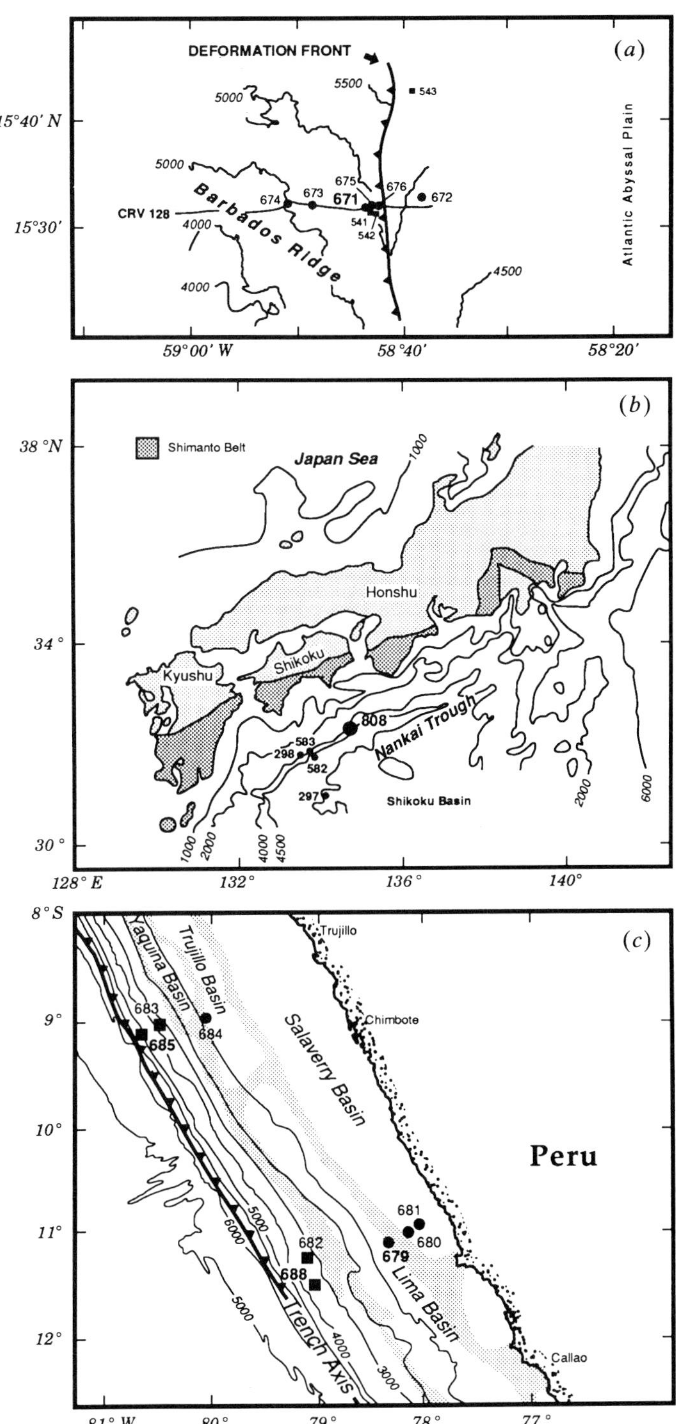

Figure 1. Schematic maps of study areas. Bathymetry is in metres. Closed circles and squares identify sites drilled. Bold numbers designate sites emphasized in this study. (*a*) At the Barbados Ridge, CVR-128 is a multichannel seismic line. (Modified from Moore *et al.* 1988.) (*b*) At the Nankai Trough. (Modified from Leg 131 ODP preliminary report no. 31, 1990, Ocean Drilling Program.) (*c*) At the Peru margin. (Modified from Kastner *et al.* 1990.)

Table 3. *The lowest Cl⁻ values as shown in figures 2 and 3*

location	Cl⁻/mM	approx. dilution of seawater (%)	reference
Barbados Ridge, site 671, at decollement	505	10	Gieskes *et al.* (1990)
Barbados Ridge, site 674, 17 km arcward from deformation front, at 448 mBSF	394	30	Gieskes *et al.* (1990)
Mud Pie, 12 km seaward of Barbados accretionary complex	250	55	Le Pichon *et al.* (1990*a*)
Nankai Trough, site 808, at 1110 mBSF	447	20	Taira *et al.* (1991); Gamo *et al.* (1991); Gieskes *et al.* (1991); Kastner *et al.* (1991)
Peru Margin, site 683, at 452 mBSF	454	20	Kastner *et al.* (1990)
Peru Margin, site 688, at Plio-Pleistocene unconformity at 365 mBSF	471	16	Kastner *et al.* (1990)
Peru Margin, site 679 at eastern edge of mid-to-upper slope, Lima Basin, 320 mBSF	339	40	Kastner *et al.* (1990)

Cl⁻ fluid along the decollement, may have occurred over 3×10^5 years ago. This is the minimum time needed to establish the long, greater than 250 m, diffusion profile observed (between 820 and 560 mBSF) in figure 2*b* (Gieskes *et al.* 1990; Gamo *et al.* 1991; Kastner *et al.* 1991; Taira *et al.* 1991).

It is worthwhile noting that, except for the high Cl⁻ fluids in the forearc basins of the Peru margin (Kastner *et al.* 1990), in the trench-fill turbidites at Nankai (Taira *et al.* 1991) and in Vanuatu, fluids with significantly higher than seawater Cl⁻ concentrations, widespread in fluid inclusions within mineralized veins (Vrolijk 1987), have not been encountered in accretionary complex sediment pore fluids as yet. Even these fluids are not as concentrated as observed in many fluid inclusions.

The origin of the ubiquitous low-Cl⁻ fluids is of great importance to the understanding of the hydrogeochemistry of convergent margins. The migration of fluids expelled by tectonic compaction, by porosity reduction processes, could not be the source for the low-Cl⁻, methane-rich, fluids that are discharging at the sea floor and supporting prolific benthic communities (Kulm *et al.* 1986, 1990; Boulègue *et al.* 1987; Ritger *et al.* 1987; Carson *et al.* 1990; Le Pichon *et al.* 1990*b*) unless they are modified by mixing with a more dilute fluid. The possible internal processes that may provide H_2O for the formation of the low-Cl⁻ fluids are: (1) dehydration or breakdown of hydrous minerals, particularly of clay minerals and opal-A; (2) dissociation of primarily methane gas hydrates (clathrates); (3) clay membrane ion filtration. Similarly, the possible sources for the associated elevated methane concentrations are: (1) biogenic or thermogenic combustion of organic matter; (2) gas hydrates dissociation; (3) decarbonation reactions plus CO_2 reduction.

The most important diagenetic dehydration reactions, the stepwise dehydration of smectite and its transformation to illite, are dependent on temperature, pore pressure and precursor composition, as well as on the availability of K^+ and Al^{3+} from other phases and on the disposability of silica by quartz precipitation, for example as a cement, or by transport in solution away from the reaction site, as indicated in the following reaction:

$$\text{smectite} + K^+ + Al^{3+} = \text{illite} + \text{silica} + H_2O$$

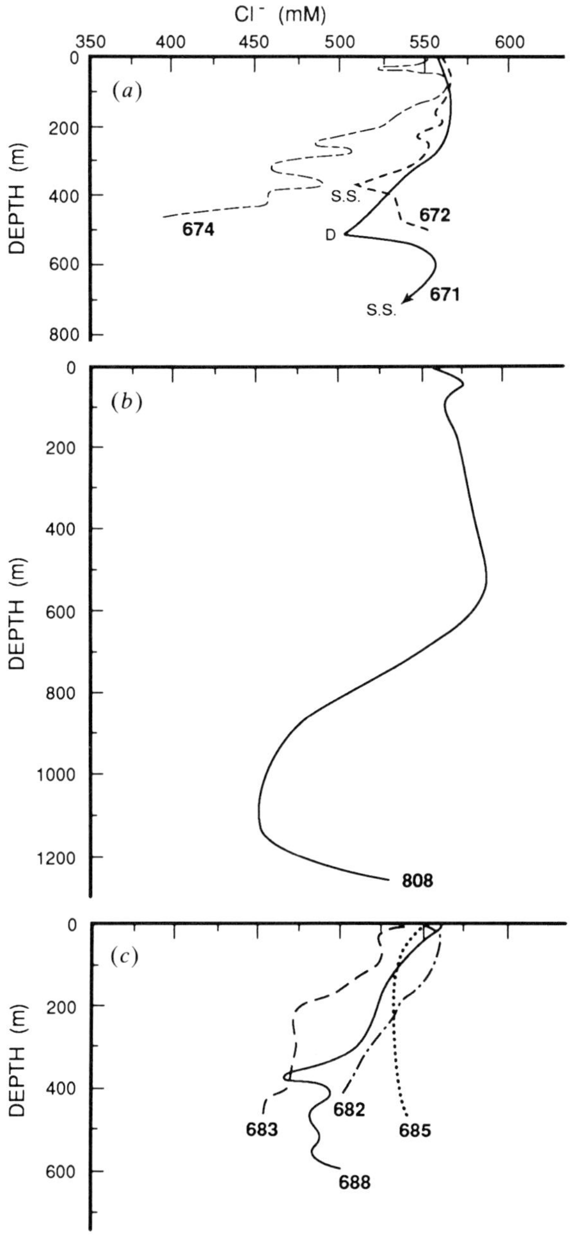

Figure 2. Depth profiles of chloride concentrations in pore fluids from the (*a*) Barbados Ridge Sites 671, 672, 674 (data from Gieskes *et al.* 1990); (*b*) Nankai Trough Site 808 (data from Taira, Hill *et al.* 1991); (*c*) Peru margin Sites 682, 683, 685, 688 (data from Kastner *et al.* 1990).

(Powers 1967; Perry & Hower 1972).

The optimal temperature range for this reaction is *ca.* 60–160 °C. These margins have geothermal gradients ranging from over 30° to 110 °C km^{-1} which correspond to a depth range of 550 m to several kilometres. Therefore, the non-steady-state features of the Cl$^-$ pore fluid profiles observed at shallower than 550 m depth in the Barbados and Peru margins (figure 2) cannot be formed in this way and must either

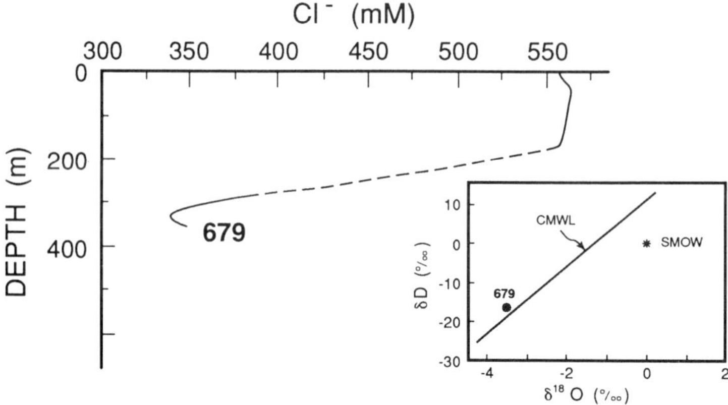

Figure 3. Depth profile of chloride concentrations in pore fluids from the Peru Margin Site 679 (data from Kastner *et al.* 1990) and a plot of the $\delta^{18}O$ and δD values of the Cl minimum fluid (339 mM) at 320 mBSF.

Table 4. *Chemical and isotopic shifts at the Barbados Ridge and Nankai Trough decollements and at the Plio-Pleistocene unconformity at* ODP *Site 688, Peru Margin*

	Barbados[a] at decollement, ODP Site 671	Nankai[b] at decollement, ODP Site 808	Peru[c] at distinct Cl^- minimum spike, ODP Site 688
Cl^-	decreases	decreases	decreases
$^{87}Sr/^{86}Sr$	decreases	increases (or no change ?)	no change
Sr^{2+}	decreases	increases	decreases
$\delta^{18}O$	increases	not available	decreases
CH_4	increases	decreases	increases
silica	increases	increases	increases
Ca^{2+}	decreases	decreases	decreases
Mg^{2+}/Ca^{2+}	increases	no change	increases

[a] Gieskes *et al.* (1990); Vrolijk *et al.* (1990).
[b] Leg 131 Shipboard Scientific Party (1990).
[c] Kastner *et al.* (1990); Elderfield *et al.* (1990).

originate from a different low Cl^- local source or be advected and/or laterally transported from other regions. Volatiles, mainly H_2O, CO_2 and CH_4 from the breakdown of clay minerals and organic matter in the subducting sediments and from hydrous silicates in the altered downgoing oceanic slab, must be migrating up and returned to the surface (Ernst 1990; Peacock 1990). Advective and/or lateral fluid flow is similarly supported by the opal-A dehydration and transformation to quartz reactions, which occur between about 30 and 40 and 80 and 100 °C (Keene 1975; Hein *et al.* 1978; Kastner 1981).

The common observation that pore fluids become progressively dilute with depth has been attributed largely to sampling artefacts or to drilling-induced pressure release that causes gas hydrate dissociation (Hesse & Harrison 1981). Indeed, gas hydrates are widespread in this environment as shown by the presence of a bottom simulating reflector (BSR) which delineates the base of the gas hydrate stability field (Kvenvolden & McMenamin 1980). Their dissociation would release both fresh water

Table 5. *Changes in oxygen and hydrogen isotopic compositions resulting from fluid–rock interactions*[a]

| | pore fluid isotopic shifts | |
reaction	$\delta^{18}O$	δD
oceanic basalt alteration		
(*a*) at $<200\ °C$, low water/rock	−	+
(*b*) at $>200\ °C$, low water/rock	+	+
hydration reactions	−	+
dehydration reactions	+	−
hydrous minerals recrystallization		
(*a*) lower temperature than precursor	−	+
(*b*) higher temperature than precursor	+	− or +[b]
clay minerals authigenesis	−	+
clay membrane ion filtration		
(*a*) filtrate	−	−
(*b*) residual fluid	+	+
meteoric water mixing	−	−
gas hydrate		
(*a*) formation	−	−
(*b*) dissociation	+	+
fluid–gas exchange reactions, with		
(*a*) CO_2	−	
(*b*) H_2S		+

[a] The table was compiled mainly from the following references: Savin & Epstein (1970); Coplen & Hanshaw (1973); Schoell (1980); Yeh (1980); Hesse & Harrison (1981); Phillips & Bentley (1987); Kvenvolden & Kastner (1990); Peacock (1990).
[b] Depending on temperature of recrystallization.

and methane. At the Peru margin, methane gas hydrates are ubiquitous and the methane source is mostly biogenic (Kvenvolden & Kastner 1990). The calculated depths of the BSRS, of 612 mBSF at Site 685 and 473 mBSF at Site 688, suggest that except for the deepest portion of the Cl^- profile at Site 688, gas hydrate dissociation could account for most of the pore fluids' Cl^- dilution seen in the figure 2*c* depth profiles. However, gas hydrates have not been observed at the Barbados drill sites nor has a BSR been identified in their vicinity although gas hydrates exist at a distance southward of the drilling region. At Site 671, the calculated BSR is at *ca.* 500 mBSF (R. Hyndman, personal communication). As seen in figure 2, low Cl^- fluids also exist below this depth and the $\delta^{13}C$ value of the methane in and below the decollement indicates a thermogenic source, implying advection from greater depths (Vrolijk *et al.* 1990). Similarly, in the Nankai Trough, at Site 808, where gas hydrates were recovered, the calculated BSR is at a considerably shallower depth, of just *ca.* 200 mBSF, than the depth range of the low Cl^- pore fluids, shown in figure 2*b* (Taira *et al.* 1991). Several additional geochemical tracers, such as oxygen and hydrogen isotopes of the pore fluids, are needed to establish the quantitative importance of gas hydrates as sources of low Cl^- fluids. The oxygen isotope profiles of figure 4*c* do not support a sole gas hydrate dissociation source, as indicated in table 5 (Hesse & Harrison 1981; Kvenvolden & Kastner 1990).

As yet the quantitative importance of the clay membrane ion filtration process in thick clay-rich sediments, originally suggested by Coplen & Hanshaw (1973), has not been evaluated. Experimental work (Haydon & Graf 1986; Phillips & Bentley 1987) indicates that the efficiency of the process is greatly pressure dependent, suggesting that in the compressional régimes of active convergent margins clay membrane ion filtration may be a more important process of low Cl^- fluid production than previously assumed.

It is important to note that the residual fluids, produced by two of the above low Cl^- fluid production processes (gas hydrate formation and clay membrane ion filtration) are 'brines', fluids of high solute concentrations and of high density. In addition, $CaCl_2$ brines form from volcanic ash and upper oceanic crust hydration reactions. Because the quantitative importance of these three processes to the overall fluid budgets in convergent margins is as yet unknown, it seems premature to evaluate the role of the residual brines in the overall hydrogeochemistry of this environment.

4. Fluid budgets

The budget of H_2O will be considered further here, the important internal sources of H_2O being: (1) porosity reduction, and (2) mineral dehydration and breakdown reactions.

Source 1. *Loss of water by porosity reduction.* If a 15% porosity reduction from 50 to 35% is assumed, in accordance with the porosity loss in the subducted sediments of the Nankai Trough (Bray & Karig 1985), 2.3×10^8 m³ H_2O would be expelled from 1 km³ sediment, or 230 m³ H_2O m⁻². The length of time over which this water is expelled differs at various margins. Assuming a 2 cm a⁻¹ subduction rate, and expulsion of water along the toe of the accretionary complex, the expected production of H_2O is 4.6 m³ a⁻¹ m⁻¹ (m³ per year per metre length of decollement or other tectonic features); 4.6 m³ H_2O a⁻¹ m⁻¹ would be expelled to the ocean. At a faster subduction rate, of 5 cm a⁻¹, 11.5 m³ H_2O a⁻¹ m⁻¹ would be expelled.

A similar calculation for 20% porosity reduction from 55 to 35% would produce 360 m³ m⁻² from 1 km³ of sediment which translates to 7.2 m³ H_2O a⁻¹ m⁻¹ for a subduction rate of 2 cm a⁻¹, and 18 m³ H_2O a⁻¹ m⁻¹ for a subduction rate of 5 cm a⁻¹.

Source 2. *Mineral dehydration and breakdown reactions.* Complete dehydration of a low porosity (30%) smectite-rich (20 wt%) sediment, with 15 wt% interlayer water in the smectite ($\rho = 2.5$ g cm⁻³), would produce 2.1×10^7 m³ H_2O from 1 km³ sediment equal to 2.1×10^{10} kg H_2O. An equivalent volume of H_2O would also be produced from a same low porosity (30%) but opal-A rich (30 wt%) sediment, with 10% water in the opal-A. This is an order of magnitude less than the 2.3×10^{11} kg H_2O produced from source 1. At a subduction rate of 2 cm a⁻¹, 4.2×10^2 kg H_2O a⁻¹ could be expelled to the ocean. The complete breakdown of the smectite to an anhydrous phase would provide an additional 0.7×10^{10} kg H_2O. In addition, based on data from the ODP on upper oceanic crust alteration and its porosity in the uppermost 1 km (Alt *et al.* 1986; Becker *et al.* 1989), a conservative value of 3 wt% volatiles, all H_2O, is assumed instead of the 2 wt% assumed by Peacock (1990). Thus a 1 km³ oceanic crust ($\rho = 3$ g cm⁻³) which contains 9×10^7 m³ H_2O could provide 9×10^{10} kg H_2O, the same order of magnitude as from smectite dehydration. At a subduction rate of 2 cm a⁻¹, 1.8 m³ H_2O a⁻¹ m⁻¹ could be expelled from an oceanic crust column of 0.02 m × 1 m × 1000 m. Similar calculations for a 5 m a⁻¹ subduction rate would provide 4.5 m³ H_2O a⁻¹ m⁻¹.

These calculations demonstrate that expulsion of water by porosity reduction is the most important internal source of volatiles in accretionary complexes. Complete smectite dehydration and oceanic crust dehydration provide an order of magnitude less volatiles.

Fluid discharge velocities of tens to over 100 m³ a⁻¹ m⁻¹ in the eastern Nankai Trough, Barbados Ridge and Cascadia (Carson *et al.* 1990; Foucher *et al.* 1990; Le Pichon *et al.* 1990 *b*) are one to two orders of magnitude greater than possible from the above internal fluid sources. Because dispersed flow is pervasive, particularly in clastic-sediment dominated margins, as manifested by the widespread occurrence of carbonate cements in Cascadia (Carson *et al.* 1990; Sample 1990), or implied by the absence of localized flow in the western Nankai Trough, Site 808 (Gieskes *et al.* 1990; Taira *et al.* 1990), this discrepancy is even larger. Localized discharge of fluids derived from large regions by non-steady-state, transient fluid flow or by the addition of large volumes of fluids from external sources, as suggested by Le Pichon *et al.* (1990 *b*), of which the most important is meteoric water, may partly explain this major discrepancy. However, hydrologic flow tends to be shallow, mostly above 1 km (Cathles 1990). The only rather clear case of long distance lateral migration of meteoric water in convergent margins is the extreme low Cl⁻ fluid observed at Site 679, Peru margin (figure 3). This site is, however, situated at just *ca.* 450 m water depth, and the Andes could have easily provided the required head. The oxygen and hydrogen isotopes of this low Cl⁻ fluid suggest a meteoric water source. The slight shift from the meteoric water line must have been caused by water–gas exchange reactions indicated in table 5. Another interesting possible solution to the problem of disparity between observed fluid fluxes and the internally available fluid, suggested by Le Pichon *et al.* (1990 *b*), is seawater mixing with a low-Cl⁻ fluid by seismic pumping or density inversion induced convection.

5. A hypothesis on the origin of the fluids

In addition to the differences in the Cl⁻ concentration depth profiles discussed already, the most distinctive features seen in the chemical and isotopic pore fluid profiles of figure 4 are: (1) the low methane concentrations in the Barbados and Nankai margins and the very high concentration in the Peru margin. This difference positively correlates with the organic C contents given in table 1; (2) the positive methane anomalies both at the Barbados decollement horizon and at the Cl⁻ minimum horizons of Peru Site 688, and the small negative anomaly at the Nankai decollement horizon; (3) a non-radiogenic low Sr concentration source, probably oceanic crust, associated with both the Barbados decollement and the distinct Cl⁻ minimum at Site 688, horizons, and a radiogenic and high Sr concentration source, most probably from clay minerals dehydration, transformation and/or breakdown at the decollement horizon at Nankai; and (4) the higher than seawater ^{87}Sr/^{86}Sr ratios of the pore fluids at Site 685, Peru margin, indicating linkage with continental crust (Elderfield *et al.* 1990; Kastner *et al.* 1990), and the apparent lack of communication between the pore fluids of adjacent thrust sheets, clearly indicated by the ^{87}Sr/^{86}Sr ratios against 1/Sr plot of Site 685 pore fluids (figure 4 *c*).

These, plus additional complexities in the pore fluids geochemical and isotopic profiles displayed in figure 4 and summarized in tables 2 and 4, together with the marked disparity between field observations of fluid flow rates at fluid discharge sites and the calculated maximum possible rates of fluid discharge from all internal sources, may be explained by the following working hypothesis.

The central fluid system consists of internally produced fluids, originally pore fluids of seawater origin. The geochemical and isotopic characteristics of this fluid system are determined by dilution through mineral dehydration reactions and by

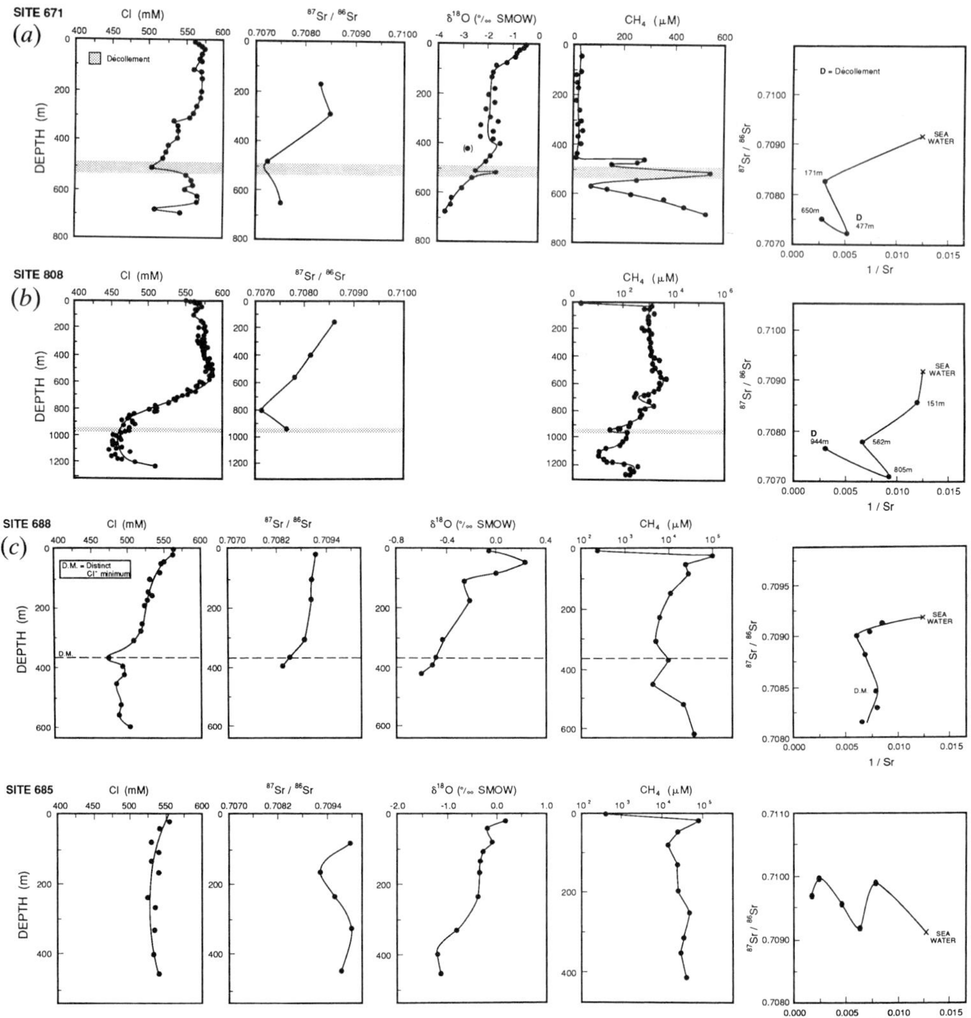

Figure 4. Depth profile of chloride and methane concentrations of $^{87}Sr/^{86}Sr$ ratios and of $\delta^{18}O$ values, and mixing relationships between $^{87}Sr/^{86}Sr$ ratios and $1/Sr^{2+}$ concentrations for (a) Site 671, Barbados Ridge (data from Gieskes et al. 1990; Vrolijk et al. 1990); (b) Site 808, Nankai Trough (data for Cl⁻ and CH₄ from Taira et al. 1991); (c) Sites 688 and 685, Peru margin (data from Kastner et al. 1990; Elderfield et al. 1990).

fluid–mineral exchange reactions plus diffusion–advection. The central fluid system is modified locally by mixing with fluids from: (1) *external sources*, e.g. meteoric water through short- or long-distance transport and possibly seawater through density inversion induced convection or seismic pumping (Le Pichon et al. 1990b). These sources may play an important role primarily in the shallower levels, within the uppermost tens to hundreds of metres, of accretionary complexes) and (2) *internal sources*, e.g. from deeper or overpressured regions of the convergent margin, transported vertically and laterally along tectonic and sedimentologic conduits, such as the decollement, faults or sand horizons. These sources may be important at all depths.

The extent of mixing with either source, whether transient or steady state, determines the intensity and nature of the geochemical and isotopic compositions at the minima and maxima horizons, observed in figure 4. The depth profiles between these 'anomalous' horizons are also affected through diffusion–advection. The minima and maxima horizons of figure 4 coincide with decollements, faults, unconformities and zones of high intergranular permeability (Gieskes *et al.* 1990; Kastner *et al.* 1990). They carry key information about the distinctive hydrology at each margin. Thus, to identify the origin of these and similar complex fluids, unique combinations of diagnostic geochemical and isotopic parameters need to be established. Clearly, as indicated in table 5, this cannot be accomplished solely by the oxygen and hydrogen isotopic compositions of the waters.

6. Diagenesis

Diagenetic reactions which are widespread in accretionary complex sediments are intimately involved in the fluid budget and fluid distribution, particularly within the central fluid system. Diagenesis significantly modifies the geochemistry and isotopic composition of the fluids as well as the porosity and permeability of the sediments and, hence, the hydrology of the complex. It also changes the mechanical properties of the sediments and even affects the thermal properties (Carson 1977; Hesse & Harrison 1981; Kulm *et al.* 1986; Ritger *et al.* 1987; Vrolijk 1987; Kastner *et al.* 1990; Kvenvolden & Kastner 1990; Sample 1990). In such rapid sedimentation régimes, diffusional communication with seawater ceases at shallow burial depths, of a maximum a few tens of metres. Therefore, the composition and reactivity of the sediments and fluid flow assume major roles in controlling the availability of solutes and consequently the nature and extent of diagenesis. The important diagenetic reactions are:

I. *Carbonate recrystallization and more importantly precipitation.* The latter reaction, which influences the physical properties of the sediments, is more likely to be limited by the availability of Ca^{2+} and/or Mg^{2+} rather than of bicarbonate, especially in margins of sediments rich in organic matter. The overall vertical distribution of the diagenetic carbonate phases is: calcite (plus Mg-calcite) kinetically precedes dolomite (or ankerite) formation. The rate of dolomite formation markedly increases with depth, with increasing Mg^{2+}/Ca^{2+} (or $Mg^{2+}+Fe^{2+}/Ca^{2+}$) ratios and bicarbonate concentrations and with decreasing sulphate concentrations. At even greater depths, coformation of dolomite and calcite or formation of only calcite may ensue. The formation of the Fe^{2+} and Mn^{2+} carbonates siderite and rhodochrosite, respectively, which depends on the availability of Fe^{2+} and Mn^{2+} relative to Ca^{2+} and Mg^{2+}, is of minor importance in most convergent margin sediments and occurs mostly at intermediate or greater depths. When dissolved (e.g. by organic or other acids) or decomposed via decarbonation, the carbonates become an important source of CO_2.

II. *Bacterial and thermal degradation of organic matter.* This is an important source of the volatiles CO_2 and CH_4 (Claypool & Kaplan 1974) as well as of some H_2O. These and the above mentioned carbonate reactions are interrelated. Furthermore, mainly CH_4, but also CO_2 and other organic volatiles, are essential for gas hydrate formation.

III. *Gas hydrate formation and dissociation.* Occurring within the uppermost 1 km of the sediment section (Kvenvolden & McMenamin 1980), gas hydrates not only modify the chemistry and oxygen and hydrogen isotopic compositions of the pore fluids (Claypool & Kaplan 1974; Hesse & Harrison 1981; Kvenvolden & Kastner 1990) but they also lead to a redistribution of the fluids involved in these reactions, in reducing the porosity and permeability of the host sediments and influencing their thermal properties. Undoubtedly, at least some of the fluids discharging at the sea-floor supporting dense benthic communities originate from gas hydrate dissociation (Kulm *et al.* 1986; Boulegue *et al.* 1987; Carson *et al.* 1990). As yet little is understood about the state of the residual 'brine' and its involvement in diagenesis and hydrology of these margins.

IV. *The dehydration and transformation of hydrous minerals, especially of clay minerals and opal-A.* The smectite to illite transformation, is one of the important reactions responsible for the evolution of low Cl^- fluids observed in all the margins studied. The physical properties of the sediments are also altered by this reaction. Based on its optimal temperature range, of *ca.* 60 to 160 °C, and dependent on the geothermal gradient, this reaction may span over a depth interval of several kilometres, starting at a burial depth of greater than 500 m. An important byproduct is dissolved silica, which is a major source of quartz cement (Perry & Hower 1972). The opal-A through opal-CT to quartz dehydration and transformation occurs between over 30 and less than 100 °C (Keene 1975; Kastner 1981) and may also, as with the clay reaction, contribute H_2O to the formation of low Cl^- fluids, over a depth interval between over 100 m and less than 2.5 km, dependent on the geothermal gradient.

These carbonate and quartz cements, forming by reactions I and IV, are extremely important phases. They include key geochemical and isotopic tracers, which could provide invaluable information about the transport of solutes and heat, by dispersed flow, during diagenesis (and metamorphism) (Bickle & McKenzie 1987).

V. *Hydrous minerals breakdown*, especially during deeper burial diagenesis and metamorphism.

VI. *Volcanic ash (and upper oceanic crust) hydration alteration to zeolites and clay minerals.* Because of the abundance of volcanic ash in convergent margin sediments these are quantitatively important hydration reactions, consuming water (plus mostly Mg^{2+} and alkalies) at low to medium temperatures and releasing it at depth, at higher temperatures, from dehydration and breakdown reactions of the authigenic zeolites and clay minerals. Dependent on the mass balance between these hydration reactions and the dehydration and transformation reactions described in IV, higher than seawater Cl^- fluids may evolve. The pore fluids of Site 808 (figure 2*b*), at under 560 m and at over 1200 m are probably good examples of such fluids.

Ion-exchange, adsorption–desorption and oxidation–reduction, and other dia-genetic reactions are of lesser importance to the fluid and geochemical budgets at convergent margins.

7. Geochemical fluxes

The volume of fluid expelled from ocean margins by subduction on a global basis has been estimated at about 1 km^3 a^{-1} (COSOD II 1987). This is similar to the values obtained in §4, considering the factor of 3 error of the COSOD II estimate. Thus, for a 30 550 km global length of subduction system, the maximum internal fluid source of 10 m^3 a^{-1} m^{-1} is equivalent to 0.3 km^3 a^{-1}. These values are very much lower than the hydrologic flow rate at margins of 100 km^3 a^{-1} estimated by COSOD II (1987). The use of values of this order taken together with data for Sr isotopic compositions and Li concentrations gives fluxes of the same order as river and hydrothermal input rates if hydrological flow is considered but of the order of 1 % of these rates for internally derived fluids (Elderfield *et al.* 1990; Martin *et al.* 1991). These estimates are typical of fluxes for several elements. This would seem to suggest that the geochemical fluxes to seawater associated with subduction processes are of minor importance to the ocean geochemical cycle but it should be remembered that the high observed fluid discharge rates have been attributed to externally derived fluids and there is, as yet, no realistic estimate of a global average flux from this source (the observed fluid discharge rates of over $100 \text{ m}^3 \text{ a}^{-1} \text{ m}^{-1}$ are equivalent to over $3 \text{ km}^3 \text{ a}^{-1}$). The high volume transport associated with such fluid sources makes them extremely important to geochemical fluxes and a detailed documentation of the hydrology of the uppermost tens to hundreds of metres of accretionary complexes is a vital prerequisite to a better definition of their importance to the oceanic budget of elements and isotopes.

We greatly appreciate the Legs 112 and 131 shipboard scientists for most stimulating dialogues, the technicians for their assistance, and the staff and drilling crews aboard the JOIDES *Resolution* for their cooperation. We are especially grateful to B. P. Holliday for help in the Leg 112 Sr isotopic analyses at Cambridge and to Dr G. Lugmair at Scripps Institution of Oceanography for the Leg 131 isotopic analysis, and value the discussions with Dr J. Natland of Scripps, on oceanic crust alteration. This research was supported through the assistance of the Ocean Drilling Program and by grants from NSF OCE (NSFOCE88-12329), United States Scientific Advisory Committee (USSAC) (M. K.) and by the NERC Grant GST/02/250 (H. E.). This is Cambridge Earth Sciences Series Contribution no. 1904.

References

Alt, J. C., Honnorez, J., Laverne, C. & Emmermann, R. 1986 Hydrothermal alteration of a 1 km section through the upper oceanic crust, Deep Sea Drilling Project Hole 504B: mineralogy, chemistry and evolution of seawater–basalt interactions. *J. geophys. Res.* **91**, 10301–10336.

Bangs, N. L. B., Westbrook, G. K., Ladd, J. W. & Buhl, P. 1990 Seismic velocities from Barbados Ridge complex: Indicators of high pore fluid pressures in an accretionary complex. *J. geophys. Res.* **95**, 8767–8782.

Becker, K. *et al.* 1989 Drilling deep into young oceanic crust, Hole 504B, Costa Rica Rift. *Rev. Geophys.* **27**, 71–102.

Bickle, M. J. & McKenzie, D. 1987 The transport of heat and matter by fluids during metamorphism. *Contrib. Mineral. Petrol.* **95**, 384–392.

Boulègue, J., Iiyama, J. T., Charleu, J. L. & Sewab, J. 1987 Nankai Trough, Japan Trench and Kuril trench: Geochemistry of fluid samples by submersible 'Nautile'. *Earth planet. Sci. Lett.* **83**, 362–375.

Bray, C. J. & Karig, E. D. 1985 Porosity of sediments in accretionary prisms and some implications for dewatering processes. *J. geophys. Res.* **90**, 768–778.

Brown, K. & Westbrook, G. K. 1988 Mud diapirism and subcretion in the Barbados Ridge accretionary complex: the role of fluids in accretionary processes. *Tectonics* **7**, 613–640.

Carson, B. 1977 Tectonically induced deformation of deep-sea sediments off Washington and northern Oregon: mechanical consolidation. *Mar. Geol.* **24**, 281–307.

Carson, B., Suess, E. & Strasser, J. C. 1990 Fluid flow and mass flux determinations at vent sites on the Cascadia margin accretionary prism. *J. geophys. Res.* **95**, 8891–8897.

Cathles, L. M. 1990 Scales and effects of fluid flow in the upper crust. *Science, Wash.* **248**, 323–329.

Claypool, G. E. & Kaplan, I. R. 1974 The origin and distribution of methane in marine sediments. In *Natural gases in marine sediments* (ed. I. R. Kaplan), pp. 91–140. New York: Plenum.

Coplen, T. B. & Hanshaw, B. B. 1973 Ultrafiltration by a compacted clay membrane. I. Oxygen and hydrogen isotopic fractionation. *Geochim. cosmochim. Acta* **37**, 2295–2310.

COSOD II 1987 Report of the Second Conference on Scientific Ocean Drilling. In COSOD II, Strasbourg, France: European Science Foundation. (142 pages.)

Davis, E. G., Hyndman, R. D. & Villinger, H. 1990 Rates of fluid expulsion across the northern

Cascadia accretionary prism: constraints from new heat flow and multichannel seismic reflection data. *J. geophys. Res.* **95**, 8861–8889.

Elderfield, H., Kastner, M. & Martin, J. B. 1990 Composition and sources of fluids in sediments of the Peru subduction zone. *J. geophys. Res.* **95**, 8811–8828.

Ernst, W. G. 1990 Thermobarometric and fluid expulsion history of subduction zones. *J. geophys. Res.* **95**, 9047–9053.

Fisher, A. T. & Hounslow, M. W. 1990 Transient fluid flow through the toe of the Barbados accretionary complex: constraints from Ocean Drilling Program Leg 110 heat flow studies and simple models. *J. geophys. Res.* **95**, 8845–8858.

Foucher, J. P., LePichon, X., Lallemant, S., Hobart, M. A., Henry, P., Benedetti, M., Westbrook, G. K. & Langseth, M. G. 1990 Heat flow, tectonics and fluid circulation at the toe of the Barbados accretionary prism. *J. geophys. Res.* **95**, 8851–8867.

Fowler, S. R., White, R. S. & Louden, K. E. 1985 Sediment dewatering in the Makran accretionary prism. *Earth planet. Sci. Lett.* **75**, 427–438.

Gamo, T., Gieskes, J. M. & Kastner, M. 1991 Geochemistry of Nankai Trough pore fluids, ODP Site 808. In *Ocean Drilling Project*, vol. 131B (ed. A. Taira *et al.*). (In preparation.)

Gieskes, J. M., Blanc, G., Vrolijk, P., Elderfield, H. & Barnes, R. 1990 Interstitial water chemistry-major constituents. In *Proc. ODP, Sci. Results*, vol. 110B (ed. J. C. Moore *et al.*), pp. 155–178. College Station, TX (Ocean Drilling Program).

Gieskes, J. M., Gamo, T. & Kastner, M. 1991 Geochemistry of Nankai Trough pore fluids, ODP Site 808. In *Ocean Drilling Project*, vol. 131B (ed. A. Taira *et al.*). (In preparation.)

Haydon, P. R. & Graf, D. L. 1986 Studies of smectite membrane behavior: temperature dependence, 20–180 °C. *Geochim. cosmochim. Acta* **48**, 723–751.

Hein, J. R., Scholl, D. W., Barron, J. A., Jones, M. G. & Mille, J. 1978 Diagenesis of late Cenozoic diatomaceous deposits and formation of bottom simulating reflector in Southern Bering Sea. *Sedimentology* **25**, 155–181.

Hesse, R. & Harrison, W. E. 1981 Gas hydrates (clathrates) causing pore-water freshening and oxygen isotope fractionation in deep-water sedimentary sections of terrigenous continental margins. *Earth planet. Sci. Lett.* **55**, 453–462.

Kastner, M. 1981 Authigenic silicates in deep-sea sediments: formation and diagenesis. In *The sea*, vol. 7 (ed. C. Emiliani), pp. 915–980. Wiley Interscience.

Kastner, M., Elderfield, H., Martin, J. B., Suess, E., Kvenvolden, K. A. & Garrison, R. E. 1990 Diagenesis and interstitial-water chemistry at the Peruvian continental margin – major constituents and strontium isotopes. In *Proc. ODP, Sci. Results*, vol. 112B (ed. E. Suess *et al.*), pp. 413–440. College Station, TX (Ocean Drilling Program).

Kastner, M., Gamo, T. & Gieskes, J. M. 1991 Geochemistry of Nankai Trough pore fluids, ODP Site 808. In *Ocean Drilling Project*, vol. 131B (ed. A. Taira *et al.*). (In preparation.)

Keene, J. B. 1975 Cherts and porcelanites from the north Pacific, Deep Sea Drilling Project leg 32. In *Initial Reports of the Deep Sea Drilling Project*, vol. 32 (ed. R. L. Larson *et al.*), pp. 421–507.

Kemp, A. E. S. 1990 Fluid flow in 'vein structures' in Peru forearc basins: evidence from backscattered electron microscope studies. In *Proc. ODP, Sci. Results*, vol. 112B (ed. E. Suess *et al.*), pp. 33–42. College Station, TX (Ocean Drilling Program).

Kulm, L. D. *et al.* 1986 Oregon subduction zone: venting, fauna and carbonates. *Science, Wash.* **231**, 561–566.

Kulm, L. D. & Suess, E. 1990 Relationship between carbonate deposits and fluid venting: Oregon accretionary prism. *J. geophys. Res.* **95**, 8891–8915.

Kvenvolden, K. A. & Kastner, M. 1990 Gas hydrates of the Peruvian outer continental margin. In *Proc. ODP, Sci. Results*, vol. 112B (ed. E. Suess *et al.*), pp. 51–526. College Station, TX (Ocean Drilling Program).

Kvenvolden, K. A. & McMenamin, M. K. 1980 Hydrates of natural gas: a review of their geological occurrence. *U.S. Geol. Survey Circ.* **815**, 11 pp.

Langseth, M. G., Westbrook, G. K. & Hobart, M. A. 1988 Geophysical survey of a mud volcano seaward of the Barbados Ridge accretionary complex. *J. geophys. Res.* **93**, 1041–1061.

Leg 131 Shipboard Scientific Party 1990 A new thrust at accretion. *Nature* **347**, 228–229.

Le Pichon, X., Foucher, J.-P., Boulègue, J., Henry, P., Lallemant, S., Benedetti, M., Avedik, F. & Mariotti, A. 1990a Mud volcano field seaward of the Barbados accretionary complex: a submersible survey. *J. geophys. Res.* **95**, 8931–8943.

Le Pichon, X., Henry, P. & Lallement, S. 1990b Water flow in the Barbados accretionary complex. *J. geophys. Res.* **95**, 8945–8967.

Lindsley-Griffin, N., Kemp, A. & Swartz, J. F. 1990 Vein structures of the Peru margin, Leg 112. In *Proc. ODP, Sci. Results*, vol. 112B (ed. E. Suess *et al.*), pp. 3–16. College Station, TX (Ocean Drilling Program).

Martin, J. B., Kastner, M. & Elderfield, H. 1991 Lithium: sources in pore fluids of Peru slope sediments and implications for oceanic fluxes. *Mar. Geol.* (In the press.)

Peacock, S. M. 1990 Fluid processes in subduction zones. *Science, Wash.* **248**, 321–337.

Perry, E. A. & Hower, J. D. 1972 Late-stage dehydration in deeply buried pelitic sediments. *Bull. Am. Ass. Petrol. Geol.* **56**, 2013–2021.

Phillips, F. M. & Bentley, H. W. 1987 Isotopic fractionation during ion filtration. I. Theory. *Geochim. cosmochim. Acta* **51**, 683–695.

Powers, M. C. 1967 Fluid-release mechanisms in compacting marine mud rocks and their importance in oil exploration. *Bull. Am. Ass. Petrol. Geol.* **51**, 1240–1254.

Ritger, S., Carson, B. & Suess, E. 1987 Methane-derived authigenic carbonates formed by subduction-induced pore-water expulsion along Oregon/Washington margin. *Geol. Soc. Am. Bull.* **98**, 147–156.

Sample, J. C. 1990 The effect of carbonate cementation of underthrust sediments on deformation styles during underplating. *J. geophys. Res.* **95**, 9111–9121.

Savin, S. M. & Epstein, S. 1970 The oxygen and hydrogen isotope geochemistry of clay minerals. *Geochim. cosmochim. Acta* **34**, 25–42.

Schoell, M. 1980 The hydrogen and carbon isotopic composition of methane from natural gases of various sources. *Geochim. cosmochim. Acta* **44**, 641–661.

Suess, E. *et al.* 1988 In *Proc. Ocean Drilling Project, Initial Reports*, Leg 112, vol. A. U.S. Government Printing Office, College Station, TX.

Taira, A. *et al.* 1991 Site 808. In *Ocean Drilling Project*, vol. 131A (ed. A. Taira *et al.*). U.S. Government Printing Office, Washington, D.C.

von Huene, R. 1984 Tectonic processes along the front of modern convergent margins – research of the past decade. *A. Rev. Earth planet. Sci.* **12**, 351–381.

Vrolijk, P. T. 1987 Tectonically-driven fluid flow in the Kodiak accretionary complex, Alaska. *Geology* **15**, 466–469.

Vrolijk, P., Chambers, S. R., Gieskes, J. M. & O'Neil, J. R. 1990 Stable isotope ratios of interstitial fluids from the Northern Barbados accretionary prism, ODP Leg 110. In *Proc. ODP, Sci. Results*, vol. 110B (ed. J. C. Moore *et al.*), pp. 181–205. College Station, TX (Ocean Drilling Program).

Westbrook, G. K. & Smith, M. J. 1983 Long decollements and mud volcanoes: evidence from the Barbados Ridge complex for the role of high pore fluid pressure in the development of an accretionary complex. *Geology* **11**, 271–283.

Yamano, M. & Uyeda, S. 1990 Heat-flow studies in the Peru trench subduction zones. In *Proc. ODP, Sci. Results*, vol. 112B (ed. E. Suess *et al.*), pp. 653–662. College Station, TX (Ocean Drilling Program).

Yeh, H.-W. 1980 D/H ratios and late-stage dehydration of shales during burial. *Geochim. cosmochim. Acta* **44**, 341–352.

The microstructural evolution of fluid flow paths in semi-lithified sediments from subduction complexes

By R. J. KNIPE[1], S. M. AGAR[2] AND D. J. PRIOR[3]

[1] Department of Earth Sciences, The University of Leeds, Leeds LS2 9JT, U.K.
[2] Department of Geological Sciences, Northwestern University, Evanston, Illinois 60208, U.S.A.
[3] Department of Earth Sciences, Liverpool University, Liverpool, U.K.

The characterization of the fluid migration pathways is essential to an understanding of the hydrodynamics of accretionary wedges located in the high levels of subduction zones. Microstructural analysis of the fluid flow pathways can provide constraints on (a) the porosity/permeability evolution of different migration pathways, (b) the deformation mechanisms histories which promote different fluid flow characteristics, (c) the interconnectivity of migration pathways. Three important fluid migration pathways can be recognized in accretionary wedges; bulk flow through the sediment, localized flow through fracture networks and localized flow along fault zones. The recent progress made towards understanding the microfabric evolution in each of these pathways in semi-lithified sediments is reviewed. The analysis highlights the role of transient deformation events in these materials and the need to characterize the detailed deformation mechanism paths and syn-deformational (dynamic) properties associated with these events.

---

1. Introduction

An important aspect of understanding the fluid flow in any sediment/rock system is the assessment of the mechanical and physical processes which control the behaviour of potential fluid flow paths. In accretionary wedges the understanding of fluid flow in sediments is critical to unravelling the wedge dynamics (Moore 1989; Langseth & Moore 1990). However, such analysis is complicated by the range of burial, compaction and deformation histories possible in this tectonic situation where the sediments experience large and rapid changes in their physical (e.g. porosity and permeability) and mechanical properties (Shephard & Bryant 1983; Bray & Karig 1985; Karig 1986, 1990). Recent research has identified the need for integrated studies which assess the detailed interaction between dewatering, lithification and deformation processes during the burial and incorporation of sediment into accretionary wedges.

Fluid flow depends upon the complex interaction between permeability and fluid potential gradients. Permeability is controlled by the microstructure and pore geometry of the sediment, hence knowledge of the microstructural evolution of sediments during burial and deformation can help constrain on the physical and mechanical properties needed to model the fluid flow. Here we review the microstructural evolution of fluid flow paths in sediments undergoing lithification, the interaction between deformation processes and fluid flow and assess the role of microstructural analysis in future fluid flow. The discussion is based on analysis of

Deep Sea or Ocean Drilling Program (DSDP/ODP) cores from active margins and of fabrics preserved in on-shore, exhumed subduction complexes where the depth of burial did not exceed 1–2 km.

2. Deformation processes and their recognition in partly lithified sediments

Data on the deformation behaviour of unlithified and partly lithified sediments is based upon experimental deformation programmes which have applied the methods developed for assessing the mechanical behaviour of soils (Lambe & Whitman 1969; Jones & Addis 1986; Maltman 1987; Karig 1990). These programmes have provided important data on the behaviour of 'soft' sediments during different stress histories and have highlighted the influence of the mean effective stress path, the differential stress magnitude, porosity and lithology in controlling the mechanical response.

The assessment of mechanical behaviour during natural deformation events in accretionary wedges requires experimental programmes designed specifically to reproduce the natural deformation conditions (Karig 1990) and confirmation that similar deformation processes have operated in the experimental and natural situations. Such a programme of comparison is still in its infancy as experiments using natural sediments are rare, and because the microstructural characterization of naturally and experimentally produced fabrics has only recently gained momentum (Knipe 1986a, b; Bennett & Hulbert 1987; Agar et al. 1989; Kemp 1990; Prior & Behrmann 1990). Recent developments in both the preparation techniques (Swartz & Lindsley-Griffin 1990) and in the improved performance of both transmission electron microscopes (TEMs) and scanning electron microscopes (SEMs) have considerably helped this research. The development of more efficient and higher resolution backscattered electron (BSE) images in SEMs (Lloyd 1985; Agar et al. 1989) have been particularly important. Backscattered images provide contrast between grains of different composition and when combined with careful preparation of polished specimens allow quantification of the microfabric elements (Prior & Behrmann 1990). The resolution of such images now provides a powerful companion to the more traditional TEM analysis (Smart & Tovey 1982; Knipe 1986a, b).

The detailed grain-scale deformation processes operating in semi-lithified sediments are still poorly understood. Thus while the general operation of particulate flow by grain-boundary (frictional) sliding is assumed to be the main mechanism of deformation (Knipe 1986c, 1989), whether the process is one of independent particulate flow (Borradaile 1981), where no grain deformation is involved, or whether dependent particulate flow, where grain (clay plate) deformation controls the strain accommodation in the aggregate needs to be resolved.

3. Deformation and fluid flow paths in accretionary prisms

Fluid flow paths available in the accretionary wedges include: (a) bulk flow through the grain framework of sediments, (b) flow through fracture networks (c) flow along fault arrays. The deformation processes which accompany the evolution of each of these paths will affect the fabric which develops and the fluid flow characteristics.

(a) Bulk flow and compaction fabrics

The pores in the grain framework of a sediment undergoing burial compaction and lithification provide a fluid flow pathway. The migration path characteristics are

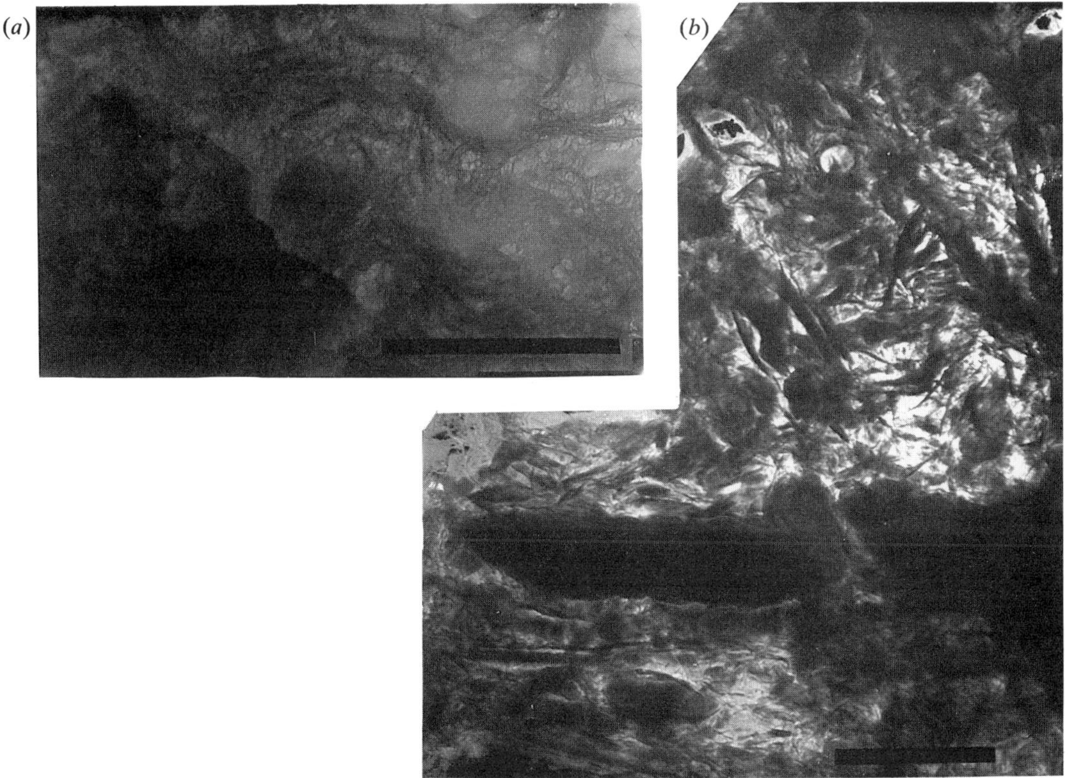

Figure 1. (*a*) Compaction fabric composed of domains of phyllosilicates aligned parallel to bedding and domains with more random and open grain framework. Scale bar 1 μm, TEM micrograph. Specimen from DSDP Leg 60 459B 46cc. (*b*) Compaction fabric containing wide domains of phyllosilicate aligned parallel to bedding (lower part of plate) and domains with a more open framework. Note the development of alignment domains at a high angle to bedding in upper part of plate. See text for discussion. Scale bar 5 μm TEM micrograph. Specimen from DSDP Leg 57.439.8.2.

known to change with time as porosity and permeability are reduced with time during burial (Bray & Karig 1985; Shephard & Byant 1983; Karig 1990). Grain collapse and cementation are known to be involved but the detailed mechanisms and the interaction of fabric collapse with fluid flow and expulsion are less well defined.

The microfabrics formed in fine-grained sediments at different stages of compaction are complex, often very heterogeneous and composed of a relatively rigid framework of grain aggregates or domains connected by weak links or chains (see review by Bennett & Hulbert 1987). The large variations in the microfabric geometries present create domains and chains with very different strengths. The complexity of the fabrics developed during compaction are illustrated in figure 1*a*, *b*. Both parts show the development of localized domains of aligned grains, in figure 1*a* they are subparallel to the mesoscopic bedding. If the alignment domains are considered to be collapse zones then the process of compaction will be heterogeneous and appears to arise from repeated, localized and transient deformation events. It is the frequency, distribution and size of these events which will control the porosity and permeability evolution. At present there is little information on these events. Analysis of the specimen shown in figure 1*a* suggests that in this case the collapse domains may involve volume losses of 25–50 μm^3 per domain. To assess the fluid flow which may arise from such compaction some estimate of the rate of collapse in the domains is needed. Such rates are unknown at present as can be illustrated by a simple calculation of the likely range of volumetric strain-rates involved. Assuming that the porosity changes in this fine-grained sediment decreases by *ca.* 1 % in *ca.* 1 Ma then the average volumetric strain rate is *ca.* 3×10^{-16} s^{-1}. However, if the volume loss per event is *ca.* 5–50 μm^3 (based on the size and density of grains in the domains shown in figure 1*a*) and events are sequential then the local strain-rate within the collapse domain may be at least 10^{-5}–10^{-6} s^{-1}. Without more detailed information on the distribution and rate of collapse, processes associated with compaction, modelling of the fluid expulsion rates and permeability properties remain incomplete.

Figure 1*b* shows a more complex microfabric where in addition to the development of domains of aligned grains parallel to the mesoscopic bedding the domain with the more open grain framework has additional alignment zones at a high angle to the bedding. The alignment zones mark domains of reduced porosity and are permeability barriers which will induce a permeability anisotropy in the bedding plane. This specimen is from DSDP Leg 57 (Japan Trench) where down slope movement and extensional faulting is present in the core (Knipe 1986*b*) and suggests that the alignment zones may represent microscopic normal faults. The compaction processes and permeability characteristics of the sediment may be dependent upon evolution of the slope instabilities and the extensional faulting as well as on the burial history.

(b) *Flow through fracture networks*

Recent work on DSDP/ODP cores and onshore studies of accretionary prisms shows that a range of fracture types may be present. At shallow depths (less than 500 m) mud-filled veins are a common feature in fine-grained slope sediments (Lundberg & Moore 1986; Knipe 1986*b*; Kemp 1990; Pickering *et al.* 1990) while web structures (anastomosing networks of low displacement, cataclastic shear zones) are common in silts/sands (reviewed in Lundberg & Moore 1986). Carbonate filled vein arrays and open fracture networks have also been reported from some active margins (Brown & Behrmann 1990; Thornburg & Suess 1990). Studies of older on-shore prisms have also revealed the ubiquitous presence of fracture systems (Agar 1990). The microstructural evolution of the mud-filled veins and web structures and their involvement in fluid flow in the shallow sections of accretionary prisms is reviewed below.

(i) *Mud-filled vein structures in fine grained sediments*

Mud-filled vein arrays are composed of sets of planar/curviplanar, dark, clay rich structures usually less than 10 cm in length and less than 1 cm wide. These structures appear to represent a common mode of deformation in partly lithified slope sediments and occur in arrays (regular or anastomosing sets) orientated at a high angle to bedding and associated with bedding parallel movement or slip zones (Lundberg & Moore 1986). They have been described from a number of the active margins drilled during the ODP (see Lundberg & Moore 1986; Knipe 1986*b*; Lindsley-Griffin *et al.* 1990 for reviews) and from onshore studies in SE Central Japan (Pickering *et al.* 1990). Vein structures do not represent simple dilational features as they contain domains of external material and microfossils, and show evidence of new mineral growth (Knipe 1986*b*; Pickering *et al.* 1990). Two main models for the vein evolution involving a sequence of disaggregation/dilation-displacement and

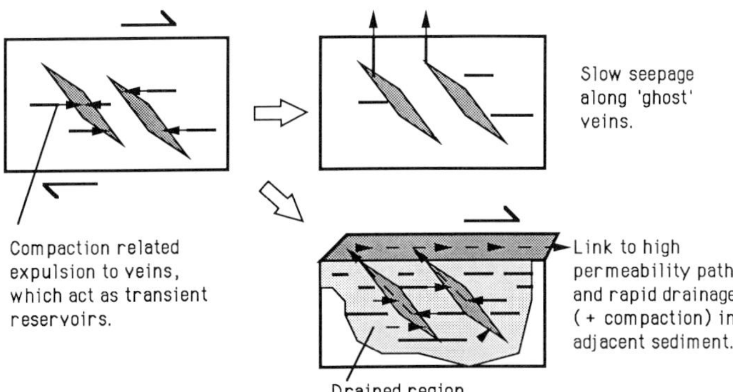

Figure 2. Review of possible interrelationships between fluid expulsion and vein evolution. The fluid ingress associated with the vein dilation may be linked to the collapse and compaction of the fabric adjacent to the vein. Fluid expulsion from the vein may either be by slow fluid flow along 'ghost' veins or by the propagation and linking of the veins to a more extensive migration pathway. Such a linking processes may induce a more extensive collapse and compaction in the adjacent sediment.

collapse were presented by Knipe (1986*b*): Model A, where the vein structures arise from fluid pressure exceeding the cohesion of the sediment and the alignment of clays within parts of the vein as the result of either rapid fluid flow or a late stage porosity collapse. Model B, where disaggregation producing the vein was associated with a shear or tensional failure not related to over-pressure but arising from slope instabilities or tectonic events. Veins described by Knipe (1986*b*) and Ritger (1985) did not show evidence of rapid dewatering. However, Kemp (1990) has reported evidence of upward fluid and mass transport in veins from Peru indicating that the model A is applicable in some cases.

The extensional geometry, evidence of disaggregation, dilation and mineral growth in veins all support the idea that fluid flow into the vein structures has occurred. The veins probably acted as small fluid reservoirs developed during deformation. There are two important aspects of vein evolution and fluid flow which require detailed assessment. Firstly, the affects of fluid movement into veins, as the disaggregation and dilation may induce enhanced compaction in the adjacent sediment (figure 2). The second important aspect of the vein evolution is the fluid escape from the veins. This escape may occur by one of two processes; (*a*) slow dewatering during continued compaction and cementation of the vein (i.e. the vein behaves in a similar fashion to the adjacent sediment), (*b*) fluid expulsion along localized channel-ways. Where vein arrays have propagated into, and are linked to, a more extensive fracture/fault network or more permeable horizons then dewatering may be rapid and a second wave of enhanced compaction may have been induced in the (over-pressured?) horizon containing the initial veins (figure 2). The magnitude of water pressure gradient in such zones and its decay rate (a function of the permeability) will control the amount of fluid released during such transient events. There are many examples of veins either connected to more extensive movement/slip surfaces or terminating at slightly different lithologies (see Lundberg & Moore 1986; Knipe 1986*b*). This simple separation of fluid expulsion implies that isolated veins with no obvious connection to high permeability pathways are confined to a slow diffuse dewatering history. However, the identification of 'ghost' veins (marked by regular to irregular networks containing a concentration of fine-

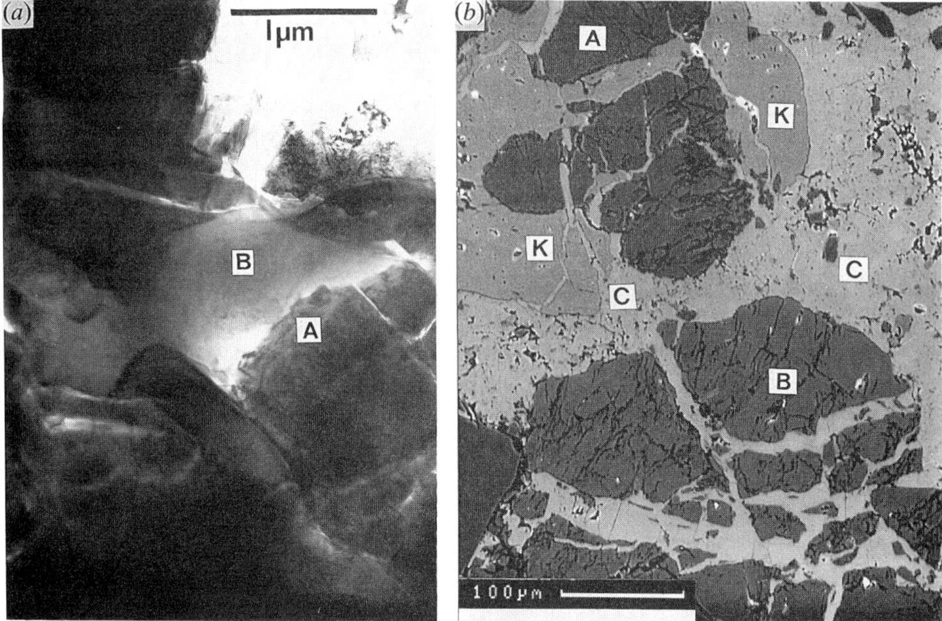

Figure 3. (*a*) Detail of multiple cementation in a web. An early pure calcite cement, A, has been fractured and a new Fe–Mn calcite cement, B, precipitated. Note the more complex defect structure in the older, deformed cement. TEM micrograph of Web from *Alvin* 2948-5V1. (*b*) Multiple fracture and cement events within a web structure. An early event has fractured Grain A and then sealed it with K-feldspar growth. K. A later event associated with calcite cementation has fractured both the original grain and the first cement. BSE image of specimen from Leg 66 493-1 100–105 cm.

grained material but not associated with a detectable modification of the sediment fabric) by Kemp (1990) during BSE analysis of sediments from the Peru forearc is extremely important as it suggests that localized fluid flow is far more extensive in sediments than previously considered.

(ii) *Web structures in silts and sands*

'Web' structures are arrays of irregular and localized cataclasite zones which commonly develop in partly cemented sandstones. The individual deformation zones are up to a few millimetres wide, usually show small displacements and form complex, intersecting arrays where the spacing is on a centimetre scale. (See reviews in Lundberg & Moore (1986) and Lucas & Moore (1986)). The zones are characterized by grain size reduction, fracture and brecciation. Lucas & Moore (1986) suggested that the process is associated with repeated strain hardening events linked to fluctuations in the pore pressure within the fractured material. Although webs are considered as important fluid migration pathways the processes involved in their evolution are poorly understood. Figure 3*a*, *b* presents evidence that the deformation in these structures is episodic and involves the introduction of fluids of different compositions. Both micrographs show the growth of cements of different compositions in dilation sites developed by different fracture events. The specimen shown in figure 3*b* is a web from a fault zone known to be an active vent and recovered by an *Alvin* dive in the frontal part of the Cascadia accretionary prism, Oregon (see Moore *et al.* (1990) for tectonic setting). The redistribution of deformation and fluid flow within the web network indicated by the multiple cements also emphasizes that different segments of the array may be active at different times during the array development and that fluid flow at any one time, or during a single deformation event, is not along all of the finite fractures preserved. This conclusion is important as the finite fracture density can not be used to estimate fluid fluxes. In addition the preferential growth of cements in and adjacent to the fracture zones leads to a change in the properties of the fracture from a transient high permeability pathway to a permeability barrier or seal.

(c) *Faulting and fluid flow*

The interaction between fluid flow and faults is well established (Carter *et al.* 1990; Sibson 1981, 1990) and a direct link between fluid flow and fault zones at active margins has been established by (*a*) the identification of biological communities and vents near faults (Suess *et al.* 1985; Carson *et al.* 1990; Moore *et al.* 1990), (*b*) the recognition of geochemical anomalies and thermal anomalies in fault zone fluids (Gieskes *et al.* 1990) and (*c*) the preferential growth of new mineral phases along fault zones (Knipe 1986*a*; Schoonmaker 1986). These associations render the assessment of the permeability properties of faults developed in semi-lithified sediment critical as these correspond to the zones where fluid expulsion is implied by calculations of heat flow and geochemical anomalies (Le Pichon *et al.* this symposium). Such an assessment requires an understanding of the mechanical properties and the relationship between fault zone fabric evolution and permeability characteristics during deformation (i.e. the dynamic properties). Experimental deformation studies of semi-lithified sediments have provided important data on the bulk mechanical behaviour of sediments with different initial physical properties and microstructural studies have already provided an insight into the deformation fabrics associated with faults.

The faults from active accretionary wedges which have been sampled by the DSDP/ODP and *Alvin* submersible operations provide an opportunity to study fabric development associated with faults in known tectonic locations within active margins where hydrogeological studies are also possible. However, only a small number of major fault zones have been sampled. Notable examples are the Barbados (Legs 78A and 110) and Nankai (Leg 131) wedges. In both cases the decollement associated with accretion was penetrated in addition to higher level faults present in the prisms. The mesoscale geometry associated with such faults range from arrays of thin (*ca.* 1 mm wide) zones to major zones (greater than 20 m wide) of scaly fabrics (defined by a mesoscopic set of anastomosing, polished and slickensided surfaces), breccias and stratal disruption/melange) (Lundberg & Moore 1986; Brown & Behrmann 1990; Lindsley-Griffin *et al.* 1990). The faults sampled exhibit a range of mesoscale fabrics and no single structure occurs on all faults.

Microstructural studies of the faults present in the semi-lithified sediments from active margins have revealed a large variety of deformation fabrics preserved. Many of the faults developed in clay rich sediments are composed of anastomosing micro-movement domains of finite porosity collapse enclosing domains/lenses of less well-aligned material (Knipe 1986*a*; Moore *et al.* 1986). The exact geometry and the spacing of the domains, varies from microscopic (less than 100 μm) in the case of mesoscopically discrete fault zones (see Knipe 1986*a*) to millimetre scale domains present in some scaly fabrics (see Moore *et al.* 1986). The deformation mechanisms involved in the fault zones are considered to be disaggregation and particulate flow

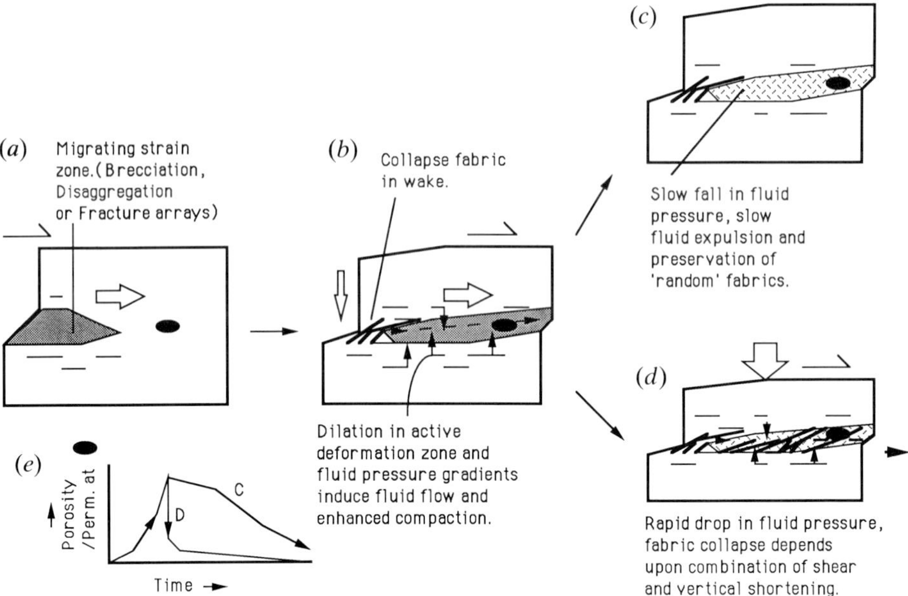

Figure 4. Review of the possible changes in the fabric and fluid flow during the migration of a strain wave in a fault zone. The diagram emphasizes the dynamic nature of the fabric, porosity and permeability associated with the migration of a strain wave and the potential influence of the decay rate of fluid pressure gradients on fabric stability and preservation.

with and without fracturing. The presence of numerous localized movement (alignment) zones suggests that new movement zones were generated during the deformation because existing ones were not able to accommodate the continuing deformation (Knipe 1986*a*). This situation may have developed either because of a hardening due to the increased coefficient of friction in alignment domains, a decrease in pore pressure or a reorientation of the slip surface relative to the stress field (Moore *et al.* 1986) or because the existing active deformation zones were incapable of accommodating an increasing strain/displacement rate (Knipe 1986*a*).

Not all the samples of fault zones studied from active margins exhibit the 'simple' two domain microfabric described above. Prior & Behrmann (1990) have reported that specimens from the decollement fault zone from Barbados (Leg 110) with a scaly clay appearance in hand specimen do not contain microdomains of aligned grains but are composed of randomly arranged aggregates less than 6 μm in size. These observations are important as they suggest that fault domains with random microfabrics do not always represent low-strain domains preserving the initial or external fabric; they may represent high strain zones where a pervasive disaggregation fabric is preserved. Prior & Behrmann (1990) emphasize the need to consider deformation by particulate flow where the grain boundary sliding is between aggregates not individual grains.

There are three aspects of the microfabric evolution in fault zones which are important to the future assessment of fluid flow (see figure 4).

(i) *Deformation mechanism paths and the transient changes in the physical properties associated with displacement*

The exact history of deformation processes, porosity, permeability and straining within the faults will control the fluid flow. Although the deformation mechanisms involved in both the experimental and the natural deformation are considered to be particulate flow and grain boundary sliding, the controls on the activity of (i) pervasive disaggregation and sliding by individual grains, or (ii) disaggregation by sliding between aggregates of grains, or (iii) by localized slip on discrete slip planes are unknown. The models of fault zone behaviour presented so far have concentrated on the fabrics where localized slip has occurred at some stage of the deformation history. Despite the lack of knowledge on the exact processes involved in fault zone fabric development the potential importance of cyclic and transient changes in the physical properties is recognized (see Karig 1986; Knipe 1986*a–c*; Moore *et al.* 1986; Moore 1989). Evaluation of the fluid flow during faulting requires knowledge of the dynamic properties (porosity and permeability) changes associated with the deformation events.

(ii) *Fabric stability*

The permeability characteristics of a material during deformation will depend upon the pore geometry and connectivity, which will be controlled by the mechanical response and deformation processes operating in the material. The changes in the effective stress, fluid pressure gradients, dilation, strain path and strain rate which accompany deformation in poorly consolidated material are likely to induce a series of changes in the microfabric and thus it is unlikely that one microfabric will be characteristic of an entire deformation event. Rather a series of microfabrics will develop during the deformation event and be modified during the late stages and/or between deformation events (Knipe 1986*a*). The fluid flow characteristics during the deformation event will be dependent on three interrelated factors of the deformation and fabric evolution.

1. *The dilation patterns induced during deformation.* A range of dilation patterns may be induced during deformation in the semi-consolidated material. The production of an interconnected fracture array which allows rapid expulsion of fluid during the production of scaly clay has been suggested by Moore (1989). At the other end of the spectrum is the pervasive disaggregation of the aggregate and dilation of the material in the fault zone during displacement under high fluid pressure, which would also enhance and increase permeability. Between these two end-member responses will be more complex patterns where dilation is linked to localized strain waves migrating through the material. The fluid flow in this case will depend upon the volume, dilation and migration velocity of these strain waves. Collapse of the material may occur in the wake of these migrating waves to produce domains of clay alignment (see figure 4). The migration of strain or creep events in over-pressured fine-grained gouges is known from the San Andres fault zones (see review by Wesson 1988) and is likely to be operational in the high level fault zones in accretionary wedges.

2. *The distribution of deformation within the fault zone.* The percentage of the fault zone volume active at different stages of the deformation event will affect the amount of dilation and the fluid flow possible. The controls on whether deformation is distributed or localized and how this pattern changes during strain/displacement

[43]

events is unknown. Initial TEM analysis of the scaly fabrics from the decollement under the Nankai wedge, sampled during ODP Leg 131, indicates that repeated localized deformation by brecciation of early fabrics has occurred.

3. *The duration of deformation activity.* This will control the time period when the different enhanced permeability paths can operate. Individual fault activity periods will depend upon the controls of fault array evolution (Platt 1990) but each fault may be associated with the generation, propagation and dissipation of large numbers of small strain events.

The above discussion raises the possibility that the variety of fabrics reported from individual fault zones may represent different types of deformation events or different stages of similar events preserved because of the exact history of conditions experienced. For example pervasive straining by disaggregation and particulate flow involving deformation throughout the entire shear zone may produce random fabrics and allow continuous and steady fluid flow. However, the stability of this fabric is low as adjustments and collapse are likely during the late stages of movement as the fluid is expelled and/or any fluid over-pressure decreased. The stability of fabric would be enhanced where collapse was prevented by a prolonged period of high fluid pressure (or a slow decay of fluid pressure) during which time cement bridges and/or a more open framework could form (figure 4c). The random fabrics (as described by Prior & Behrmann 1990) may be characteristic of such a situation and reflect one type of dynamic (syn-deformation) fabric. The development of domains of clay alignment may be more characteristic of the end of a deformation event where the displacement rate, fluid pressure and permeability changes induce localized grain-framework collapse (figure 4d). It is interesting to note that the production of such alignment zones may well change the material response to the next deformation event as a new pre-strain fabric is present. These points also emphasize that the fabrics preserved (and physical properties measured) are the finite features which are likely to have been modified during deformation.

(iii) *Cyclic versus continuous deformation to fluid flow*

Different fluid flow characteristics are likely to be associated with each of the deformation processes and fabrics discussed above. Three end-member modes of flow may be recognized:

(1) fluid flow associated with pervasive, continuous creep in fault zone;

(2) transient events of increased deformation and fluid flow;

(3) flow along faults during periods of low or no strain between deformation events where the permeability anisotropy of the grain alignment, generated during deformation, may create localized flow along the fault but act as a seal to cross fault flow. Arch & Maltman (1990) have discussed the possible role of permeability anisotropy in fault zones.

It is clear from the foregoing discussion that evidence for cyclic deformation is common in semi-lithified sediment. What is uncertain is the cause of the cyclic deformation and fluid flow. The possible causes which require future assessment may be divided into two groups: (1) *internally generated* events caused, for example by the linking of localized initial fabric failures, generated by fluid over-pressuring or by high differential stresses and (2) *externally imposed* cyclic straining caused by deformation events migrating into the volume considered. An example of the latter would be the seaward migration of the wedge deformation front (Platt 1990). A numerical analysis of the behaviour of event clustering applied to fluid flow along faults by Stark & Stark (1991) has demonstrated the important application of percolation theory to this problem. Establishing the exact nucleation, propagation and clustering processes as well as the size (volume affected), duration and migration distances of such events are all important to understanding the associated permeability and fluid flow characteristics and will require a combination of microstructural and numerical modelling studies.

4. Conclusions

The review of microstructural research presented above demonstrates the role that such studies have in assessing the migration of fluids in semi-lithified sediments. As permeability is controlled by the detailed microstructure of the grain framework characterizing the evolution of fabrics generated in experimental and natural deformation events can provide important information needed for understanding the hydrology of accretionary wedges. The discussion has identified a number of research areas where future analysis is particularly important. These include the assessment of; (*a*) the microfabrics associated with different fluid flow paths and their stability during deformation under different conditions, (*b*) the characterization of transient deformation events in terms of the deformation mechanisms paths, the volumes affected, the dilation histories and the propagation processes and (*c*) the interaction and interdependence of the different fluid migration paths.

The authors acknowledge support from the NERC, Deutsche Forschungs Gemeinschaft and the ODP. We thank Alex Maltman, Kevin Pickering and John Tarney for comments on a first draft and Casey Moore for providing ALVIN specimens.

References

Agar, S. M. 1990 The interaction of fluid processes and progressive deformation during shallow level accretion: examples from the Shimanto Belt of SW Japan. *J. geophys. Res.* **95**, 9133–9148.

Agar, S. M., Prior, D. J. & Behrmann, J. H. 1989 Back-scattered electron imagery of the tectonic fabrics of some fine-grained sediments: implications for fabric nomenclature and deformation processes. *Geology* **17**, 901–904.

Arch, J. & Maltman, A. 1990 Anisotropic permeability and tortuosity in deformed wet sediments. *J. geophys. Res.* **95**, 9035–9046.

Bennett, R. H. & Hulbert, M. H. 1987 Clay microstructure. Boston: IHRDCC.

Borradaile, G. J. 1981 Particulate flow and the generation of cleavage. *Tectonophys.* **72**, 306–321.

Bray, C. J. & Karig, D. E. 1985 Porosity of sediments in accretionary prisms and some implications for dewatering processes. *J. geophys. Res.* **90**, 768–778.

Brown, K. M. & Behrmann, J. H. 1990 Genesis and evolution of small scale structures in the toe of the Barbados Ridge accretionary wedge. In *Ocean Drilling Programme, Scientific Results*, vol. 110, pp. 229–243.

Carson, B., Suess, E. & Strasser, J. C. 1990 Fluid flow and mass flux determinations at vent sites on the Cascadia margin accretionary prism. *J. geophys. Res.* **95**, 8891–8899.

Carter, N. L., Kronenburg, A. K., Ross, J. V. & Wiltschko, D. V. 1990 Control of fluids on deformation in rocks. In *Deformation mechanisms, rheology and tectonics* (ed. R. J. Knipe & E. H. Rutter). *Geol. Soc. Lond. Spec. Publ.* no. 54, pp. 1–13.

Gieskes, J. M., Vrolijk, P. & Blanc, G. 1990 Hydrogeochemistry of the Northern Barbados accretionary complex transect: ODP Leg 110. *J. geophys. Res.* **95**, 8809–8818.

Jones, M. E. & Addis, M. A. 1986 The application of stress paths and critical state analysis to sediment deformation. *J. struct. Geology* **8**, 575–580.

Karig, D. E. 1986 Physical properties and mechanical state of accreted sediments in the Nankai Trough, Southwest Japan Arc. In *Structural fabric in deep sea drilling project cores from forearcs* (ed. J. C. Moore). *Geol. Soc. Am. Mem.* **166**, 117–136.

Karig, D. E. 1990 Experimental and observational constraints on the mechanical behaviour in the toes of accretionary prisms. In *Deformation mechanisms rheology and tectonics* (ed. R. J. Knipe & E. H. Rutter). *Geol. Soc. Lond. Spec. Publ.* no. 54, pp. 383–398.

Kemp, A. E. S. 1990 Fluid flow in 'vein structures' in Peru forearc basins: Evidence from Backscattered electron microscope studies. In *Proc. Ocean Drilling Program Scientific Results*, vol. 112, pp. 33–41.

Knipe, R. J. 1986*a* Faulting mechanisms in slope sediments: examples from deep sea drilling project cores. In *Structural fabric in deep sea drilling project cores from forearcs* (ed. J. C. Moore). *Geol. Soc. Am. Mem.* **166**, 45–54.

Knipe, R. J. 1986*b* Microstructural evolution of vein arrays preserved in deep sea drilling project cores from the Japan Trench, Leg 57. In *Structural fabric in deep sea drilling project cores from forearcs* (ed. J. C. Moore). *Geol. Soc. Am. Mem.* **166**, 75–88.

Knipe, R. J. 1986*c* Deformation mechanism path diagrams for sediments undergoing lithification. In *Structural fabric in deep sea drilling project cores from forearcs* (ed. J. C. Moore). *Geol. Soc. Am. Mem.* **166**, 151–160.

Knipe, R. J. 1989 Deformation mechanisms – recognition from natural tectonites. *J. struct. Geol.* **11**, 127–146.

Lambe, T. W. & Whitman, R. V. 1969 *Soil mechanics.* New York: Wiley.

Langseth, M. & Moore, J. C. 1990 Introduction to special section on the role of fluids in sediment accretion, deformation, diagenesis and metamorphism in subduction zones. *J. geophys. Res.* **95**, 8737–8742.

Lindsley-Griffin, N., Kemp, A. E. S. & Swartz, J. F. 1990 Vein structures of the Peru margin, Leg 112. In *Proc. Ocean Drilling Program Scientific Results*, vol. 112, pp. 3–16.

Lloyd, G. E. 1985 Review of instrumentation, techniques and applications of SEM in mineralogy. In *Applications of electron microscopy in the Earth sciences* (ed. J. C. White). Mineralogical Society of Canada Short Course **11**, 151–188.

Lucas, S. E. & Moore, J. C. 1986 Cataclastic deformation in accretionary wedges; DSDP Leg 66, and on-land examples from Barbados and Kodiak Islands. In *Structural fabric in deep sea drilling project cores from forearcs. Geol. Soc. Am. Mem.* **166**, 89–104.

Lundberg, N. & Moore, J. C. 1986 Macroscopic structural features in DSDP cores from forearc regions. In *Structural fabric in deep sea drilling project cores from forearcs* (ed. J. C. Moore). *Geol Soc. Am. Mem.* **166**, 13–44.

Maltman, A. J. 1987 Shear zones in argillaceous sediments – an experimental study. In *Deformation of sediments and sedimentary rocks. Spec. Publ. Geol. Soc. Lond.* **29**, 71–76.

Moore, J. C. 1989 Tectonics and hydrogeology of accretionary prisms: role of the decollemont zone. *J. struct. Geol.* **11**, 95–106.

Moore, J. C., Orange, D. & Kulm, La V. D. 1990 Interelationship of fluid venting and structural evolution: *Alvin* observations from the frontal accretionary prism. *Oregon. J. geophys. Res.* **95**, 8795–8809.

Moore, J. C., Roeske, S., Lundberg, N., Schoonmaker, J., Cowan, D., Gonzales, E. & Lucas, S. 1986 Scaly fabrics from the deep sea drilling project cores from forearcs. In *Structural fabric in deep sea drilling project cores from forearcs* (ed. J. C. Moore). *Geol. Soc. Am. Mem.* **166**, 55–74.

Pickering, K. T., Agar, S. M. & Prior, D. J. 1990 Vein structure and the role of pore fluids in early wet sediment deformation, Late Miocene volcanoclastic rocks, Muira group, SE Japan. In *Deformation mechanisms, rheology and tectonics* (ed. R. J. Knipe & E. H. Rutter). *Geol. Soc. Lond. Spec. Publ.* no. 54, pp. 417–430.

Platt, J. 1990 Thrust mechanics in highly overpressured accretionary wedges. *J. geophys. Res.* **95**, 9025–9034.

Prior, D. J. & Behrmann, J. H. 1990 Thrust related mudstone fabrics from the Barbados forearc: a backscattered electron microscope study. *J. geophys. Res.* **85**, 9055–9067.

Ritger, S. D. 1985 Origin of vein structures in slope deposits of modern accretionary wedges. *Geology* **13**, 437–439.

Schoonmaker, J. 1986 Clay mineralogy and diagenesis of sediments from deformation zones in the Barbados accretionary wedge. In *Structural fabric in deep sea drilling project cores from forearcs* (ed. J. C. Moore). *Geol. Soc. Am. Mem.* **166**, 105–116.

Shephard, L. E. & Bryant, W. R. 1983 Geotechnical properties of the lower trench inner slope sediments. *Tectonophys.* **99**, 279–312.

Sibson, R. H. 1981 Fluid flow accompanying faulting: field evidence and models. In *Earthquake prediction; an international review* (ed. D. W. Simpson & P. G. Richards), pp. 593–603. Am. Geophys. Union M. Ewing Series. no. 4.

Sibson, R. H. 1990 Conditions of fault-value behaviour. In *Deformation mechanisms, rheology and tectonics* (ed. R. J. Knipe & E. H. Rutter). Geol. Soc. Spec. Publ. no. 54, pp. 15–28.

Smart, P. & Tovey, K. 1982 *Electron microscopy of soils and sediments*. Oxford University Press.

Stark, C. P. & Stark, J. A. 1991 Seismic fluids and percolation theory. *J. geophys. Res.* (In the press.)

Suess, E., Carson, B., Ritger, S., Moore, J. C., Jones, M. L., Kulm, L. D. & Cochrane, G. R. 1985 Biological communities at vent sites along the subduction zone off Oregon. The hydrothermal vents of the eastern Pacific: an overview (ed. M. L. Jones). *Bull. Biol. Soc. Wash.* **6**, 475–484.

Swartz, J. F. & Lindsley-Griffin, N. 1990 An improved impregnation technique for studying structure of unlithified cohesive sediments. In *Proc. Ocean Drilling Program Scientific Results*, vol. 112, pp. 87–91.

Thornburg, T. M. & Suess, E. 1990 Carbonate cementation of granular and fracture porosity. In *Proc. Ocean Drilling Program Scientific Results*, vol. 112, pp. 95–109.

Wesson, R. L. 1988 Dynamics of fault creep. *J. geophys. Res.* **93**, 8929–8951.

Plumbing accretionary prisms: effects of permeability variations

By J. Casey Moore[1], Kevin M. Brown[1], Frank Horath[1],
Guy Cochrane[1], Mary MacKay[2] and Greg Moore[2]

[1] Earth Sciences, University of California at Santa Cruz, California 95064, U.S.A.
[2] SOEST, University of Hawaii, Manoa, Honolulu, Hawaii 96822, U.S.A.

Fault zones focus fluid expulsion in the muddy northern Barbados Ridge accretionary prism with fault-parallel permeabilities about 1000 times greater than intergranular permeabilities in the adjacent sediment. In the Oregon prism the low bedding-perpendicular permeability (due to mudstones) inhibits intergranular dewatering; however, intergranular flow is concentrated where submarine erosion breaches high permeability sandy layers. Even so, faults can capture fluid flow from these exposed sandy layers suggesting the faults have a still higher permeability. Such observations coupled with laboratory measurements permeabilities suggest that faults off Oregon may have fault-parallel permeabilities at least 10–10000 times greater than the adjacent sediments.

Results from Barbados and Oregon suggest fluid flow is concentrated along the most active faults. At the toe of prisms the fault zones are being progressively loaded by the thickening wedge and are undergoing compaction. Preliminary experiments show that permeability decreases relative to the surrounding wall rocks along faults within this compactive deformation régime; we believe that these faults must undergo dilation, perhaps linked to transient increases in pore pressure if they are to be preferential fluid conduits. Farther upslope erosion exposes rocks that are more consolidated, commonly more cemented, and generally of lower intergranular permeability than rocks of equivalent burial further seaward. Because of their lithification and overconsolidation these rocks dilate during faulting, locally enhancing fracture permeability. In such dilative régimes, faults become evermore focused zones of fluid expulsion relative to occluded intergranular pathways.

1. Introduction

Permeability is the most important factor controlling fluid flow in accretionary prisms. Fluid flow rate depends principally on permeability and hydraulic gradient. In accretionary prisms, permeability variations due to lithology can range over six to seven orders of magnitude (Horath 1989; Taylor & Leonard 1990); whereas, hydraulic gradients change by less than an order of magnitude (Moore & von Huene 1980); thus, permeability is at least one million times more variable than hydraulic gradient in governing fluid flow in accretionary prisms. In addition to its lithologic variability, permeability changes with structural evolution, diagenesis and metamorphism; permeability may also change during short-term deformational events on the timescale of the earthquake cycle (Sibson 1981).

Because fractures can radically affect permeability and accretionary prisms are highly faulted, large-scale field determination of permeability would be most effective. Unfortunately such measurements have yet to be made; permeability data from accretionary prisms is limited to measurements from small samples. Inferences from these few data are based on geologic reasoning, experiments, and numerical modelling. Here we review information on the permeability structure of hydro-geologic end-members of accretionary prisms, present some new observational and preliminary experimental results, and attempt to conceptualize possible permeability variations in time and space.

We principally use units of m^2 for permeability but also scale diagrams in hydraulic conductivity with units of $cm\ s^{-1}$. Hydraulic conductivity is the proportionality constant in Darcy's law, sometimes erroneously referred to permeability; hydraulic conductivity equals the product of permeability, density, and the acceleration of gravity divided by viscosity (Freeze & Cherry 1979, p. 27). Equivalent permeability is used to describe an average or effective bulk measure of permeability in various lithologies or in various directions in units of anisotropic permeability structure (Freeze & Cherry 1979, pp. 32–34).

2. Permeability variation in muddy accretionary prism: Northern Barbados ridge

Long-distance lateral flow of fluids in subduction zones was first explicitly recognized through Ocean Drilling Program (ODP) investigations of the Northern Barbados Ridge (Moore *et al.* 1987, fig. 1). Here, the evidence for active fluid flow, principally along faults, is from geochemical (Gieskes *et al.* 1990) and temperature anomalies (Fisher & Hounslow 1990). These geochemical and temperature anomalies must be periodically replenished by fluid flow or they would be obliterated through diffusion. The geochemical and temperature data suggest fluid moves along the decollement zone with minor leakage into the imbricate thrusts, and negligible flow into the surrounding low permeability (Taylor & Leonard 1990), muddy sedimentary rock matrix (figure 1). Because the methane from the decollement zone has a thermogenic component, deep sources are necessary (Vrolijk *et al.* 1990). For fluids to be channelled for great distance along a fault zone and not be depleted by flow into the wall rocks, the fault zone permeabilities must be much higher than the wall rock. In the absence of *in situ* measurements, numerical models provide the only constraints on the relative magnitudes of wall rock and fault zone permeabilities.

To simulate the pore pressures of 60–100 % of lithostatic near the base of the prism, the numerical model of Screaton *et al.* (1990) requires the overall equivalent permeability of the prism to lie between 10^{-16} and $10^{-18}\ m^2$ over the depth range penetrated by Site 671 of Leg 110. Because the permeability measurements on samples (Taylor & Leonard 1990) are so close to modelled equivalent permeability estimates, it appears that the permeability of the prism is relatively uniform and not controlled by fractures outside the local environment of certain of fault zones. To focus flow in the decollement the model of Screaton *et al.* requires that the decollement zone be at least three orders of magnitude more permeable than the overlying prim. Because an equivalent permeability of $10^{-14}\ m^2$ gives the most realistic values of the magnitude and distribution of pore pressure relative to the structure features, Screaton *et al.* believe this value is the most representative permeability for fluid flow along the decollement fault zone (figure 1). The decollement zone is described as having an equivalent permeability because the fluid

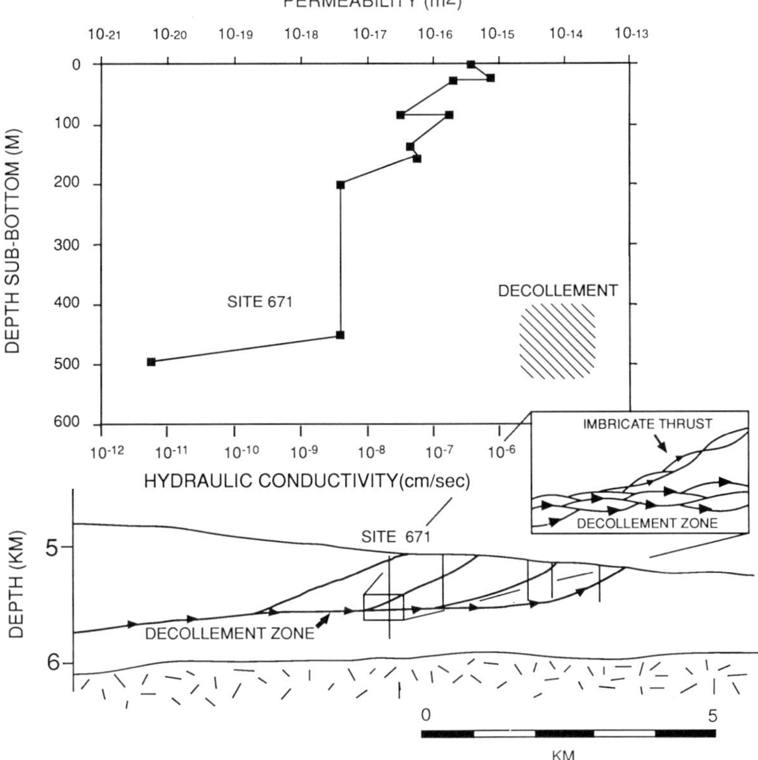

Figure 1. Permeability measurements for Site 671 penetrating the Northern Barbados Ridge (Taylor & Leonard 1990). Cross section and inset schematically depicts active fluid flow along decollement with minor leakage into imbricate thrust. Decollement zone permeability based on flow modelled parallel to this fault zone (Screaton *et al.* 1990).

would flow through both the intergranular matrix and fractures. The observation of veined fractures in fault zones from the Leg 110 area (Vrolijk & Shepard 1991) suggests that any significant permeability increase is due to dilation of fractures.

Overall Screaton *et al.* (1990) illustrate how powerful modelling can be in providing estimates of equivalent permeability and other types of data from the broad constraints of fluid pressure, flow paths, and prism geometry. Their particular approach involved steady-state deformation and fluid flow and hence inferences of permeability are time-averaged; episodic flow is likely and should be investigated through field monitoring and modelling.

3. Permeability variation in a sandy accretionary prism: Oregon margin

Although faults are important fluid conduits in muddy accretionary prisms, how important are faults in focusing fluid flow in sandy accretionary prisms, composed of thick sequences of siliclastic turbidites? Here, we review the observed flow paths and infer the permeability structure of the clastic-dominated Oregon margin. Individual permeability measurements are available from samples located with respect to structural features; seep distribution is established from more than 40 *Alvin* dives; the geometry of potential fluid conduits is constrained by an extensive multichannel seismic reflection grid.

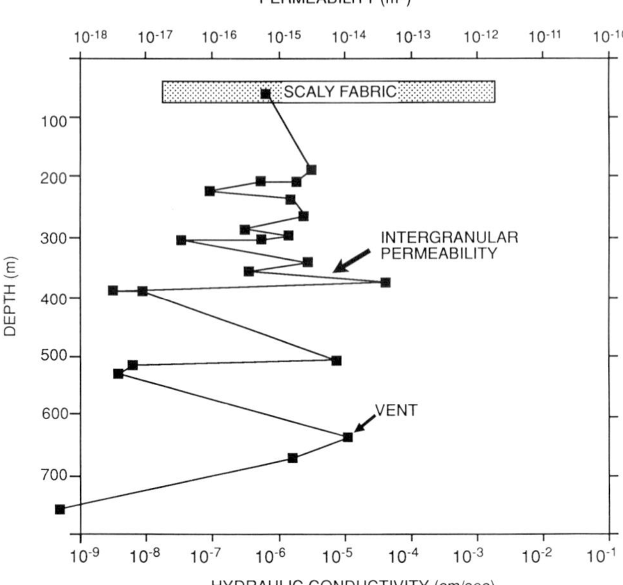

Figure 2. Permeability measurements from (1) submarine canyon exposure through ramp anticline at base of slope, Oregon subduction zone, and (2) related scaly fabrics from onshore fault zones (Horath 1989). Sample depth for intergranular permeabilities from Oregon margin is based on structural projection of samples to a single profile (figure 5*b*) with an allowance for erosion. Horizontal bar shows range of permeabilities from scaly fabrics from Neogene on-land exposures of the Cascadia accretionary prism. Burial depth for the scaly fabric measurements represents effective stress at measurement, assuming hydrostatic pressure. Higher values are for samples with the scaly fabric oriented parallel to flow direction; lower values are for samples with scaly fabric perpendicular to flow direction.

Intergranular permeability: directional heterogeneity

Permeability determinations were made on samples collected by *Alvin* from a submarine canyon exposure using a constant flow rate apparatus under isotropic confining stress of 0.3–0.6 MPa. Measurements of intergranular permeability (figure 2) were either measured at the effective stress of sampling (assuming hydrostatic pressure gradient) or restored to this pressure using various permeability-effective stress functions (Horath 1989). Because fluid pressures in the accretionary prism are likely to exceed hydrostatic, the reported permeabilities are minimum values.

A vertical permeability profile through a ramp anticline at the marginal ridge of the Oregon margin shows about four orders of magnitude variation in intergranular permeability, due to variations from the more muddy to the more sandy and silty lithologies (figure 2). This profile of intergranular permeabilities allows evaluation of the permeability of various structural features based on how they interact with the stratigraphic section. Equivalent permeabilities calculated for flow parallel and perpendicular to bedding (Freeze & Cherry 1979, p. 34) indicate a permeability of 3×10^{-15} m² along bedding surfaces and 7×10^{-18} m² across bedding surfaces. Thus, although sandy Oregon prism may have a higher mean intergranular permeability than the Northern Barbados Ridge, its large-scale equivalent permeability structure shows about a 400-fold anisotropy.

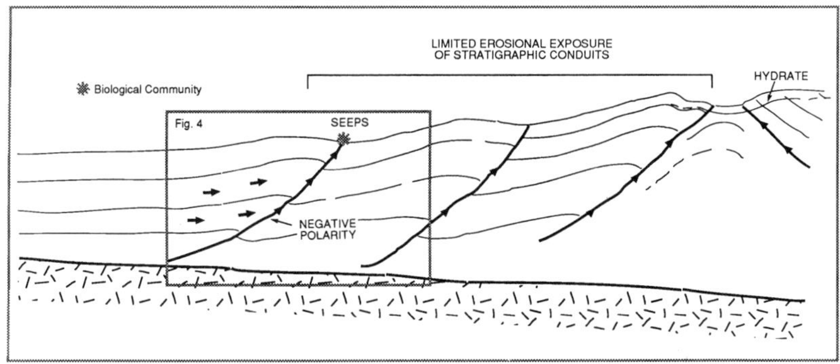

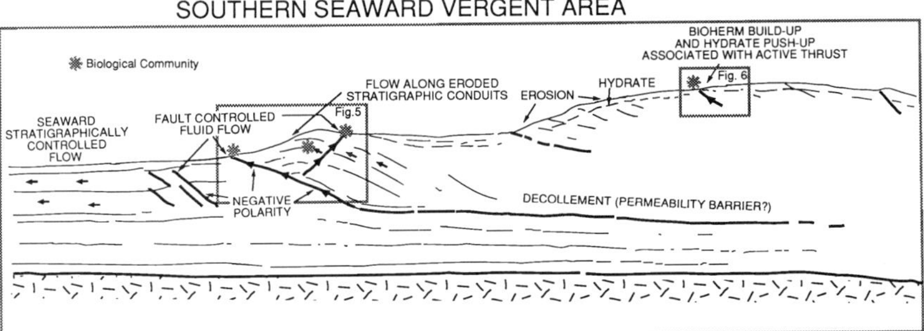

Figure 3. Generalized cross sections showing contrasting structural styles, geomorphic features, and seep geometries of northern landward and southern seaward vergent areas of Oregon margin.

Fluid flow paths in a faulted and uneroded stratigraphic section

In the landward vergent area offshore central Oregon minimal erosion has occurred off the hanging wall of the frontal thrust and the bedding parallel, high permeability conduits are not exposed. Flow to the surface must be either constrained to cut across bedding along flow paths with low equivalent permeabilities or to utilize faults (figures 3 and 4).

Biological evidence suggests the frontal thrust in the landward vergent area off Oregon is a conduit of fluid expulsion and therefore that it has significantly increased permeabilities. Concentrations of *Solemya sp.* clams suggest active flow along this fault zone (Kulm *et al.* 1986; Lewis & Cochrane 1990). The clam concentrations tend to occur in isolated groups along linear zones in a low area marking the surface intersection of the frontal thrust (figure 4).

The frontal thrust on seismic line OR-22 shows a high-amplitude, negative polarity reflection at about 1.75–2 s subbottom (about 2–2.5 km depth) (figure 4). We interpret this type of reflection as being due to a decrease in density and velocity in the fault zone due to an enhanced fluid content. The displacement along the fault is only in the region of a few hundred metres; this displacement, insufficient to cause an impedance contrast that would produce such a strong reflection, given the limited rate of change in density and velocity at this depth (Gregory 1977; Hamilton 1978). A reduced density in the fault zone is consistent with the fault zone having dilated during its formation (Gregory 1977). High fluid pressures developed as a consequence of rapid sediment accumulation in the fan sequence (Wang *et al.* 1990) many have contributed to this dilative behaviour.

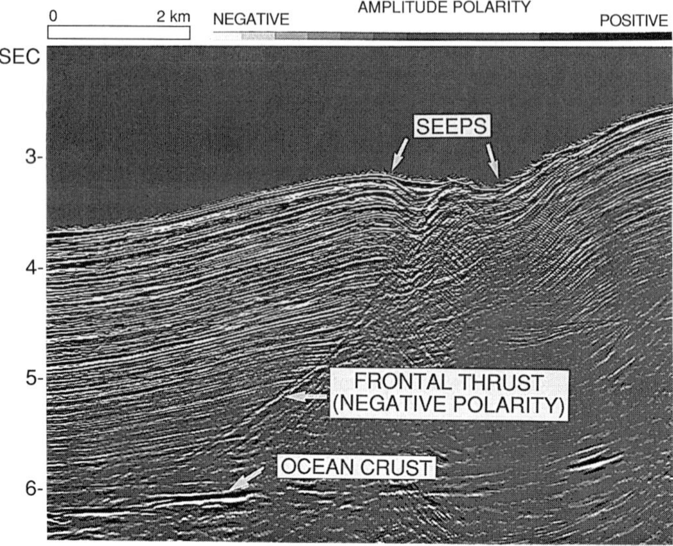

Figure 4. Detail of seismic reflection line OR-22 from deformation front of landward vergent region, Oregon subduction zone. Note the frontal thrust roots very near oceanic crust and shows a high-amplitude negative polarity reflection at depth. True amplitude display. Seep at surface marked by concentrations *Solemya* clams.

In the landward vergent area off Oregon faults apparently provide the highest permeability conduit for fluid expulsion. The existence of overpressures of depositional origin combined with any loading of the footwall due to overthrusting provides the pressure gradient to drive flow. The permeability of the fault must be higher than that of the equivalent permeability for flow perpendicular to the stratigraphic section or higher than about 10^{-17} m².

Exposed stratigraphic conduits and faults: seaward vergent area

In the southern part of the Oregon margin survey area, seaward verging structures have developed seaward facing erosional scarps, commonly on hanging walls of thrusts. Erosion cuts through sections of interbedded mudstone and sandstone-siltstone. Biological communities, marking vents, occur in the sandier and siltier layers at the crest of the breached anticlines indicating stratigraphic control on vent location (Moore *et al.* 1990). Concretions are common near the biological communities providing additional evidence for enhanced fluid flow. The single permeability measurement from one of the vent sites is amongst the highest measured from any of the sandy and silty lithologies (figure 2). The pressure gradient for fluid flow up landward tilted layers is caused by thrust imbrication (Wang *et al.* 1990).

The ramp anticline underlying the marginal ridge off Oregon is cut by a back thrust demarked by surface vents (Moore *et al.* 1990). Excepting the seep at the crest of the breached anticline, biological communities were not observed on the exposed bedding outcrops in a large submarine canyon up dip of the regions in which the back thrust is developed. Thus, this fault must either seal and/or capture lateral flow. Because the biological communities above the backthrust are more vigorous than those associated with the stratigraphic vents, we believe that flow is substantially captured due to a significantly higher permeability in the fault zone. A high permeability for scaly fabrics from fault zones supports this supposition (figure 2).

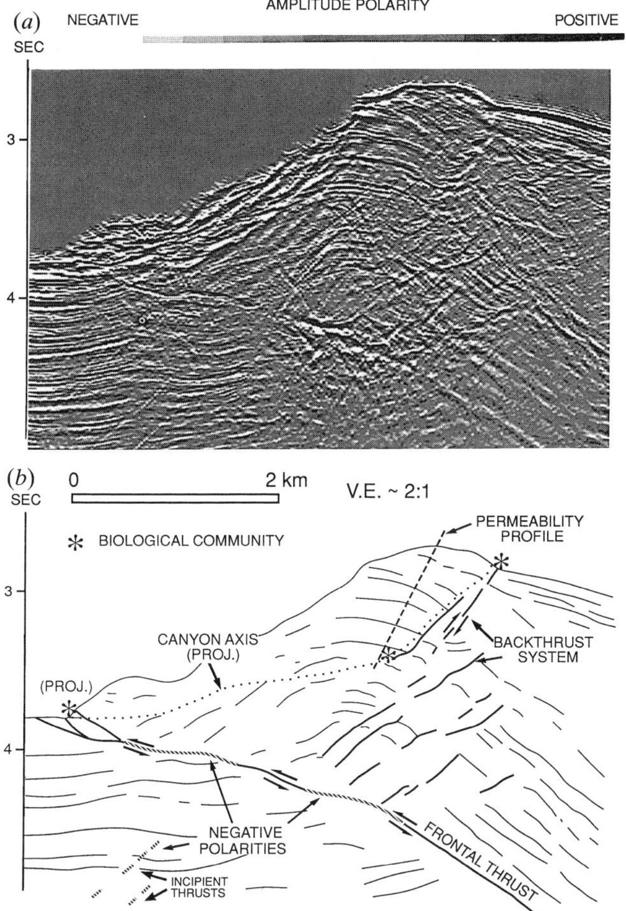

Figure 5. (*a*) Detail of seismic reflection line OR-9 from deformation front of seaward vergent area, Oregon subduction zone. True amplitude display. (*b*) Line drawing of line OR-9. Note regions of high-amplitude negative polarity reflections, associated with faults. Submarine canyon cuts ramp anticline along strike and allowed sampling and structural measurements within interior of fold. Canyon axis is projected from 1 to 2 km north. Biological community marking seep at frontal thrust is projected from 3 km to south.

Minor thrusts beneath the frontal thrust zone of the ramp anticline show amplitude anomalies with negative polarities (figure 5). These anomalies are unlikely to be due to displacement of rocks of significantly differing impedance, and probably result from fluid-rich low-density–low-velocity fault zones. Significant amplitude anomalies of negative polarity are also observed along the frontal thrust. The displacement on this fault is more than 1 km and may juxtapose sedimentary rocks of differing physical properties, creating the amplitude anomalies. In this area, however, a total of eight lines cross the frontal thrust, all with similar displacement but differing patterns of amplitude anomalies. The anomalies may be in part due juxtaposition of rocks of originally differing burial depths, but dilation along the fault zone is probably an important factor. The decreasing amplitude anomaly up the fault dip (figure 5) may be due to collapse of the fault resulting from lateral leakage of fluid to layers that are exposed in the canyon. This leakage also may explain *Alvin*

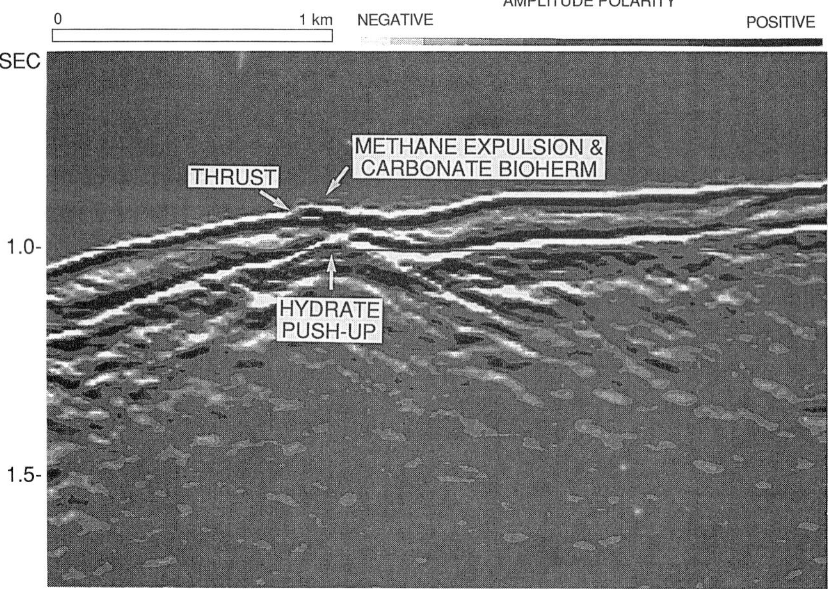

Figure 6. Detail of seismic reflection line OR-9 (see figure 3 for location). True amplitude display. Hyrate reflector is first prominent negative polarity reflector below water bottom. Hydrate reflector rises beneath trace of landward dipping thrust fault and is interpreted as being pushed up by flow of warm fluids. Submersible observations indicate active methane venting at surface and carbonate build-ups of *Calyptogena* clams and inorganically precipitated carbonate.

observations indicating a lack of vents up dip from the negative polarity anomaly on the seismic line in figure 5, but the presence of vents at the frontal thrust 3 km to the south where erosion of the ramp anticline is less severe.

Upslope venting and permeability evolution off Oregon

The evidence for continuing fluid seeps is apparent off Oregon both at a midslope level and beyond to shelf depths (Kulm & Suess 1990). In the region with seaward vergent structures discussed above, the midslope region is characterized by late thrusts that cut a large bathymetric high. One of these faults shows an upward deflection in the gas hydrate reflector that could be caused by enhanced flow and locally higher geothermal gradient (figure 6). Recent submersible observations at the surface trace of this fault indicate that it is venting methane bubbles, and that it supports the most concentrated chemosynthetic biological community yet seen off Oregon. Precipitation and biological action has led to the build up of a carbonate ridge many metres high. The focused flow indicates a substantial permeability contrast between the country rock and the fault zone in this region of the prism. One measurement of permeability from a cemented sandstone in this midslope here is 10^{-18} m². Because the faults are cutting cemented and substantially consolidated rocks, deformation would be dilatant with permeability enhancement in fault zones.

Discussion of Oregon accretionary prism observations

Faults that transmit fluids up through uneroded sections must have minimum permeabilities larger than the equivalent permeability necessary for flow perpendicular to the layering which corresponds to values greater than 10^{-16}–10^{-17} m². Apparently some faults have a higher permeability along their extent than the more permeable sandy layers (greater than 10^{-13}–10^{-14} m^2), because these faults can capture flow from the sandy layers.

Although no fault zone material was collected during the *Alvin* program off Oregon, measurements of Neogene on-land equivalents suggest permeability values as high as 10^{-12} m^2 along scaly fabrics of faults and as low as 10^{-16}–10^{-17} m^2 across the scaly fabric of the fault rock (Horath 1989). Even though the sample scale is an order of magnitude greater than the mean fracture spacing, the measurements probably are minimum values (Sowers 1981); Because the measurements were made at confining pressures of 0.3–0.6 MPa and we have no applicable permeability-depth function, the measurements can only be compared with intergranular permeability measurements from shallow depths or low effective stresses. Nevertheless, the measurements indicate high permeabilities along flow paths parallel to the scaly fabric, permeabilities that would be sufficient to capture flow from the stratigraphic section at under low effective stress conditions. Moreover, modelling of the fluid expulsion off Oregon using Horath's (1989) values for intergranular permeability and fault permeabilities reproduce the observed flow paths well (Henry & Wang 1991).

4. Inferences on evolution and transience of permeability in accretionary prisms

Results outlined above from the Barbados Ridge and oregon accretionary prisms indicate that some faults, particularly active ones, localize fluid flow. However, not all faults encountered during Barbados ODP drilling conduct fluid preferentially. Apparently any permeability enhancement of fault zones is governed by the stress path of the fault zone material and cementation.

The compactive path: permeability evolution of faults being progressively loaded in a thickening wedge

Where erosion at the surface of the prism is minimal, most entering sediments will follow trajectories where they become progressively buried by sedimentary and tectonic thickening. We would expect these sediments, in both the wall rocks and fault zones, to compact (Karig 1986; Taylor & Leonard 1990). As permeability is strongly dependent on the number and degree of interconnection between pore voids (Freeze & Cherry 1979), higher degrees of compaction in fault zones (Atkinson & Bransby 1978; Karig 1986) should result in fault permeabilities being reduced relative to the wall rocks; such faults should not form fluid conduits.

Preliminary experimental results support the hypothesis that faults undergoing compaction and failure will have reduced fault-parallel permeabilities, in comparison with adjacent wall-rock (Brown & Moore 1990). In these experiments a mud sample (50% quartz, 30% illite, and 20% montmorillonite) 3 mm thick was placed along a 30° saw cut in a jacketed sliding block with permeability and diffusivity measured by a pulse test (Morin & Olsen 1987) along a flow path parallel to the sample extent. The deformation of the sample was incrementally stopped for permeability and diffusivity measurements. The axial load and confining pressure remained constant at 0.6 MPa during the permeability determinations so that the stress conditions correspond closely to the points on the failure curve (figure 7).

Our experiments suggest that faults deforming under compactive stress conditions will not increase in permeability but instead will become aquitards. As active faults in accretionary prisms can transmit fluids for at least portions of their histories, we

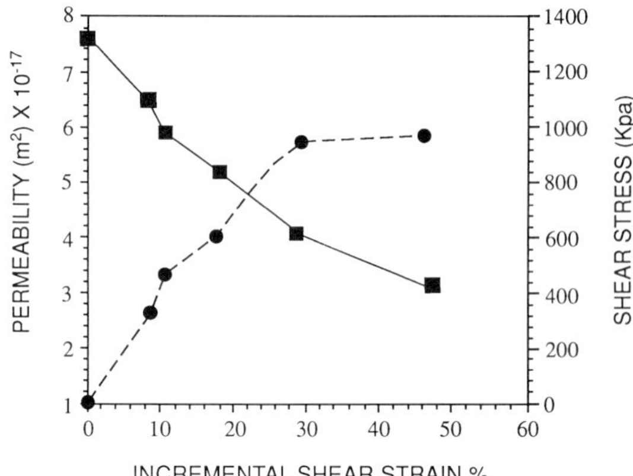

Figure 7. Permeability against shear strain and shear stress (o_1-o_3) against strain curves. Shear strain equals displacement divided by shear zone thickness. The progressive decrease in shear zone thickness (due to compaction) is accounted for in both the permeability and shear strain determinations. ■, Permeability; ●, failure curve.

believe that at these times these faults are not undergoing compactive failure but are dilated. The correlation of active venting with the apparently dilated faults associated with the negative polarity anomalies off Oregon (figures 4 and 5) supports this view. Moreover, Karig (1990) has argued for intermittent dilative deformation and high fluid pressure to account for the development of scaly fabrics in accretionary fault zones otherwise on a compactive strain path. Such a deformation scheme would also permit intermittent fluid flow.

In contrast to our experimental results, Arch & Maltman (1990) found that permeability increased parallel to shear zones in muddy sediments. However, their permeability measurements were made after unloading of the sample, reducing the effective stress to zero. The increased permeability is explicable by sample dilation. A combined interpretation of the experimental results of Brown & Moore (1990) and Arch & Maltman (1990) suggests that faults in unlithified sediments can only become fluid conduits when failure occurs under episodically reduced effective stresses conditions (figure 8). Increased fault-parallel permeabilities are probably associated with mechanical dilation of the sediments, probably along weak failure surfaces; typical fault-parallel orientation of the dilated failure surfaces leads to the focusing of flow along the fault plane. Arch & Maltman (1990) showed that the clay fabric alignment does produce a secondary anisotropy in permeability that would also assist in the channelling of flow along the fault zone.

The dilative path: permeability evolution during erosion

Significant mass wasting is common on the slopes of accretionary prisms (figures 3 and 9). Sediments uplifted and unloaded by erosion or faulting (Platt 1986) will tend to be overconsolidated and better lithified than sediments at equivalent depth in uneroded sections. Due to the overconsolidated and cemented nature of the sediments, regions undergoing active deformation will dilate (Atkinson & Bransby 1978) and increase in permeability. Undeformed regions will retain their low permeabilities. In active accretionary prisms, sites of significant erosion will always tend to be associated with active deformation because the prism equilibrium requires the erosion to be balanced by a similar amount of uplift (Davis *et al.* 1983). In the

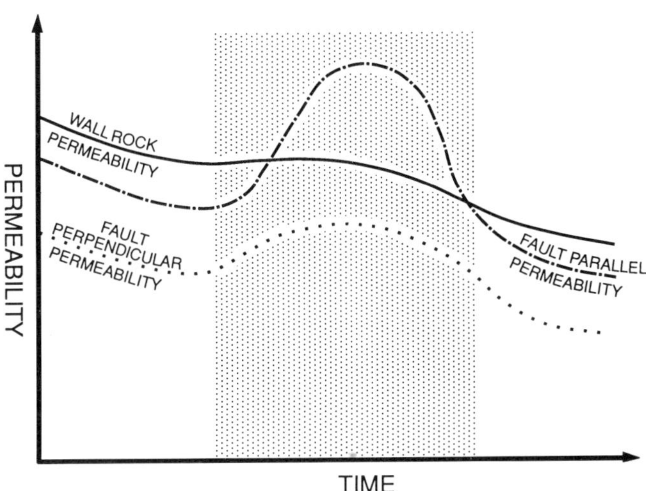

Figure 8. Conceptual evolution of fault permeability during compactive loading. Relative permeability trajectories of a segment of fault zone and the surrounding wall rocks at similar depths as they are progressively loaded during thrusting. Permeabilities in the fault zone are anisotropic but generally below the wall rock permeabilities during periods when the effective confining stresses are high. Transiently raised pore fluid pressures result in the episodic dilation of the shear surfaces and the fault parallel permeability exceeds the wall rock permeabilities. The higher fault-parallel permeabilities allow the excess pore pressure to diffuse along the fault zone and propagate fluid flow and displacement. Drainage of excess pressures results fracture closure and fault sealing. Stipple represents dilation and elevated pore pressure.

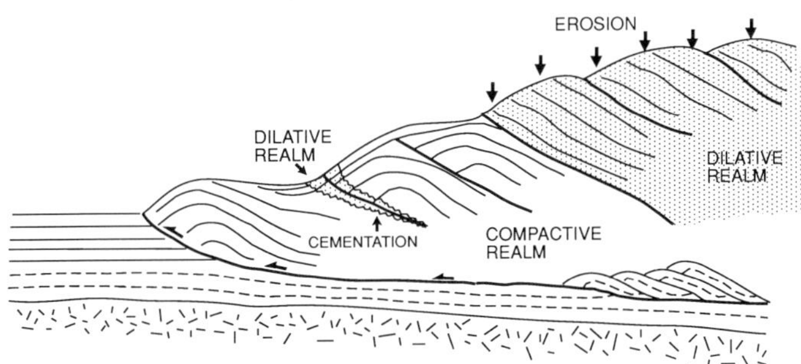

Figure 9. Schematic depiction of compactive and dilative realms of potential deformation in an accretionary prism. In the compactive realm sediments are under to normally consolidated, generally undergoing burial, and would suffer a decrease in volume and permeability during deformation. In the dilative realm sediments are over consolidated, due to erosion or cementation, and would increase in volume and permeability during deformation. In the generally compactive realm, fluid pressure transients could result in episodes of dilative deformation and permeability increase. As sediments are accreted and uplifted they would travel on deformation paths that are in general initially compactive (with possible dilative transients) followed by a generally dilative paths.

older regions of accretionary prisms, this uplift will commonly be accommodated on out-of-sequence thrusts; at the toe frontal imbricate thrusts produce the uplift. Because of the tendency of the uplifted rocks to dilate, these active structures in regions of erosion will become sights of considerable focused flow. For example, the active late thrust in the more interior regions of the Oregon prism is actively flowing fluid (figure 6).

Lithification, metamorphism, and permeability evolution

At deep levels in the prism (3–10 km burial) diagenesis and metamorphism apparently continue to reduce the intergranular permeability of low strain regions. Eventually only active deformation or hydrofracture development will allow further significant through-flow of fluid and the hydrogeologic system will become dominated by fracture flow (Vrolijk *et al.* 1988; Speed 1990). In addition crack seal textures and fluid inclusion studies (Vrolijk 1987; Fisher & Byrne 1990; Byrne & Fisher 1990), indicate that fluid flow in the deeper regions of accretionary prisms is associated with episodic phases of vein opening and mineral infilling. We would expect permeabilities of such fractured systems to fluctuate greatly with time.

5. Conclusions

1. Fault-parallel permeabilities in both in muddy and sandy accretionary prisms may be 1000 times higher than the average intergranular permeability of the adjacent sediment.

2. Transverse permeability in some fault zones may be as low as some of the lower permeability shales.

3. Sandy prisms vary in intergranular permeability over four orders of magnitude at a given depth of burial and show bulk anisotropy due to layering.

4. The more permeable layers in sandy accretionary prisms approach the fault-parallel permeability of the faults that actively leak fluid.

5. The magnitude of surface erosion and fault-sandy layer geometry significantly affects the geometry of fluid expulsion.

6. During compactive strain paths the permeability of fault zones generally decreases except during periods of dilative deformation, permeability increase, and fluid flow. During dilative strain paths, overconsolidation due to uplift and erosion, and cementation result in dilative deformation and permeability increase in fault zones; however, the adjacent wall rocks retain their low permeability caused by cementation and consolidation. Overall, a dilative strain path results in a more significant contrast between fault zone and wall rock permeability than a compactive strain path.

We thank the Royal Society for support to participate in the Discussion Meeting that fostered this paper. Research supported by NSF grants OCE-8609965, OCE-8912272, OCE-8813907, OCE-8917705 (tp C. M.) and OCE-8821577 (to G. M.). We thank Alex Maltman and Dave Prior for review of the typescript and John Tarney for editorial assistance.

References

Arch, J. & Maltman, A. 1990 Anisotropic permeability and tortuosity in deformed wet sediments. *J. geophys. Res.* **95**, 9035–9045.

Atkinson, J. H. & Bransby, P. L. 1978 *The mechanics of soils: an introduction to critical state soil mechanics.* London: McGraw-Hill.

Brown, K. M. & Moore, J. C. 1990 Deformation related permeability changes in muddy rocks. *Int. Conf. Fluids in Subduction Zones Abstr. vol.*, p. 27, Paris.

Byrne, T. & Fisher, D. 1990 Evidence for a weak and overpressured decollement beneath sediment-dominated accretionary prisms. *J. geophys. Res.* **95**, 9081–9097.

Davis, D., Suppe, J. & Dahlen, F. A. 1983 The mechanics of fold-and-thrust belts. *J. geophys. Res.* **88**, 1153–1172.

Fisher, D. & Byrne, T. 1990 The character and distribution of mineralized fractures in the Kodiak

Formation, Alaska: implications for fluid flow in an underthrust sequence. *J. geophys. Res.* **95**, 9069–9080.

Fisher, A. T. & Hounslow, M. 1990 Heat flow through the toe of the Barbados accretionary complex. In *Proc. Initial Reports (Part B) Ocean Drilling Project*, vol. 110, pp. 345–363.

Freeze, R. A. & Cherry, J. A. 1979 *Groundwater.* Englewood Cliffs, New Jersey: Prentice-Hall.

Gieskes, J. M., Vrolijk, P. & Blanc, G. 1990 Hydrogeochemistry of the Northern Barbados Accretionary complex transect: Ocean Drilling Project Leg 110. *J. geophys. Res.* **95**, 8809–8818.

Gregory, A. R. 1987 Aspects of rock physics from laboratory and log data that are important to seismic interpretation. In Seismic stratigraphy – applications to hydrocarbon exploration (ed. C. E. Payton). *Am. Ass. Petr. Geologists Mem.* **26**, 15–46.

Hamilton, E. L. 1978 Sound velocity-density relations in sea-floor sediments and rocks. *J. acoust. Soc. Am.* **63**, 366–377.

Henry, P. & Wang, C.-Y. 1991 Modeling of fluid flow and pore pressure at the toe of the Oregon and Barbados accretionary wedges. *J. geophys. Res.* (In the press.)

Horath, F. 1989 Permeability evolution in the Cascadia accretionary prism: examples from the Oregon prism and Olympic Peninsula melanges. M.Sc. thesis, University of California, Santa Cruz.

Karig, D. E. 1986 Physical properties and mechanical state of accreted sediments in the Nankai Trough, S. W. Japan. In Structural fabrics in Deep Sea Drilling Project cores from forearcs. *Geol. Soc. Am. Mem.* **166**, 117–133.

Karig, D. E. 1990 Experimental and observational constraints on the mechanical behavior in the toes of accretionary prisms. *Geol. Soc. Lond. Spec. Publ.* **54**, 383–398.

Kulm, L. D. & Suess, E. 1990 The relation of carbonate deposits to fluid venting processes: Oregon accretionary prism. *J. geophys. Res.* **95**, 8899–8915.

Kulm, L. D., Suess, E., Moore, J. C., Carson, B., Lewis, B. T., Ritger, S. D., Kadko, D. C., Thornberg, T. M., Embley, R. W., Rugh, W. D., Massoth, G. J., Langseth, M. G., Cochran, G. R. & Scamman, R. L. 1986 Oregon subduction zone: venting, fauna, and carbonates. *Science, Wash.* **231**, 561–566.

Lewis, B. T. R. & Cochrane, G. C. 1990 Relationship between chemosynthetic benthic communities and geologic structure on the Cascadia subduction zones. *J. geophys. Res.* **95**, 8783–8793.

Moore, J. C. & von Huene, R. 1980 Abnormal pore pressure and hole instability in forearc regions: a preliminary report. Ocean Margins Drilling Program, Washington, D.C.

Moore, J. C., Orange, D. L. & Kulm, L. D. 1991 Interrelationship of fluid venting and structural evolution, Oregon margin. *J. geophys. Res.* **95**, 8795–8808.

Moore, J. C. *et al.* 1987 Expulsion of fluids from depth along a subduction-zone decollement horizon. *Nature* **326**, 785–788.

Morin, R. H. & Olsen, H. W. 1987 Theoretical analysis of the transient pressure response from a constant flow rate hydraulic conductivity test. *Water Resources Res.* **23**, 1461–1470.

Platt, J. P. 1986 Dynamics of orogenic wedges and the uplift of high-pressure metamorphic rocks. *Geol. Soc. Am. Bull.* **97**, 1037–1053.

Screaton, E. J., Wuthrich, D. R. & Dreiss, S. J. 1990 Permeabilities, fluid pressures, and flow rates in the Barbados Ridge Complex. *J. geophys. Res.* **95**, 8997–9007.

Sibson, R. H. 1981 Fluid flow accompanying faulting: field evidence and models. In *Earthquake prediction: an international review.* American Geophysical Union, Maurice Ewing Series vol. 4, pp. 593–603.

Sowers, G. F. 1981 Rock permeability or hydraulic conductivity – an overview. In *Permeability and groundwater contaminant transport.* ASTM STP 746, pp. 65–83.

Speed, R. 1990 Volume loss and defluidization history of Barbados. *J. geophys. Res.* **95**, 8983–8996.

Taylor, E. & Leonard, J. 1990 Sediment consolidation and permeability at the Barbados forearc. In *Proc. Ocean Drilling Project, Scientific Results*, vol. 110, pp. 289–308.

Vrolijk, P. J. 1987 Tectonically-driven fluid flow in the Kodiak accretionary complex, Alaska. *Geology* **15**, 466–469.

Vrolijk, P. & Shepard, S. M. F. 1991 Syntectonic carbonate veins from the Barbados accretionary prism (ODP Leg 110) – record of paleohydrology. *Sedimentology.* (In the press.)

Vrolijk, P., Myers, G. & Moore, J. C. 1987 Warm fluid migration along tectonic melanges in the Kodiak accretionary complex, Alaska. *J. geophys. Res.* **93**, 10313–10324.

Vrolijk, P., Chambers, S., Gieskes, J. & O'Neil, J. 1990 Stable isotope ratios of interstitial fluids from the northern Barbados accretionary prism, ODP Leg 110. In *Proc. Ocean Drilling Program, Scientific Results*, vol. 110: College Station TX (Ocean Drilling Program), pp. 189–205.

Wang, C. Y., Shi, Y., Hwang, W. & Chen, H. 1990 Hydrologic processes in the Oregon–Washington accretionary complex – seaward vergent versus landward vergent margins. *J. geophys. Res.* **95**, 9009–9023.

Sediment deformation and fluid activity in the Nankai, Izu-Bonin and Japan forearc slopes and trenches

By A. Taira[1] and K. T. Pickering[2]

[1] *Ocean Research Institute, University of Tokyo, 1-15-1 Minamidai, Nakano-ku, Tokyo 164, Japan*

[2] *Department of Geology, University of Leicester, Leicester LE1 7RH, U.K.*

A variety of sediment and crustal deformation, associated with fluid activity, has been observed in the present forearc slopes, trenches and ancient onland outcrops in the Japanese island arcs. The Nankai forearc represents a typical clastic-dominated accretionary prism where the expulsion of pore fluids from sediments seems to have occurred intermittently, through both channelized and diffusive mechanisms, some of which appears to be pulsed. Mud diapirs occur within the trench, near the toe and upper slope of the accretionary prism. The widespread development of a gas-hydrate phase boundary (bottom simulating reflector (BSR)) may indicate pervasive fluid advection throughout the prism. The Japan Trench forearc is characterized by tectono-gravity slope instability and collapse, resulting in zones of fluid escape along normal faults. No BSR is observed in the Japan forearc despite the presence of organic-rich sediments. The Izu-Bonin forearc, apparently unlike the Japan and Nankai Trench forearcs, contains recently discovered serpentinite diapirs in which there are blocks of basic-ultrabasic rocks, together with blueschist. Fossil *Calyptogena* beds in the Pliocene basins of the Izu Collision Zone provide good examples of fluid activity associated with venting in an ancient accretionary prism now exposed onland.

Contrasting tectonic processes in the three forearcs may be explained by differences in the plumbing systems and hydrogeologic characteristics of the forearc basements: dewatering sediments in the Nankai Trough prism; impermeable and consolidated sediments in the Japan Trench prism, and the very permeable pillow-lava-volcaniclastic complex of the Izu-Bonin Trench forearc.

1. Introduction

The Japanese arc–trench systems contain the Kuril, Japan, Nankai, Izu-Bonin (Izu-Ogasawara) and Ryukyu trenches (figure 1). Among these, the Japan, Nankai and Izu-Bonin forearc regions have been most intensively studied using a panoply of techniques from shallow and deep seismic profiling, earthquake studies, deep-sea drilling, sampling and submersible dives.

The forearc to the Nankai Trough is associated with the subduction of the Philippine Sea plate. Active sediment accretion and the formation of a classic thrust and fold belt characterizes this region (Nasu *et al.* 1982; Aoki *et al.* 1983; Moore *et al.* 1990). North of the present forearc of southern Japan, the geology charts a long

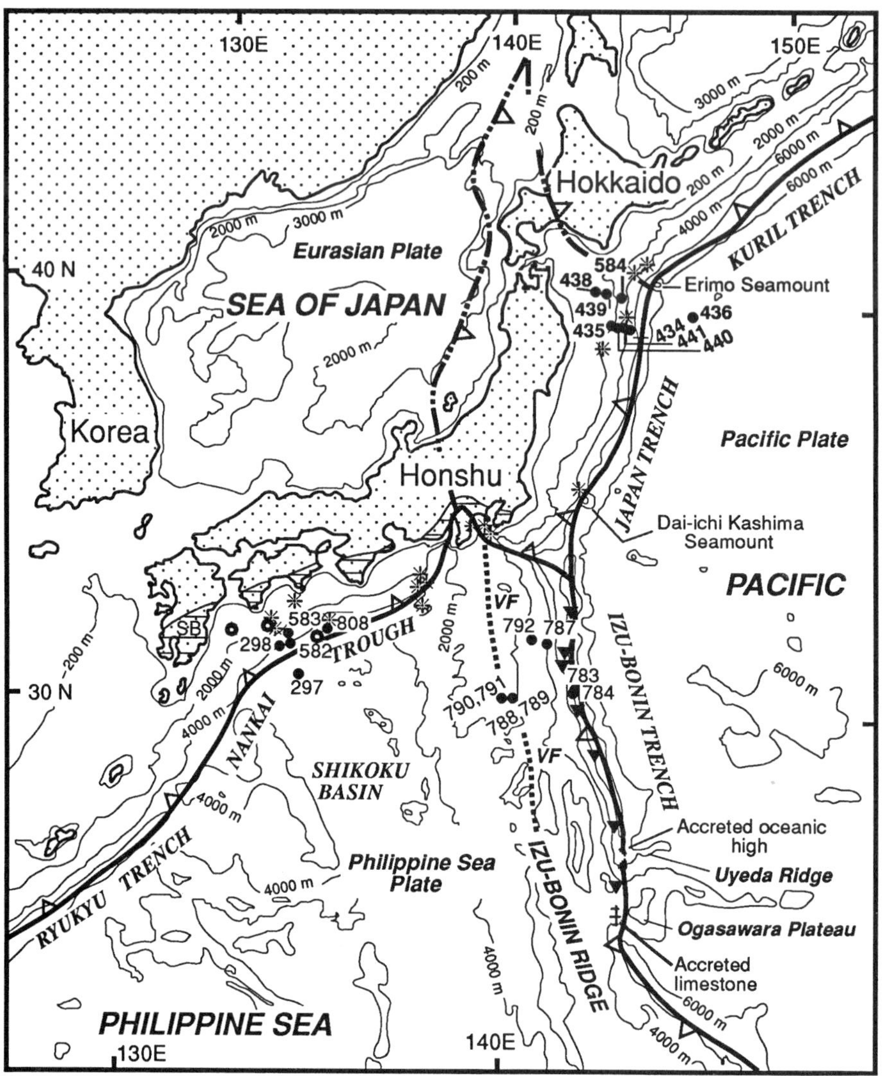

Figure 1. ⎯△⎯, Japanese subduction zones; ⎯...△..., plate boundaries; ●, IPOD drilling sites (DSDP and ODP); volcanic front of Izu-Bonin Arc (dash line labelled VF); *, probable fluid-venting-related biological communities; ▼, serpentinite diapirs; ○, location of selected mud volcanoes referred to in this paper. SB, Shimanto Belt.

history of subduction–accretion, at least since the Late Palaeozoic times (Taira *et al.* 1989).

The Japan Trench is the plate boundary where the Pacific plate is being subducted. The age of the Pacific plate here is about 130–140 Ma and comprises magnetic lineations trending ENE–WSW (Nakanishi *et al.* 1989). The forearc basement is interpreted as mainly Cretaceous accretionary prisms and to the more landward, a Cretaceous forearc basin (Yezo Basin, von Huene *et al.* 1982). The topography of the trench landward slope is characterized by large-scale gravity tectonics, including massive failure of the forearc slope (Cadet *et al.* 1987; von Huene & Culotta 1989).

Phil. Trans. R. Soc. Lond. A (1991)

The forearc to the Izu-Bonin Trench is associated with subduction of the slightly older Pacific plate where the oceanic crust ranges in age from 150–130 Ma (Nakanishi *et al.* 1989). The arc is mainly composed of a Cenozoic volcano–plutonic complex. The trench landward slope is characterized by two discrete slope segments separated by a terrace at water depths of 5–6 km (Honza & Tamaki 1985). Serpentinite diapirism is a prominent feature of this terrace (Fryer *et al.* 1985; Fryer & Fryer 1987).

The collision zone between the mainland southeastern central Japanese, or Honshu, arc and the Izu-Bonin arc records Neogene–Recent uplift and incremental accretion by crustal imbrication (Taira *et al.* 1989). The Neogene accreted material of the ancient Izu-Bonin forearc is exposed in Honshu, immediately north of the arc–arc collision zone. This collisional zone is called the Izu Collision Zone (ICZ).

There are various hydrogeological features associated with the three forearc regions and the ICZ. Although the type, coverage and quality of data from these modern forearc regions differs considerably, we believe that there are now sufficient data to critically review the hydrogeological features in relation to the morpho-tectonic aspects in these forearc regions.

Our approach in this paper is to firstly review the morpho-tectonics of the three forearcs and to describe the important hydrogeological aspects. The distribution of fluid-venting related biological communities and diapiric structures as surface manifestations of fluid flow provide one of the major ingredients in delineating recent to present flow paths. We then provide an overview of the heat flow, location of gas-hydrate (mainly methane-hydrate) related reflectors or bottom simulating reflector (BSR), and deep-sea drilling results as tools for evaluating fluid flow pathways. Finally, we assess the implications for the nature of fluid flow within the forearcs, particularly the accretionary prisms. We aim to show just how pervasive and important such processes are in generating forearc geology.

2. Nankai Trough forearc

(a) *Morpho-tectonics*

The Nankai Trough (figure 1) is the topographic expression of the plate boundary between the Philippine Sea plate and the Eurasian plate, with a convergence rate south of the island of Shikoku of about 3–4 cm a^{-1}, estimated by seismic slip data (Seno 1977), and 2–3 cm a^{-1} using geological constraints (Karig & Angevine 1986). The Philippine Sea plate in this region is composed of the Oligocene–Miocene Shikoku Basin (Kobayashi & Nakada 1978; Shih 1980; Chamot-Rooke *et al.* 1987). The Nankai Trough is relatively shallow (4.8 km maximum water depth) and filled by a 1–2 km thick pile of sediments (Le Pichon *et al.* 1987 *a, b*).

There are basically two layers of sediments subducting below the forearc at the Nankai Trough: an upper turbiditic layer, and a lower hemipelagic stratigraphic succession (Kagami *et al.* 1986; Taira *et al.* 1991). The turbidites in Nankai Trough were supplied mainly from the Suruga Trough drainage area to the east, especially from the Fuji River (Taira & Niitsuma 1986). The sedimentation rate in the turbidite succession reached 2000 m Ma^{-1}, or more, comparable with present rates found in coastal-fluvial deposition around Japan. Taira & Niitsuma (1986) have shown that this enormous sediment flux, mainly from the Fuji River drainage basin to Nankai, occurred because of the arc–arc collision between the Izu-Bonin island arc and the Honshu (mainland Japanese) arc.

The structure of the forearc to the Nankai Trough can be summarized as a trench

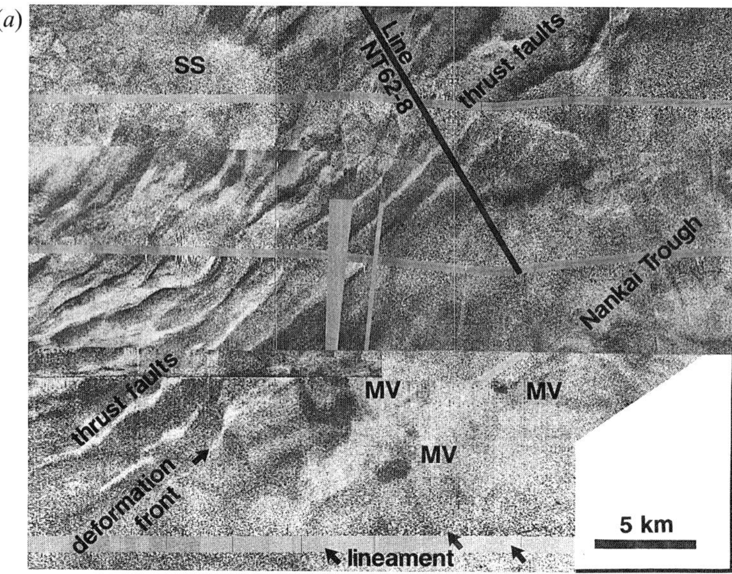

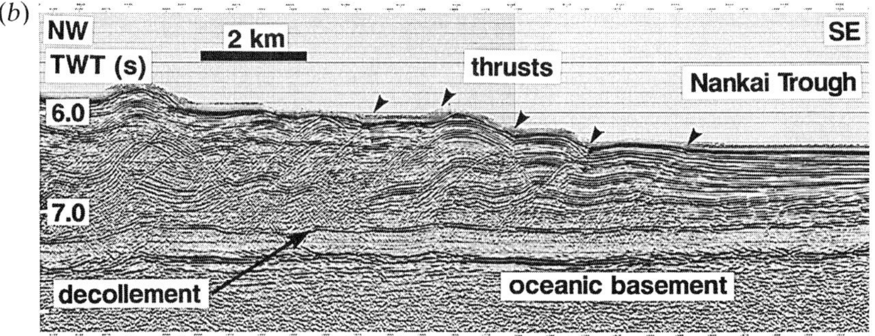

Figure 2. (a) IZANAGI sidescan seafloor image of the Nankai forearc near toe of accretionary prism at ODP Leg 131, Site. MV, mud volcano; SS, submarine slide. Lower slope thrusts intersecting seafloor visible. Lineations on outer trench slope. Site survey seismic line marked. (b) The seismic line NT62-8 shown in (a).

landward slope, comprising an active accretionary prism partly covered by slope basins, a trench-slope break and forearc basins. Deformation of the trench sediments initiates a 'proto-thrust zone', where progressive thickening of the trench wedge can be observed on seismic reflection profiles (Moore *et al.* 1989) as well as core analysis (Karig & Lundberg 1990). In 3.5 kHz sub-bottom profiling records, faults are predominantly high-angle reverse faults with spacing on a scale of about 0.5–1 km and with vertical offsets of 5–20 m.

The frontal thrust zone shows a series of regularly spaced active thrusts spaced at 1–2 km apart (figure 2), and with landward dips that converge to a detachment zone (decollement) which, in some seismic lines, can be traced to about 30 km landward from the trench area (Moore *et al.* 1990). The decollement zone occurs at a particular horizon in the upper Shikoku Basin hemipelagic stratigraphy, but appears unrelated to any obvious lithological change (Taira *et al.* 1991).

Farther landward, the internal image of the accretionary prism is marked by a loss of the typical seismic signal and, instead, the presence of vague landward-dipping reflectors. The decollement seems to change elevation to a lower level and the detachment of the hemipelagic layer from the oceanic basement appears to occur at 4.5 s TWT below the sea floor (Ashi 1991). Normal faults perpendicular, as well as oblique, to the frontal thrust trend begin to develop in this zone, suggesting a possible change in the relative magnitude of the principal stress direction. Taira *et al.* (1988) interpreted this zone to coincide with the transition from frontal accretion to underplating.

Landward of the active accretionary prism, a structural high, the trench-slope break, composed of deformed slope basin sediments and older accretionary prism material, occurs typically at water depths of 0.5–1.5 km. To the arcward side of the trench-slope break, forearc basins are developed with 1–2 km thickness of Neogene sedimentary successions (Okamura 1990), underlain by older accretionary prism material, referred to as the Shimanto Belt (Taira *et al.* 1988).

(b) *Hydrogeologic features*

Fluid-venting related biological communities have been observed at several locations along the Nankai forearc through submersible dives, deep-sea photography and dredging. The best studied areas are located in the eastern Nankai Trough where Kaiko and Kaiko–Nankai expeditions revealed the extensive distribution of biological communities (Le Pichon *et al.* 1987 c). They occur basically in two regions: the frontal part of the accretionary prism, and the upper trench landward slope, and we suggest that both these regions are zones of preferred fluid venting in the accretionary prism.

The frontal part of the accretionary prism seems to be a zone of preferred fluid expulsion (Brown & Westbrook 1988; Moore *et al.* 1988; Moore *et al.* 1990), and they have been interpreted as sites of rapid tectonic dewatering and/or associated with fluid flow from the decollement (Moore *et al.* 1990). Henry *et al.* (1989) showed that the inferred rate of fluid flow associated with the biological communities is large and suggest that fluid venting is rapid (100 m a^{-1}) and transient.

In the upper slope region, several locations are identified as sites of fluid venting, including both the eastern and western Nankai Trough regions (Okamura *et al.* 1986). The presence of biological communities in these regions appears to be associated with zones of active faulting and erosion. As the biological communities generally prefer coarse-grained substrata for growth, wherever turbidity currents and/or deep semi-permanent currents, like the Kuroshio current, have eroded or winnowed out a sandy seafloor, there exists a greater propensity for the establishment of such communities. We also suspect a possible link between gas-hydrate development at many of these sites (see below).

Other evidence for fluid migration in the Nankai forearc has been obtained from the distribution of mud volcanoes revealed by IZANAGI sidescan sonar. The region mapped by IZANAGI sidescan is located in the western Nankai Trough, including the ODP Leg 131 sites (figure 2). Some volcanoes have diameters up to 2 km and relief above the surrounding sea floor of 200 m. The mud volcanoes occur in three areas: (*a*) trench floor adjacent to the toe of the accretionary prism, with an alignment at high angles to the deformation front, interpreted as associated with wrench faults in the prism but controlled by basement structures (Ashi 1991); (*b*) upper trench landward slope, and (*c*) at the base of the landward slope of the forearc basins. The mud volcanoes in the upper slope region occur within a large sediment slide approximately 20×30 km. The sediment slide was a consequence of seismo-tsunamigenic thrust faulting in this region of the prism (Ando 1975). The relationship between sediment sliding and mud diapirism remains unclear, but we suggest that many slides are possibly related to the presence of the extensive, shallow, BSR development.

The nature of the fluid pathways within the accretionary prism has been constrained by surface heat flow measurements, the distribution of BSRs, and drilling results from various DSDP and ODP sites.

Studies of the thermal structure of the prism, based on surface heat flow measurements (Kinoshita & Yamano 1985; Yamano *et al.* 1988; Taira *et al.* 1988*b*), and the location of the BSR (Yamano *et al.* 1982; Ashi 1991), indicate that anomalously high heat flow occurs at the toe of the prism which deviates from the predicted trend based on conductive flux of heat from the oceanic crust. In the Nankai prism, the BSR shows several features (Ashi 1991): (1) the seaward limit of the BSR lies near the toe of the prism, generally only a few thrust-sheet packages landward of the frontal thrust; (2) the BSR does not extend through to the undeformed slope-basin successions; (3) the BSR is well developed in the upper slope region extending through the deformed slope and older accretionary prism deposits, and (4) the BSR is not detected in the forearc basins. The common occurrence of the BSR in the accretionary prism also suggests the widespread advection of fluids throughout much of the forearc.

Hyndman *et al.* (1991) proposed a new mechanism for the formation of gas-hydrate in which, essentially, the gas-hydrate forms by the percolation of methane undersaturated fluid, in contrast to previously suggested processes involving generation from a gas-saturated fluid (Miller 1974). The observed distribution of the BSR in the Nankai accretionary prism, undeformed slope basins and forearc basins, suggests that locally the volume of fluid migration has not been substantial enough to form a gas-hydrate layer, due to fast rates of sediment accumulation and/or low permeability.

In the Nankai prism, in the upper slope region there is a well-developed shallow gas-hydrate phase boundary approximately 200 m below the seafloor (mBSF) which, potentially, could be a source for fluid venting. Sediment sliding and faulting in this region may trigger accelerated rates of fluid migration to release trapped fluid beneath the gas-hydrate phase boundary. The distribution of vent-related bio-communities and mud diapirs (volcanoes) in this region may be related to the development of the shallow BSR.

ODP Leg 131 drilling at Site 808 penetrated 1327 m of stratigraphy below the toe of the prism, from the upper 20 m thick slope veneer, through the frontal thrust and decollement, to the *ca.* 15 Ma subducting oceanic basement (figure 3). Shipboard results show no unequivocal mineralogical or geochemical reason for present active, concentrated, fluid flow within the sedimentary column, even at the thrust front and the decollement, the latter being within monotonous hemipelagic sediments of the palaeo-Shikoku Basin, although there is a slightly more carbonate-rich interval from within the decollement (960 mBSF) to about 25 m below (1000 mBSF). There are some geochemical anomalies associated with the decollement, such as a chloride anomaly in the pore fluids (figure 3), but there is no observable mineralization that could be ascribed to channelized or diffusive fluid flow within the prism, for example calcite-filled microstructures. There is, however, a broad low-chloride pore-water zone below 550 mBSF (figure 3). The start of the chloride anomaly appears to coincide exactly

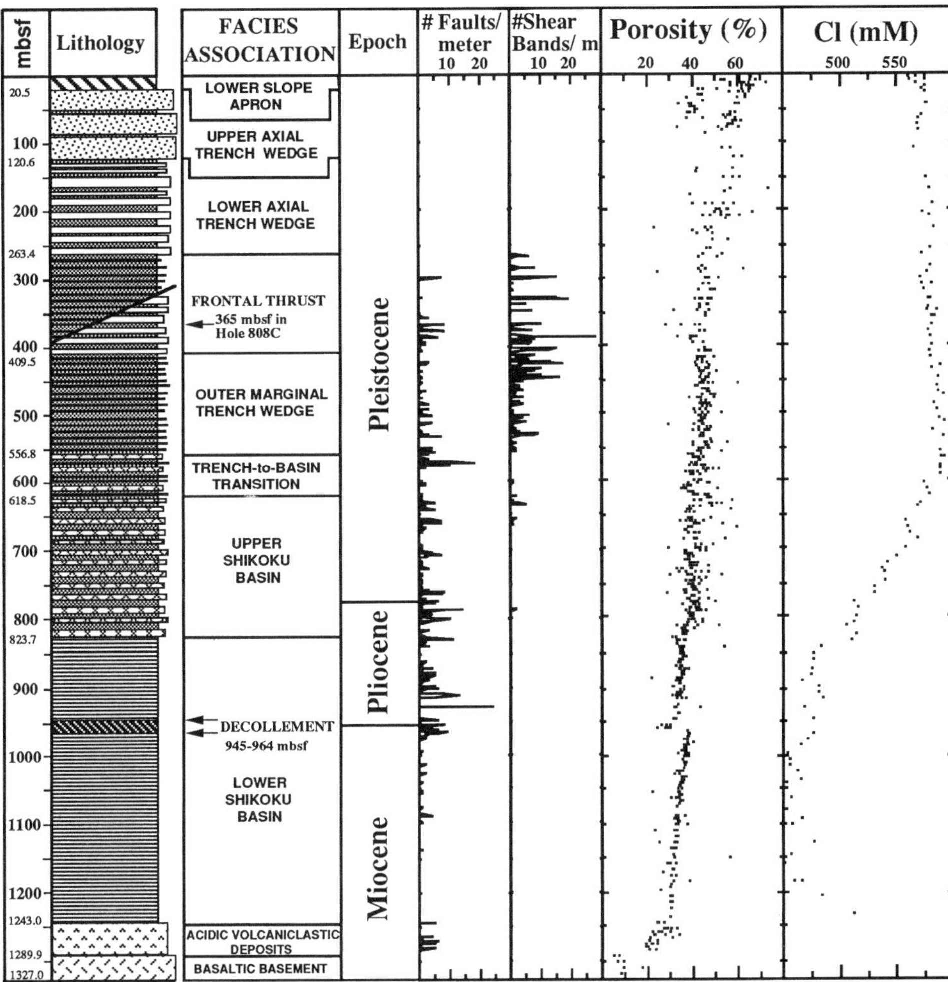

Figure 3. ODP Leg 131 results (from Shipboard Scientific Party, ODP Leg 131, 1990).

with the onset of the illitization of smectite at about 530 MBSF, therefore at least some component of the chloride anomaly could be due to the release of clay mineral bound water during early diagenesis. The overall magnitude of the anomaly, however, would require at least some pulse of deeper fluid flow in the past. Thus the results from ODP Leg 131 contrast with the enhanced methane concentration and low-chloride spike detected in pore fluids from the decollement and fault zone in the Barbados accretionary prism (Moore *et al.* 1988). The origin of the low-chloride fluid remains enigmatic.

There is a sharp contrast in the intensity of deformation above and below the decollement. Sediments above the decollement are intensely deformed, whereas below there is virtually no deformation. The porosity data (figure 3) suggest that the decollement is an overpressured zone separating normally consolidated from underlying underconsolidated muddy sediments. Permissible interpretations of the ODP Leg 131 data-set are: (*a*) fluid flow is pulsed and channelized; or (*b*) there is present channelized fluid flow along the decollement, but the shipboard measurements were unable to detect this and the fluids have not produced any clearly

visible mineralization, for reasons that we do not yet understand. In addition, the
BSR and the actual presence of an overpressured decollement demand continuous
diffusive flow throughout the accretionary prism. The decollement is overpressured
and also acts as a seal to inhibit the dewatering of underthrust sediments. The
mechanism by which the decollement maintains both the characteristics of an
overpressured zone and seal at the same time remains unclear.

3. Japan Trench forearc

(a) *Morpho-tectonics*

The Japan Trench (figure 1) is 7–8 km deep and sediment starved, except in the
area of the trench triple junction where sediments are supplied from the ICZ (Taira
et al. 1989). The trench seaward slope is characterized by well-developed half-graben
structures on the subducting oceanic plate (Kaiko 1 Research Group 1986; Cadet
et al. 1987). The trench landward slope is steep, with a mean slope angle of 7°,
commonly forming a steeper upper slope with benched topography. From the
coastline of the subaerial forearc massif to the trench-slope break at a depth of
2.5 km, there is a broad 150 km wide terrace underlain by a Cenozoic succession
(forearc basin, about 2 km thick) (Nasu *et al.* 1979).

Seismic profiles from the forearc basins to the upper trench landward slope are
characterized by landward-dipping reflectors which are unconformably overlain by
a latest Palaeogene to Neogene sedimentary succession in which normal faults
predominate (Nasu *et al.* 1979). IPOD results and industrial drilling showed that the
landward-dipping reflectors correlate with Cretaceous sedimentary rocks (von Huene
et al. 1982) which probably equate with the Yezo-Sorachi Belt of Hokkaido
(ophiolitic forearc basin sequence). The seaward portion of the basement rocks below
the trench-slope break to trench landward slope region may represent late Cretaceous
to early Palaeogene accretionary prism material (Hidaka Belt). The unconformable
contact between the sedimentary cover and subjacent landward-dipping reflectors
generally cannot be traced seaward beyond the region of the trench landward slope.
The lower portion of the trench landward slope is characterized by discontinuous,
high-frequency reflections. Below the trench landward slope, a detachment surface
(decollement) can be traced for about 15 km landward (Nasu *et al.* 1979). Extensive
horst and graben features are developed on the subducting oceanic plate and are
partly to completely covered by sediment slides and debris from the trench landward
slope (von Huene & Culotta 1989).

IPOD drilling at Sites 438 and 439, from the forearc terrace, recovered a
Pleistocene to Lower Miocene sequence of interbedded mudstone, diatomaceous
mudstone and siltstone above a succession of sandy turbidities (von Huene *et al.*
1982). Below the latter deposits, Oligocene shallow-marine sandstones and volcanic
breccias occur above an unconformity dated at about 40 Ma, and below which there
are Cretaceous claystones. DSDP sites in the lower and mid-slope region are
dominated by Quaternary to Middle Miocene mudstones, some of which are
diatomaceous (von Huene *et al.* 1982).

SeaBeam topographic mapping shows that the trench landward slope of the Japan
Trench is dominated by large-scale gravity collapse (Cadet *et al.* 1987*a*). The
topography is characterized by steep concave scours, large blocks of slope material
and a debris apron on the trench floor. The most spectacular examples are associated

with seamount subduction. Near the junction of the Japan and Kuril Trenches, the Cretaceous Erimo Seamount, part of a seamount chain, is approaching the trench axis. The inner trench wall in this area shows signs of seamount indentation and gravity failure (Cadet *et al.* 1987; Fujioka & Taira 1989). A large embayment with enormous scours is observed to the north of the Erimo Seamount. Yamasaki & Okamura (1988) investigated this area further using magnetic anomalies, and suggested that a seamount of about the same size as Erimo Seamount is buried beneath the irregular topography partly created by slope failure. Southeast of Erimo Seamount, the inner wall of the Japan Trench comprises a series of crescent-shaped escarpments, and an intervening hummocky terrain, with lobate topography at the trench. Fujioka & Taira (1989) interpreted these features as the expression of a large submarine landslide complex. The slide, 50×40 km, probably resulted from seamount subduction in which oversteepened escarpments eventually collapsed.

One of the principal conclusions from a study of the Japan Trench forearc is that the lower portion of the trench landward slope is mechanically very weak and, therefore, subject to frequent massive gravity failure.

(b) Hydrologic features

In the Japan Trench forearc, a number of fluid-venting related bio-communities have been observed, mainly consisting of the giant clam *Calyptogena*, suggesting pervasive fluid venting from the inner trench landward slope and forearc terrace.

Observations from submersible dives show that the trench landward slopes comprise highly sheared mudstones and talus breccias forming a step-like topography (Pautot *et al.* 1987; Cadet *et al.* 1987*b*; Kaiko II Research Group 1987; Lallemant 1989). Vent-related bio-communities are commonly located on the coarse-grained talus veneer and on sheared mudstones associated with normal faulting (Fujioka & Taira 1989; Henry *et al.* 1989).

Recently, several specimens of *Calyptogena* and other vent-related clams, have been obtained by beam trawling from the seafloor of the upper terrace in water depths of *ca.* 1730 m. Although there is no direct visual observation, the existence of fluid-seepage has been suggested from the same area (Ohta 1990, personal communication), this area being within part of a canyon system with a large embayment along the margin of a probable large normal fault block on an anticlinal ridge. Inspection of the nearest seismic profiles to this area, *ca.* 50 km to the north and to the south, reveal that the deep-sea terrace at depths of 1500–2000 km generally coincides with an anticlinal ridge in which intensely faulted Miocene–Pliocene sediments are recorded. The origin of the venting fluids remains uncertain.

Mud diapirism has not been identified in the Japan Trench forearc. This is probably due to the lack of suitable sidescan sonar imaging of the area.

Two lines of evidence exist to constrain the internal fluid pathways in the Japan Trench forearc, i.e. heat flow and drilling data There are relatively few heat flow measurements (less than 20) available for the entire Japan Trench forearc (Yamano & Uyeda 1988). It is therefore premature to infer any systematic variation of heat flow in this region unlike in the case of the Nankai Trough forearc. To date, however, the heat flow data suggest: (1) in contrast to the Nankai prism, no heat flow maxima occurs at the toe of the landward slope of the prism, and (2) heat flow values appear much lower than for Nankai, being about 30 mW m^2.

The downhole temperature measurements at IPOD Site 440 revealed non-linear thermal profiles, from which Burch & Langseth (1981) suggested a vertical fluid flux of approximately 1.4 cm a^{-1}. As no further augmentation of data has been made in this area, these data seem to be the only indication of possible fluid advection in the Japan Trench forearc.

There is no gas hydrate related BSR identified in the Japan Trench forearc despite the presence of organic-rich diatomaceous sediments.

Drilling cores from IPOD sites in the forearc basin and upper trench landward slope yielded abundant evidence of small-scale deformation structures, particularly from the diatomaceous mudstones and include post-depositional veins, healed fractures and microfaults (von Huene 1980; Carson *et al.* 1982). Beneath the forearc basin, these small-scale structures occur only at depths greater than 620 m (Miocene) associated with a zone of normal faulting. Beneath the landward slope of the trench, however, they occur at depths shallower than 250 m (Pliocene). Bulk densities in the landward trench sites indicate a zone of over-pressure at *ca.* 200–700 mBSF. Carson *et al.* (1982) interpreted this zone as associated with tectonic dewatering of the underlying sediments and the upward migration of fluids resulting in an excess pressure horizon above. Such advection of fluids should provide favourable conditions for the development of a gas-hydrate phase boundary. No BSR, however, has been detected to date in the Japan Trench forearc. The reasons for this are unclear, but it may be that there is only about 1 km of unconsolidated sediments (above the Cretaceous basement) which may not generate sufficient gas hydrate to replenish any potential incipient hydrate layer. Pressure–temperature conditions may also be inadequate. In the Peru forearc, a BSR is documented from a comparable geotectonic setting to that in the Japan Trench forearc, in an area dominated by large normal faults (von Huene *et al.* 1988). We speculate that the BSR is absent because there is not sufficient pervasive advection of fluids in the Japan Trench forearc compared with many other forearcs. Further detailed study is required to resolve this problem.

4. Izu-Bonin Trench forearc

(a) *Morpho-tectonics*

The evolution of the Izu-Bonin arc system started with the initiation of westward subduction of the Pacific Plate in the early–middle-Eocene (Uyeda & Ben-Avraham 1972). An episode of Middle Oligocene to early Miocene rifting and seafloor spreading generated the Shikoku Basin and divided the arc in two, with the present Izu-Bonin arc and the remnant, submerged, Kyushu-Palau Ridge. The present incipient rifting of the Izu-Ogasawara arc commenced in the late Pliocene to early Pleistocene.

The present configuration of the Izu-Bonin arc comprises, from east to west, a 7–9 km deep prominent trench, forearc region including a trench landward slope, an outer-arc high, a volcanic arc, an active backarc rift, and the Shikoku marginal basin (Honza & Tomaki 1985) (figure 1). The forearc basin is filled by volcaniclastic and hemipelagic sediments banked behind the outer-arc high. In the southern part of the Izu-Bonin forearc, the outer-arc high is called the Ogasawara Ridge and includes the Ogasawara Islands where the forearc basement, Eocene volcanics and shallow-marine limestones, are exposed (Hanzawa 1947). Several mature dendritic submarine canyon systems have developed across the Izu-Bonin forearc basins to the trench. These canyons have incised as much as 1.5 km into the surrounding sedimentary successions (Taylor & Smoot 1984). The Izu-Bonin forearc thus shows more canyon-eroded morphology than either the Nankai Trough or Japan Trench forearcs.

The lower trench landward slope of the Izu-Bonin Trench is characterized by the offscraping of an oceanic high. SeaBeam mapping by the hydrographic department of the Japanese Maritime Safety Agency (Iwabuchi *et al.* 1988) of the Uyeda Ridge (Smoot & Heflner 1986), shows a linear, E–W trending, topographic high about 150 km in length, 18 km wide, and with a maximum elevation of 4.2 km, to the north of the Ogasawara Plateau. The ridge is being subducted at the Izu-Bonin Trench where water depths are in excess of 9 km. At the toe of the landward wall, there is a prominent 8 × 4 km high at about 5 km landward from the trench axis. This topographic feature is oblique to the linear and smooth N–S trending cliff line of the trench landward slope. It also shows a continuous magnetic anomaly from the ocean side and no associated gravity anomaly, suggesting that this portion is a continuation of the oceanic crust. Iwabuchi *et al.* (1988) interpreted the linear high as a probable offscraped segment of the Uyeda Ridge at the toe of the inner trench wall.

Part of the Ogasawara Plateau is subducting at the southern Izu-Bonin Trench: a prominent structural high occurs in the toe of the trench landward slope at this boundary, as the rectangular-shaped, 50 × 20 km, Hahajima Seamount, with about 2 km of relief from the surrounding oceanic floor. Dredge samples from this seamount are ultramafics. To the southeast of this seamount, there is a steep, N–S trending, cliff interpreted as the toe of the trench landward slope. At a depth of 3.3 km, which is about 1 km above the trench axis, a deep-tow survey by JAMSTEC (Momma *et al.* 1990) revealed exposures of white-coloured rocks, which upon dredging were shown to include late Cretaceous (Turonian) nannofossil limestones. Micritic limestones of this age have not previously been recorded from the Izu-Bonin forearc, but they are comparable with the limestones that veneer the Ogasawara Plateau. It seems likely that at least part of this steep trench landward slope may comprise accreted Ogasawara Plateau crust.

The above evidence leads us to speculate that at least part of the forearc crust to the Izu-Bonin arc consists of accreted oceanic material from the Pacific plate. This, then, suggests different types of accretionary processes operate in the Izu-Bonin Trench forearc compared to the Nankai Trough and Japan Trench forearcs.

(b) *Hydrogeologic features*

On the mid-slope bench of the trench landward slope of the Izu-Bonin Trench to the Mariana Trench, there are a number of serpentinite seamounts ranging from 5 km to more than 20 km in diameter, and from several hundred to more than 1400 m in height (Ishii 1985). These seamounts are spaced at intervals of 15–60 km along the bench situated less than 50 km from the trench axis.

Two sites on ODP Leg 125 were drilled on the forearc seamounts within the Izu-Bonin forearc (Torishima Seamount), and three sites into Conical Seamount in the Mariana forearc. Primary mantle material was recovered as predominantly highly depleted, tectonized, harzburgite which has experienced medium-grade metamorphism. The associated fluids contain significant amounts of hydrocarbons, with methane concentrations of up to 400 µl l^{-1} of sediments.

One of the major findings from ODP Leg 125 is the recovery of blueschist facies metamorphic rocks from the Conical Seamount of the Mariana Trench (Maekawa *et al.* 1991). Conical Seamount is situated about 30 km above the subducting slab, and the blueschist metamorphism probably occurred at *ca.* 150–250 °C and 4–6 kbar (0.4–0.6 GPa) (13–18 km deep).

An *Alvin* diving survey revealed the presence of chemical chimneys associated with fluid venting (Fryer *et al.* 1990). These chimneys are made of carbonate, calcite and

[73]

aragonite, together with silicate (a new Mg-silicate) (Haggerty 1985). Fryer *et al.* (1990) considered that a biogenic source in the Mariana forearc is unlikely to have been responsible for producing these chimneys. Instead, they suggest that methane was generated at depth beneath the forearc wedge, possibly by the interaction between mantle material and water from the subducting slab, including both sediment-hosted water and slab-hosted water.

Thermally, the forearc region is characterized by an arcward increasing trend in heat flow ranging from 20 to 100 mW m^{-2}. Heat flow values greater than 50 mW m^{-2} pose a problem in accounting for the thermal structure of the forearc region: the forearc temperatures become too high at the appropriate depths for generating blueschist metamorphism. Although the heat flow measurements from the Izu-Bonin forearc are rather few in number to satisfactorily resolve this paradox, we discuss some possible solutions in a later section of this paper.

ODP Legs 125 and 126 (Fryer *et al.* 1990; Taylor *et al.* 1990) encountered abundant dark vein-like structures in the sediments, typically 3–5 cm long, subvertical to bedding, in some cases S-shaped, and with bifurcating terminations. Sediment dykes and other fluidization structures were also reported from the forearc sediments. Some vein structure occurs at depths of less than 50 m below the seafloor, suggesting a phase of early fluid migration. Recently, Kimura *et al.* (1989) documented vein structure farther south from within 50 cm of the seafloor, in unconsolidated volcaniclastic muds, near the junction of the Mariana and Yap Trenches. The vein structure occurs widely throughout the accreted forearc sediments exposed in onland sections. The role of vein structure in fluid migration and expulsion is not yet fully understood, but their abundance suggests that vein structure may be an important indication of early porosity reduction processes within some forearcs. Vein structure is discussed further in a later section.

5. Fluid activity in the arc-arc Izu Collision Zone

The Izu Collision Zone (ICZ) encompasses the region of Neogene-Recent arc-arc collision between the mainland Japanese, or Honshu, arc and the Izu-Bonin island arc immediately west of the trench–trench–trench triple junction off SE Japan (figure 4). The present collision boundary extends from west to east through the Suruga Trough, below Mt Fuji, and through Sagami Trough to the triple junction area.

Reconstructing plate boundaries around the ICZ suggests that accretion of successive segments of the Izu-Bonin (Izu-Ogasawara) arc crust has occurred by crustal delamination and incremental accretion (Taira *et al.* 1989; Soh *et al.* 1991).

The Miura Group (Eto *et al.* 1987) ranges from late Miocene–Pliocene (10–3 Ma) and comprises deep shelf to basinal sediments, mainly as scoriaceous to pumiceous volcaniclastics derived from the Izu-Bonin arc. These sediments, now exposed onland, represent part of the Neogene accretionary complex resulting from arc–arc collision in the ICZ. The sediments accumulated in relatively short-lived foreland basins, filled by overall coarsening and shallowing upward sequences developed along the major basin-bounding thrust faults. The basin fills tend to become younger in age from north to south, interpreted as a consequence of the incremental migration of the plate boundary caused by the incorporation of segments of the Izu-Bonin arc onto the Honshu arc (Taira *et al.* 1988).

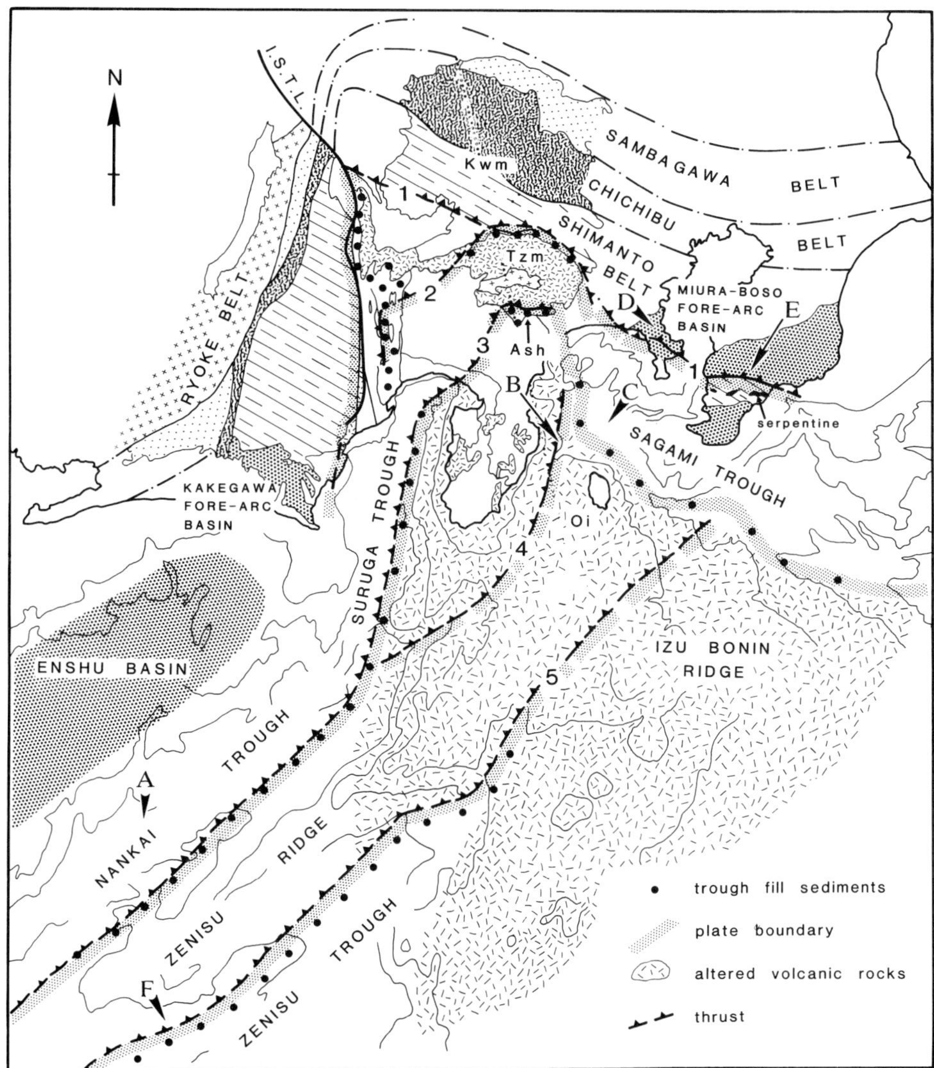

Figure 4. Tectonic framework of the Izu collision zone and location of related seep communities. A, Tenryu Community; B, Hatsushima Community; C, Okinoyama Community; D, Ikego Formation; E, Kurotaki Formation; F, Community at the Zenisu Ridge. Arrow to bottom-right is the direction of plate motion. Nos 1–5 refer to successive plate boundaries caused by the incremental accretion of Izu-Bonin are segments; 4 and 5 are incipient plate boundaries. Oi, Oshima Islands.

The main purpose of introducing the ICZ here is to discuss recent offshore fluid venting along the Sagami Trough and fossil fluid activity in the Miura Group.

(a) *Fluid venting in the Sagami Trough*

In the Sagami Trough, a collisional plate boundary between the forearc segment of the Izu-Bonin arc send Honshu arc, several fluid-vent-related bio-communities have been identified (figure 4) (Ishii *et al.* 1988; Hashimoto *et al.* 1989). The bio-communities occur in three tectonic zones: (1) Hatsushima zone of the Izu Bonin arc; (2) along the frontal thrust of the accretionary prism (Sagami accretionary prism) in

Figure 5. Vein structure, Miocene–Pliocene Misaki Formation, Miura Peninsula, SE Japan. These structures represent the earliest post-depositional expression of pore-fluid activity which generates a systematic fabric in muddy lithologies.

which the forearc sediments of the Izu-Bonin arc and trough-fill sediments have been accreted, and (3) the landward margin of the mid-slope terrace and head of submarine canyons.

The Hatsushima fluid-venting zone is situated along the base of the N–S trending fault scarp to the east of Izu peninsula. Submersible dives have located dense populations of *Calyptogena* communities (Hashimoto *et al.* 1989), and by the remote operating vehicle *Dolphine* 3K (Hattori 1989). Kinoshita & Yamano (1990) measured very high heat flow, up to more than 190 mW m^{-2}, and showed that these values occur close to the volcanic front, and suggests that this unique area is a fluid venting site. Furthermore, this data shows that volcanic heat sources must influence fluid circulation.

The fluid-venting-related bio-communities in the Sagami accretionary prism provide an important comparison with the Nankai accretionary prism farther west. In both prism sites, fluid venting is apparently related to: (1) thrust faults and the hinge regions of anticlines, and (2) canyon systems, again probably linked to tectonic features. In the Sagami Trough, the two fluid-venting sites are located either on anticlinal ridges or along thrust faults, both constituting the wall of a submarine canyon. This setting is similar to ones found in the Kaiko–Nankai sites. Dredge samples from the mid-slope show clam colonies on a thrust fault eroded by a submarine canyon. These morpho-tectonic settings indicate that fluid venting is currently taking place along active thrust faults in the Sagami Trough.

(b) *Fossil fluid activity in the Miura Group, SE Japan*

Evidence for fluid activity in the Miocene–Pliocene Miura Group, SE Japan, is grouped into two different types: (1) fossil *Calyptogena* beds in comparable tectonic positions compared to those from the present Sagami Trough, and (2) vein structures, sediment injections and related phenomena in the palaeo-Izu-Bonin forearc sediments developed prior to, and during, accretion of the 'Miura block' onto the Honshu arc.

On Miura peninsula, the deep-marine volcanic sandstones and conglomerates of the Ikego Formation, deposited in upper bathyal depths (inferred from benthic foraminifera), contains a giant clam colony of *Calyptogena* (Niitsuma *et al.* 1989). Our onland fieldwork indicates localized concentration of these clams, many of which are preserved in life position, suggesting their preservation close to fossil fluid-venting sites. Abundant evidence of sediment dykes and injection-related brecciation, suggests the possibility of high pore-pressure within the dewatering and lithifying sedimentary pile. Locally, there is extensive cementation, some of which developed in the shape of circular conduits and chimneys, reminiscent of the modern fluid-venting chimneys found in the Japanese forearcs. Calcite cements obtained from the shell-bearing volcanic conglomerates, within the Ikego Formation, have extremely light carbon isotopic ratios ($\delta^{13}C = -19.7\%_{oo} - 49.2\%_{oo}$), suggesting an origin from biogenic methane due to fluid seepage. Detailed mapping shows that the *Calyptogena* colony in the Ikego Formation probably formed on fault-related talus (chaotic) deposits along a canyon wall trending NNW–SSE. The reconstructed sedimentary environment suggests that this onland ancient setting is similar to the mid-slope site in the Sagami accretionary prism.

Vein structure is abundant in the Miura Group (figure 5). The vein structure is generally restricted to layer-parallel zones 1–10 cm thick, in which individual veins are approximately equidistant, ranging from 1–30 mm apart (Pickering *et al.* 1990). They tend to be planar and typically up to 0.1–2.0 mm in width in the central parts. There are, however, several generations of vein structure, with later sets tending to be larger and more widely spaced, even with sigmoidal geometries.

In general, the vein structure in the Miura Group predates virtually all of the faulting, although there are a few rare cases where the vein structure may cut layer-parallel thrusts, which may also be very early structures. Vein structure occurs within blocks of sediment caught up in early, chaotic, sediment slide deposits. Our unpublished X-ray tomographic analysis of such vein structure suggests that the veins apparently contain more dense interiors than the surrounding material and host sediments, indicating that they represent a porosity reduction. Such data suggest that the vein structure formed early in the deformational history of the dewatering sediments, and probably at rather shallow depths of burial.

Vein structure has been recognized from various forearc trench-slope settings from DSDP and ODP sites (see summary by Lundberg & Moore 1986). Recently, Kimura (1989) documented vein structure from within a few metres of the seafloor in unconsolidated volcaniclastic muds near the junction of the Mariana and Yap Trenches. Drilling in the Mariana and Izu-Bonin forearcs also revealed the shallow occurrence of vein structure, from within 100 mBSF (Fryer *et al.* 1990; Taylor *et al.* 1990).

There have been many interpretations of vein structure (Knipe 1986; Lundberg & Moore 1986). Cowan (1982), who was the first to describe vein structure from the Middle American Trench, interpreted it as extensional fractures. Knipe (1986) interpreted the structure as forming in response to gravity-induced downslope failure of sediment. Knipe (1986) also noted the possibility that veins may develop or be modified to a sigmoidal geometry as a result of shear parallel to bedding in unconsolidated slope sediments (cf. Kimura *et al.* 1989). Pickering *et al.* (1990) interpreted vein structure in the Miura Group as forming in an extensional setting under gravity-controlled creep processes.

The evidence for a porosity reduction within the vein structures suggests that they provided effective and pervasive fluid expulsion conduits. Although the overall significance of vein structure in the consolidation history of forearc sediments

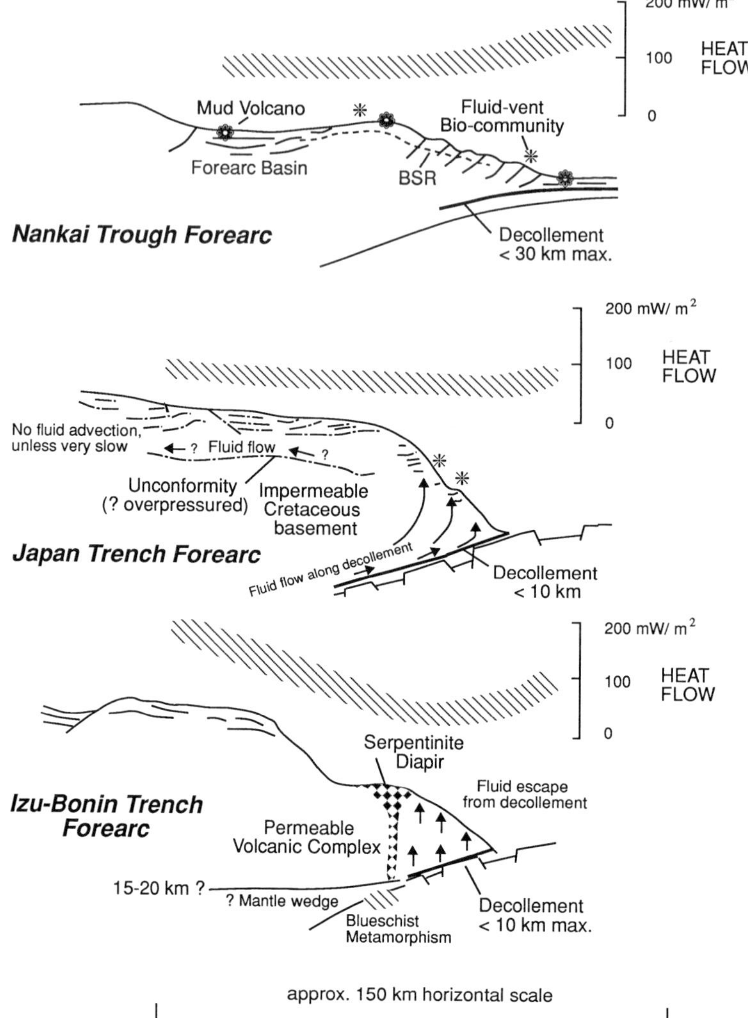

Figure 6. Synthesis of forearc structure and hydrogeologic features in the Nankai, Japan and Izu-Bonin Trench forearcs. Schematic only.

remains uncertain, their abundance and orientation within the Miura Group suggests that they acted as important local conduits for fluid flow in certain finer-grained lithologies.

7. Discussion

The three forearc regions described in this paper show a wide variety of tectonic style and fluid activity which is summarized in figure 6 and table 1. The three fundamental questions to address are: (1) the source of the various fluids in forearcs; (2) the hydrogeologic framework and fluid pathways (mechanisms of fluid migration, circulation patterns, fluid fluxes, etc.), and (3) the relationships between tectonics, diagenesis-metamorphism and fluid circulation. Although we stand a considerable distance away from solving all these problems, we can assess at least some of the important implications bearing on these problems from the available data.

Table 1. *Comparison of morpho-tectonics and hydrogeology*

	Nankai	Japan	Izu-Bonin
trench sediments and tectonics (subduction rate)	1–2 km thick turbididites and hemipelagites, no horst and graben on outer swell (3 cm a^{-1})	500 m pelagites and hemipelagites, horst and graben (10 cm a^{-1})	500 m thick pelagites and ashes, horst and graben (6 cm a^{-1})
shape of trench landward slope	single slope or gentle lower slope and steep upper slope	very steep lower slope and gentle upper slope	steep lower slope and gentle upper slope
tectonics of trench landward slope	sediment accretion	gravity failure	partial accretion of oceanic highs
forearc rocks	accretionary prism (mostly turbidites and hemipelagites)	Cretaceous sedimentary rocks overlain by Cenozoic sequence	Eo–Oligocene volcanic complex overlain by Neogene volcaniclastics
vent-related biological communities	occur at the toe region and upper slope	occur at the lower trench landward slope and trench-slope break	not known
diapiric features	mud volcanoes at the toe region, upper slope and forearc basin	not known	serpentinite diapirs at the mid-slope bench
bottom simulating reflectors	well developed in the trench landward slope	no development	no development
heat flow	landward decrease (200–500 mW m^{-2}) high at the toe	uniform (20–50 mW m^{-2})	landward increase (20–80 mW m^{-2})
vein structure	not known	present	present
hydrogeologic interpretation	concentrated flow at the toe and upper slope; slow advection in entire prism; overpressured decollement	overpressured decollement with forward fluid flow; overpressured lower slope	less overpressured decollement; slow advection through forearc basement; migration of deep-seated fluids through serpentinite diapir

(a) *Fluid sources*

In subduction settings, fluid sources include the subducting slab and sediments, accreted and subcreted sediments, cover sequences, basement magma-related fluids and meteoric water. At present, we cannot quantify the relative amounts of fluid generated by these sources at different levels in the forearcs. The following discussion of fluid sources, therefore, tends to be related to the amounts of sediments being subducted.

It is possible to estimate the volume of sediments that are incorporated into the subduction zone in the three forearc regions discussed in this paper. The parameters used are: (1) the thickness of sediments incorporated into the subduction zone, and (2) the rate of subduction. In the Nankai forearc, the mean thickness of sediments is about 1000 m, including trench wedge and hemipelagites. The rate of subduction off Shikoku is about 3 cm a^{-1}. In the Japan Trench, the estimate is not so simple due to the development of horst and graben structure. We estimate a mean vertical displacement of graben blocks of about 300 m, based on seismic profiles (Kaiko I Research Group 1986). The thickness of the pelagic and hemipelagic cover on the Pacific seafloor is about 600 m. Because the horsts and grabens are about equally developed, the mean thickness of sediments subducted is *ca.* 750 m, assuming that the graben are filled by debris at the trench. The subduction rate is *ca.* 10 cm a^{-1}. In the Izu-Bonin forearc, the sediment thickness on the oceanic crust is about 400 m. The mean displacement of normal fault is similar to those in the Japan Trench, i.e. *ca.* 300 m. This gives a mean sediment thickness of 550 m of sediment that potentially can be subducted. The subduction rate is 6 cm a^{-1}. These estimates, although they are first approximations and ignore the porosity variations, suggest that the annual rate of sediment being subducted in the three forearcs, expressed in cubic metres per metre width are as follows:

Nankai: 30 m^3,   *Japan*: 75 m^3,   *Izu-Bonin*: 33 m^3.

These values show that the volumes are the same order of magnitude in the three forearcs: the greater value for the Japan Trench forearc possibly indicates that it receives more water than the other forearcs. The sediment volumes involved in the Japan Trench and Izu-Bonin Trench forearcs, where sediments are subducted directly underneath the trench landward slope without frontal accretion, are roughly comparable with, or probably larger than, that for the Nankai accretionary prism in which frontal accretion is a major process. These contrasting accretionary processes of predominantly frontal accretion (Nankai) and subcretion (Japan and Izu-Bonin) should provide very different mechanisms for sediment dewatering and fluid migration paths.

One of the very important sources of fluids in the accretionary complex is the release of loosely, and more tightly, bound structural water from sediments, especially clay minerals, during compaction and early diagenesis. The general decrease in porosity with depth, as seen in ODP Leg 131 Site 808 (figure 3), provides an indication of the volumes of trapped seawater that are released during the early phases of compaction when primary porosity is lost. The illitization of smectites, for example, will release relatively large volumes of low-chloride fluids. In ODP Leg 131 Site 808 near the toe of the Nankai accretionary prism, the decollement is at a temperature of about 110 °C, which is well within the smectite–illite transformation field, therefore a component of the low-chloride anomaly probably owes its origin to this early diagenetic mineralogical change, from about 530 MBSF.

Among the possible fluid sources, geochemical data suggest that magma-related fluids are probably relatively insignificant in the case of the Japanese forearcs. An input from meteoric water may be important, as for example is the case in the Peru margin, but little is known of its significance in the Japanese forearcs.

(b) Fluid Pathways

The Japanese forearcs and trenches exhibit considerable evidence for diffusive and channelized, focused, fluid flow and venting. There are three different modes of fluid migration in forearcs: (1) diffusive advection through porous media; (2) channelized flow in some permeable conduits, and (3) migration of fluids related to host material transfer such as diaprism. Again, the relative importance of these mechanisms and temporal nature (transient against steady-state), remains unclear in the three forearcs. We suggest that the fluid pathways will, to a large extent, be controlled by the nature of the forearc basement and the volume of subducted sediments.

Probably, the best evidence for diffusive fluid flow is the presence of a BSR. The widespread distribution of a BSR in the Nankai Trough forearc suggests that fluid migration occurs throughout most of the landward trench slope region. The absence of a BSR in the Japan and Izu-Bonin Trench forearcs does not preclude fluid migration throughout the landward slope. However, we suggest the possibility of very different fluid circulation patterns in the lower landward slope region of the Japan and Izu-Bonin forearcs.

The Japan Trench forearc is underlain by Cretaceous accretionary prism and forearc sediments. The outcrops of equivalent rocks in Hokkaido suggests that the concealed prism is composed of highly consolidated and impermeable rocks, mainly of deformed flysch deposits. Such strata may act as an impermeable seal on any subducted water-rich sediments: the reason why there is no BSR in the Japan Trench forearc may be a consequence of insufficient water being supplied to the overlying sediments to cause pervasive fluid advection. As pointed out by von Huene & Culotta (1989), the existence of a long and continuous seismic image of the decollement in the Japan Trench forearc suggests an overpressured zone. Thus, the decollement with an upper seal, should be a conduit for fluid flow. As there is insufficient material to accrete above the decollement, no internal deformation related to accretion occurs. Instead, the forward migration of fluids should produce an overpressured toe in the lower landward slope. This may be the reason why the lower trench slope is so weak in the Japan Trench prism, which is subject to repeated sediment mass failure.

In the Izu-Bonin arc, the basement of the trench landward slope is mainly composed of Eocene–Oligocene volcanic rocks, including volcaniclastics, hyaloclastic breccias, and pillow and sheet lavas (Taylor *et al.* 1990); locally, there is accreted oceanic material. We believe that the Izu-Bonin forearc has essentially achieved a steady state, with neither voluminous accretion nor major gravity tectonics to modify the volume of the forearc. Furthermore, it is reasonable to assume that the Izu-Bonin forearc basement comprises very permeable materials. In such a setting, subducted sediments would be expected to dewater rapidly. We therefore speculate that in the Izu-Bonin forearc, a water-saturated decollement is even less likely to develop than in the Japan Trench forearc. As material in the Izu-Bonin forearc is generally permeable, massive sediment failure, as slides, is uncommon. A corollary of these arguments is that the decollement in the Izu-Bonin forearc has a greater frictional resistance compared with that in the Nankai and Japan Trench forearcs. The absence of a BSR in the Izu-Bonin forearc may be due to lithologic control: organic-poor volcaniclastic rocks, which may explain why the accretion of an oceanic high is taking place at the Izu-Bonin forearc.

(c) *Implications for forearc seismicity and thermal structure*

The foregoing discussion leads to two broad implications for the deeper seismicity and thermal conditions in the forearcs.

It is a striking feature of subduction zones in general that there is virtually no seismic activity along the subduction zone to depths of 20–30 km (Yoshii 1979; Fukao 1979). Such seismically quiet zones suggest that the zone is subject to smaller frictional effects due to the presence of fluids (Shimamoto 1985). As suggested earlier in this paper, the amount of subducted fluids may, at least in part, be a function of the rate of dewatering at the shallow part of the prism, at depths of about 5 km. The linkage between shallow dewatering processes and the availability of fluids from greater depths is unclear. It is therefore important to investigate the width and depth of the aseismic slip zone in relation to the shallow level processes operating towards the toe region of a prism. Very little data is currently available on this topic.

Reck (1987) pointed out that heat flow in the Japan Trench forearc is apparently too high to produce blueschist metamorphism at depth. He suggested that the high heat flux in the forearc is a manifestation of extensive fluid migration to the surface.

The calculated temperature at a depth of 30 km in the three forearc case studies is: 600 °C in Nankai, 300 °C in the Japan Trench forearc, and 700 °C in the Izu-Bonin forearc (Uyeda *et al.*, unpublished data). Although the surface heat flow measurements in the forearcs are too few to make any conclusive statement, these temperatures suggest that Reck's hypothesis may be valid. The recent recovery of blueschist from the Izu-Bonin forearc suggests that blueschist metamorphism is probably taking place at depth. The Izu-Bonin forearc, therefore, could provide a unique opportunity to investigate the role of fluid migration linked to blueschist metamorphism, i.e. the question of the depths from which fluid migration in the forearc can affect the thermal structure of the subduction zone. To date, the temperature of subduction zones is primary determined by the age of oceanic lithosphere and calculations of shear heating along the slip zones, whereas the energy transfer by fluid flux has not been fully evaluated.

8. Conclusions

In this paper, we have stressed the pervasiveness of fluid activity in the Nankai Trough, Japan and Izu-Bonin Trench forearcs, and shown that the fluid flux can play an important role in shallow to deep level tectonics, and the thermal structure of the forearc regions. We have reviewed a variety of fluid activity in three, morpho-tectonically, contrasting forearcs.

The main morpho-tectonic characteristics of the Nankai Trough, Japan Trench, and Izu-Bonin Trench forearcs are:

1. The Nankai Trough forearc is an accretionary system dominated by frontal accretion, and with a long history of development as a fold-thrust belt.

2. The Japan Trench forearc is predominantly erosional, mainly by slope failure and sliding, and has a pronounced subsidence history. The basement of the trench landward slope is composed of Cretaceous consolidated sediments.

3. The Izu-Bonin Trench forearc appears to be more stable than the other two forearcs, with evidence of local oceanic crustal accretion. The basement is composed of an older lava-volcaniclastic complex.

We suggest that the fluid pathways and permeability variations within these three forearcs are quite different. In the Nankai forearc, concentrated fluid expulsion occurs at the toe of the prism and upper trench landward slope; the rest of the prism is dominated by slow fluid advection as sediments gradually dewater. In the Japan Trench forearc, concentrated fluid flows occur in relation to normal faulting, and there appears to be no pervasive advection from depth. The decollement could be water-saturated and subject to relatively lower frictional forces. The forward migration of fluids along the decollement produces overpressuring at the lower trench landward slope, which triggers massive sediment failure. In the Izu-Bonin forearc, slow fluid advection may occur throughout the permeable forearc basement. The decollement, however, may not be water-saturated. The rapid loss of water from the decollement, through permeable arc-basement, could explain the reason for the accretion of oceanic highs in the Izu-Bonin forearc, and also provide an explanation for the relatively steep lower trench landward slope, assuming the Coulomb wedge model (Davis *et al.* 1983) is applicable.

We thank T. Byrne, J. Ashi, K. Fujioka for useful discussion. S. Ohta provided us with information on vent-related bio-communities in the Japan Trench forearc upper slope. T. Barber and A. Robertson critically read an early version of the typescript and made many useful comments.

References

Ando, M. 1975 Source mechanisms and tectonic significance of historical earthquakes along the Nankai Trough, Japan. *Tectonophysics* **27**, 119–140.

Aoki, Y., Tamano, T. & Kato, S. 1933 Detail structure of the inner trench slope of the Nankai Trough, from migrated seismic sections. *Am. Ass. petrol. Geol. Mem.* **34**, 309–324.

Arthur, M. A. & Von Huene, R. 1980 Sedimentary evolution of the Japan forearc region off northern Honshu (DSDP Legs 56 and 57). In *Initial Rep. DSDP* **56**, **57**, 521–568. Washington, D.C.

Ashi, J. 1991 Structure and hydrogeology of the Nankai accretionary prism. D.Sc. thesis, Dept of Geology, University of Tokyo.

Brown, K. & Westbrook, G. K. 1988 Mud diapir and subduction in the Barbados Ridge accretionary prism complex: the role of fluids in accretionary processes. *Tectonics* **7**, 613–640.

Burch, T. K. & Langseth, M. 1981 Heat-flow determinations in three DSDP boreholes near the Japan trench. *J. geophys. Res.* **86**, 9411–9419.

Cadet, J.-P., Kobayashi, K., Lallemant, S., Jolivet, L., Aubouin, J., Boulegue, J., Dubois, Hotta, H., Ishii, T., Konishi, K., Niitsuma, N. & Shimamura, H. 1987 Deep scientific dives in the Japan and Kuril Trenches. *Earth planet. Sci. Lett.* **83**, 313–328.

Cadet, J.-P., Kobayashi, K., Aubouin, J., Boulegue, J., Deplus, C., Dubois, J., von Huene, R., Jolivet, L., Kanazawa, T., Kasahara, J., Koizumi, K., Lallemant, S., Nakamura, Y., Pautot, G., Suyehiro, K., Tani, S., Tokuyama, H. and Yamazaki, T. 1987 The Japan Trench and its juncture with the Kuril Trench: cruise results of the Kaiko project, Leg 3. *Earth planet. Sci. Lett.* **83**, 267–284.

Carson, B., Von Huene, R. & Arthur, M. 1982 Small-scale deformation structures and physical properties related to convergence in Japan Trench slope sediments. *Tectonics* **1**, 277–302.

Chamot-Rooke, N., Renard, V. & Le Pichon, X. 1987 Magnetic anomalies in the Shikoku Basin: a new interpretation. *Earth planet Sci. Lett.* **83**, 214–228.

Davis, D. M., Suppe, J. & Dahlen, F. A. 1983 Mechanics of fold-and-thrust belts and accretionary wedges. *J. geophys. Res.* **88**, 1153–1172.

Davis, E. E., Hyndman, R. D. & Villinger, H. 1990 Rates of fluid expulsion across the northern Cascadia accretionary prism: constraints form new heat flow and multichannel seismic reflection data. *J. geophys. Res.* **95**, 8869–8889.

Eto, T., Oda, M., Hasegawa, S., Honda, N. & Funayama, M. 1987 Geologic age and paleoenvironment based upon microfossils of the Cenozoic sequence in the middle and northern parts of the Miura Peninsula. *Sci. Rep. Yokohama National University, Ser.* II **34**, 41–57. In Japanese with English abstract.)

Fryer, P. & Fryer, G. J. 1987 Origin of nonvolcanic seamounts in a forearc environment. In *Seamounts, islands and atolls* (ed. B. H. Keating, P. Fryer, R. Batiza & G. W. Boehlert), *Geophysical Monograph* **43**, 61–69. American Geophysical Union, Washington, D.C.

Fryer, P., Ambos, E. L. & Hussong, D. M. 1985 Origin and emplacement of Mariana forearc seamounts. *Geology* **13**, 774–777.

Fryer, P. *et al.* 1990 *Proc. ODP, Initial Rep.*, **125**: College Station, TX (Ocean Drilling Program).

Fryer, P., Saboda, K., Johnson, L., Mackay, M. E., Moore, G. F. & Stoffer, P. 1990 Conical Seamount: SeaMARK II, Alvin submersible and seismic-reflection studies. In *Proc. ODP, Initial Rep.* **125**: College Station, TX (Ocean Drilling Program), 69–80.

Fujioka, K. & Taira, A. 1989 Tectono-sedimentary settings of seep biological communities – a synthesis from the Japanese subduction zones. In *Sedimentary facies in the active plate margin* pp. 577–602. Tokyo: Terra Scientific.

Fukao, Y. 1979 Tsunami earthquakes and subduction processes near deep-sea trenches. *J. geophys. Res.* **84**, 2303–2314.

Haggerty, J. A. 1987 Petrology and geochemistry of Neogene sedimentary rocks from Mariana forearc seamounts: implications for emplacement of the seamounts. In *Seamounts, islands and atolls, Geophysical Monograph* **43**, 175–185.

Hanzawa, S. 1947 Eocene foraminiferal from Hahajima. *J. Paleont.* **21**, 254–259.

Hashimoto, J., Ohta, S., Tanaka, T., Hotta, H., Matsuzawa, S. & Sakai, H. 1989 Deep-sea communities dominated by the giant clam, *Calyptogena soyoae*, along the slope foot of Hatsushima Island, Sagami Bay, central Japan. *Paleogeography Paleoclimatology Paleoecology* **71**, 179–192.

Henry, P., Lallemant, S. J., Le Pichon, X. & Lallemant, S. E. 1989 Fluid venting along Japanese trenches: tectonic context and thermal modelling. *Tectonophysics* **160**, 277–291.

Hilde, T. W. C. 1983 Sediment subduction versus accretion around the Pacific. *Tectonophysics* **99**, 381–397.

Honda, S. 1985 Thermal structure beneath Tohoku, Northeastern Japan – a case study for understanding the detailed thermal structure of the subduction zone. *Tectonophysics* **112**, 69–102.

Honda, S. & Uyeda, S. 1983 Thermal process in subduction zones – a review and preliminary approach on the origin of arc volcanism. In *Arc volcanism – physics and tectonics* (eds. D. Shimozuru & I. Yokoyama), pp. 117–140. Tokyo: Terra Scientific.

Honza, E. & Tamaki, K. 1985 Bonin Arc. In *The ocean basins and margins the Pacific Ocean:* (vol. 7), pp. 459–499. New York: Plenum Press.

Hyndman, R. D., Davis, E. E. & Bornhold, B. 1991 A mechanism for the formation of methane hydrate and seafloor bottom simulating reflectors by vertical fluid expulsion. *J. geophys. Res.*

Ishii, T. 1985 Dredged samples from the Ogasawara fore-arc seamount or 'Ogasawara Paleoland' forearc ophiolite. In *Formation of active ocean margins*, pp. 307–342. Tokyo: Terra Scientific.

Ishii, T., Watanabe, M., Ishizuka, T., Ohta, S. Sakai, H., Haramura, H., Shikazono, N., Togashi, K., Minai, Y., Tominaga, T., Chinsei, K., Horikoshi, M. & Matsumoto, E. 1988 Geological study with the 'SHINKAI 2000' in the West Sagami Bay including Calyptogena colonies. *JAMSTEC Deepsea Res.* 1988, pp. 189–218. (In Japanese with English abstract.)

Iwabuchi, Y., Kato, S. & Kasuga, S. 1987 A Pacific ridge accretion onto the Philippine Sea Plate across the Izu-Ogasawara Trench. *Modern Sea Bottom Res.* **7**, 121–126. Japan Hydrographic Association. (In Japanese with English abstract.)

Kagami, H. *et al.* 1985 *Initial Rep. DSDP* **87**. U.S. Government Printing Office, Washington, D.C.

Kaiko I Research Group 1986 Topography and structure of trenches around Japan. *Data Atlas of Franco-Japanese Kaiko Project, Phase I* (ed. A. Taira & H. Tokuyama). University of Tokyo Press.

Kaiko II Research Group 1987 *6000 meters deep: a trip to the Japanese trenches*. University of Tokyo Press.

Karig, D. E. & Angevine, C. L. 1986 Geologic constrains on subduction rates in the Nankai Trough. In *Initial Rep. DSDP* **87**, 789–796.

Karig, D. E. & Lundberg, N. 1990 Deformation bands from the toe of the Nankai accretionary prism. *J. geophys. Res.* **95**, 9099–9109.

Kimura, G., Koga, K. & Fujoka, K. 1989 Deformed soft sediments at the junction between the Mariana and Yap Trenches. *J. struct. Geol.* **11**, 463–472.

Kinoshita, H. & Yamano, M. 1985 The heat flow anomaly in the Nankai Trough area. In *Initial Rep. DSDP* **87**, 737–743.

Kinoshita, M. & Yamano, M. 1990 Heat flow anomaly associated with pore fluid circulation at the Izu Collision Zone, Japan. In *Int. Conf. Fluids in Subduction Zones*. Abstract.

Knipe, R. J. 1986 Microstructural evolution of vein arrays preserved in Deep Sea Drilling Project cores from the Japan Trench, Leg 57. In *Structural fabrics in Deep Sea Drilling cores from forearcs. Geol. Soc. Am. Mem.* **166**, 75–87.

Kobayashi, K. & Nakada, M. 1978 Magnetic anomalies and tectonic evolution of the Shikoku interarc basin. *J. Phys. Earth* **26**, 391–402.

Lallemant, S. (ed.) 1989 The Japanese Trenches – Kaiko Program 'Nautile' Submersible Cruise. IFREMER, *Campagnes Oceanographiques Françaises* **no. 10**, pp. 274.

Lallemant, S., Culotta, R. & von Huene, R. 1989 Subduction of the Daiichi Kashima Seamount in the Japan Trench. *Tectonophysics* **160**, 231–247.

Lallemant, S., Chamot-Rooke, N., Le Pichon, X. & Rangin, C. 1989 Zenisu Ridge: a deep intraoceanic thrust related to subduction, off Southwest Japan. *Tectonophysics* **160**, 151–174.

Le Pichon, X., Iiyama, T., Chamley, H., Charvet, J., Faure, M., Fujimoto, H., Furuta, T., Ida, Y., Kagami, H., Lallement, S., Leggett, J., Murata, A., Okada, H., Rangin, C., Renard, V., Taira, A. & Tokuyama, H. 1987*a* Nankai Trough and the fossil Shikoku Ridge: results of Box 6 Kaiko survey. *Earth planet. Sci. Lett.* **83**, 186–198.

Le Pichon, X., Iiyama, T., Chamley, H., Charvet, J., Faure, M., Fujimoto, H., Furuta, T., Ida, Y., Kagami, H., Lallement, S., Leggett, J., Murata, A., Okada, H., Rangin, C., Renard, V., Taira, A. & Tokuyama, H. 1987*b* The eastern and western ends of Nankai Trough: results of Box 5 and Box 7 Kaiko survey. *Earth planet. Sci. Lett.* **83**, 199–213.

Le Pichon, X., Iiyama, T., Boulegue, J., Charvet, J., Faure, M., Kano, K., Lallemant, S., Okada, H., Rangin, C., Taira, A., Urabe, T. & Uyeda, S. 1987*c* Nankai Trough and Zenisu Ridge: a deep-sea submersible survey. *Earth Planet. Sci. Lett.* **83**, 285–299.

Lundberg, N. & Moore, J. C. 1986 Macroscopic structural features in Deep Sea Drilling Project cores from forearc regions. In *Structural fabrics in Deep Sea Drilling Project cores from forearcs*. *Geol. Soc. Am. Mem.* **166**, 13–44.

Maekawa, H., Shozui, M., Ishii, T., Saboda, K., Ogawa, Y. 1991 Metamorphic rocks from the serpentinite seamounts in the Marian and Izu-Ogasawara forearcs. In *Proc. ODP, Sci. Results* **125**: College Station, TX (Ocean Drilling Program). (In press.)

Mascle, A. & Moore, J. C. 1990 ODP Leg 110: Tectonic and hydrogeologic synthesis. *Proc. ODP, Initial Rep.* **110**: College Station, TX (Ocean Drilling Program), pp. 409–422.

Miller, S. L. 1974 The nature and occurrence of clathrate hydrates. In *Natural gases in marine sediments*, pp. 151–178. New York: Plenum Press.

Momma, H., Naka, J. & Matsumoto, T. 1990 Preliminary report of deep tow surveys in the Hachijo depression, Kaikato Seamount and Hahajima Seamount (DK 88-3-1ZU). *Tech. Rep. Japan mar. Sci. Technol. Center (JAMSTEC)* **23**, 219–236. (In Japanese with English Abstract.)

Moore, G. F., Shipley, T. H., Stoffa, P. L., Karig, D. E., Taira, A., Kuramoto, S., Tokuyama, H. & Suyehiro, K. 1990 Structure of the Nankai Trough accretionary zone from multichannel seismic reflection data. *J. geophys. Res.* **95**, 8753–8765.

Moore, J. C. 1989 Tectonics and hydrogeology of accretionary prisms: role of the decollement zone. *J. struct. Geol.* **11**, 95–106.

Moore, J. C. & Scientific Party of ODP Leg 110 1988 Tectonics and hydrogeology of the northern Barbados: Results from Ocean Drilling Program leg 110. *Geol. Soc. Am. Bull.* **100**, 1578–1593.

Moore, J. C., Orange, D. & Kulm, L. 1990 Interrelationship of fluid venting and structural evolution: Alvin observations from the frontal accretionary prism, Oregon. *J. geophys. Res.* **95**, 8795–8808.

Nakanishi, M., Tamaki, K. & Kobayashi, K. 1989 Mesozoic magnetic anomaly lineation and seafloor spreading history of the northwestern Pacific. *J. geophys. Res.* **94**, 15437–15462.

Nasu, N. *et al.* 1979 Multichannel seismic reflection data across the Japan Trench: IPOD-Japan. *Basic Data Series no. 3. Ocean Research Institute, University of Tokyo.*

Nasu, N. *et al.* 1982 Multichannel seismic reflection data across Nankai Trough: IPOD-Japan. *Basic Data Series no. 4. Ocean Research Institute, University of Tokyo.*

Niitsuma, N., Matsushima, Y. & Hirata, D. 1989 Abyssal molluscan colony of Calyptogena in the Pliocene strata of the Miura Peninsula, central Japan. *Paleogeography Paleoclimatology Paleoecology* **71**, 193–203.

Niitsuma, N., Matsushima, Y. & Hirata, D. 1989 Abyssal molluscan colony of Calyptogena in the Pliocene strata of the Miura Peninsula, central Japan. *Paleogeography Paleoclimatology Paleoecology* **71**, 193–203.

Ogawa, Y. & Miyata, Y. 1985 Vein structure and its deformational history in the sedimentary rocks of the Middle America Trench slope off Guatemala, Deep Sea Drilling Project Leg 84. In *Initial Rep. Deep Sea Drilling Project* **84**, 811–829. U.S. Government Printing Office, Washington, D.C.

Okamura, Y. 1990 Geologic structure of the upper continental slope off Shikoku and Quaternary tectonic movement of the outer zone of southwest Japan. *J. geol. Soc. Japan* **96**, 223–237. (In Japanese with English abstract.)

Okamura, Y., Tanaka, T. & Nakamura, K. 1986 Diving survey of the knolls on the trench slope break off Kochi, Southwest Japan. *Tech. Rep. Japan Marine Science and Technology Center (JAMSTEC), Special Issue on the 2nd Symposium on Deepsea Research Using the Submersible 'SHINKAI 2000' System*, pp. 173–189. (In Japanese with English abstract.)

Pautot, G., Nakamura, K., Huchon, P., Angelier, J., Bourgois, J., Fujioka, K., Kanazawa, T., Nakamura, Y., Ogawa, Y., Seguret, M. & Takeuchi, A. 1987 Deep-sea submersible survey in the Suruga, Sagami and Japan Trenches: preliminary results of the 1985 Kaiko cruise, Leg 2. *Earth planet. Sci. Lett.* **83**, 300–312.

Pickering, K. T., Agar, S. M. & Prior, D. J. 1990 Vein structure and the role of pore fluids in early wet-sediment deformation, Late Miocene volcaniclastics, Miura Group, SE Japan. In *Deformation mechanisms, rheology and tectonics. Spec. Publ. geol. Soc. Lond.* **54**, 417–430.

Reck, B. H. 1987 Implications of measured thermal gradients for water movement through the northeast Japan accretionary prism. *J. Geophys. Res.* **92**, 3683–3690.

Sakai, H., Gamo, T., Endo, K., Ishibashi, J., Ishizuka, T., Yanagisawa, F., Kusakabe, M., Akagi, T., Igarashi, G. & Ohta, S. 1987 Geochemical study of the bathyal seep communities at the Hatsushima site, Sagami Bay, Central Japan. *Geochem. J.* **21**, 227–236.

Seno, T. 1977 The instantaneous rotation vector of the Philippine Sea plate related to the Eurasian plate. *Tectonophysics* **102**, 209–226.

Shih, T. C. 1980 Magnetic lineations in the Shikoku Basin. In *Initial Rep. DSDP* **58**, 783–788.

Shimamoto, T. 1985 The origin of large or great thrust-type earthquakes along subducting plate boundaries. *Tectonophysics* **119**, 37–65.

Shimamura, K. 1988 Sedimentation and tectonics of Zenisu Ridge, eastern Nankai Trough and Suruga Trough regions. *Sci. Rep. Tohoku University, Ser. 2*, **58**, 107–167.

Smoot, N. C. & Heflner, K. J. 1989 Bathymetry and possible tectonic interaction of the Uyeda Ridge with its environment. *Tectonophysics* **124**, 23–36.

Soh, W., Pickering, K. T., Taira, A. & Tokuyama, H. 1991 Basin evolution in the arc–arc collision zone, Mio–Pliocene Miura Group, central Japan. *J. geol. Soc. Lond.* **147**, 317–330.

Taira, A. & Niitsuma, N. 1986 Turbidite sedimentation in the Nankai Trough as interpreted from magnetic fabric, grain size, and detrital modal analyses. In *Initial Rep. DSDP* **87**, 611–632.

Taira, A. *et al.* 1991 *Proc. ODP, Initial Rep.* **131**: College Station, TX (Ocean Drilling Program). (In the press.)

Taira, A., Katto, J., Tashiro, M., Okamura, M. & Kodama, K. 1988 The Shimanto Belt in Shikoku, Japan–Evolution of Cretaceous to Miocene accretionary prism. *Modern Geol.* **12**, 5–46.

Taira, A., Tokuyama, H. & Soh, W. 1989 Accretion tectonics and evolution of Japan. In *The evolution of the Pacific Ocean Margins*, pp. 100–123. Oxford University Press.

Taylor, B. *et al.* 1990 *Proc. ODP, Initial Rep.* **126**: College Station, TX (Ocean Drilling Program).

Taylor, B. & Smoot, N. C. 1984 Morphology of Bonin fore-arc submarine canyons. *Geology* **12**, 724–727.

Uyeda, S. & Ben-Avraham, Z. 1972 Origin and development of the Philippine Sea. *Nature* **240**, 176–178.

Von Huene, R. & Culotta, R. 1989 Tectonic erosion at the front of the Japan Trench convergent margin. *Tectonophysics* **160**, 75–90.

Von Huene, R., Langseth, M., Nasu, N. & Okada, H. 1982 A summary of Cenozoic tectonic history along the IPOD Japan Trench transect. *Geol. Soc. Am. Bull.* **93**, 829–846.

Von Huene, R. *et al.* 1988 Results of Leg 112 drilling, Peru continental margin, part 1, Tectonic history. *Geology* **16**, 934–943.

Yamano, M. & Uyeda, S. 1988 Heat flow. In *The ocean basins and margins*, vol. 7B, pp. 523–557. *The Pacific Ocean*. New York: Plenum Press.

Yamano, M., Uyeda, S., Aoki, Y. & Shipley, T. H. 1982 Estimates of heat flow derived from gas hydrates. *Geology* **10**, 339–343.

Yamano, M., Honda, S. & Uyeda, S. 1984 Nankai Trough: A hot trench? *Mar. geophys. Res.* **6**, 187–203.

Yamazaki, T. & Okamura, Y. 1989 Subducting seamounts and deformation of overriding forearc wedges around Japan. *Tectonophysics* **160**, 207–229.

Yoshii, T. 1977 Crustal and mantle structure of Northeast Japan. *Kagaku* **47**, 170–176. (In Japanese.)

Yoshii, T. 1979 Compilation of geophysical data around the Japanese Islands (1). *Bull. Earthquake Res. Inst. Univ. Tokyo* **54**, 75–117.

Yoshikawa, T., Kaizuka, S. & Ohta, Y. 1981 *The landforms of Japan.* University of Tokyo Press.

Water budgets in accretionary wedges: a comparison

By X. Le Pichon[1], P. Henry[1] and the Kaiko–Nankai Scientific Crew†

[1]*Laboratoire de Géologie, Ecole Normale Supérieure, 24 rue Lhomond, 75231 Paris, Cedex 05, France*

Direct or indirect measurements of fluid flow out of the toe of accretionary wedges have now been made in the Barbados, Central Oregon, Northern Cascadia and Nankai subduction zones. The steady-state local compaction model predicts velocities of fractions of a millimetre per year and total outflows from the toe of a few cubic metres per year per metre of length along strike of the subduction zone ($m^2 a^{-1}$). Sea bottom measurements reveal channellized flows at velocities of hundreds of metres per year and total outflows from the toes of a few hundreds of square metres per year. Thermal arguments show in the Nankai area that all of this large surface flow cannot come from as deep as the decollement and that consequently a significant dilution by shallow sea water convection must be present. We propose that this convection is driven by the reduced density of less saline fluid of deep origin. Thus the outflow of water of deep origin may be only a few tens of square metres per year. We note, however, that there are indications in the Barbados area of massive flow of low salinity water at depth along the decollement. These flows imply the existence of large scale non-steady-state lateral transport and require the existence of sources of low salinity water. These might include the smectite–illite transformation and the fluid contained in the oceanic crust. However, these sources are limited and the possibility exist that other more important sources may be required such as long distance transport of fresh water from adjacent sedimentary basins and (or) recharge mechanisms such as the seismic pumping one.

1. Introduction

We compare quantitative estimates of fluid outflow from the toes of accretionary wedges in subduction zones, based on various direct or indirect measurements, with steady-state flow due to compaction of the sediment incorporated within the wedge. We use recent indirect measurements of flow made on the Eastern Nankai wedge (Le Pichon *et al.* 1990*a*; Foucher *et al.* 1990*a*; Henry *et al.* 1990), direct (Carson *et al.* 1990) and indirect (Davis *et al.* 1990) measurements made on the Oregon wedge and indirect measurements made on the Barbados wedge (Foucher *et al.* 1990*b*; Fisher & Hounslow 1990; Le Pichon *et al.* 1990*b, c*; Langseth *et al.* 1990). We conclude that in all these cases the outflow is significantly larger than the volume of fluid released by compaction processes.

† K. Kobayashi, J. P. Cadet, J. Ashi, J. Boulègue, H. Cambray, N. Chamot-Rooke, A. Fiala-Medioni, J. P. Foucher, T. Furuta, T. Gamo, J. T. Ilyama, H. Kinoshita, S. Lallemand, S. Lallemant, M. Nakanishi, Y. Ogawa, S. Ohta, H. Sakai, P. Schultheiss, J. Segawa, K. Shitashima, M. Sibuet, A. Taira, A. Takeuchi, P. Tarits, H. Toh and M. Watanabe.

2. Fluid outflow due to steady-state compaction of sediments

Typically, the sediments entering subduction zones have a high porosity, especially if they consist of recently deposited terrigenous sediments. Tectonic stacking of thrust packages at the toe of an accretionary wedge results in rapid compaction and expulsion of most of the interstitial water. Thus, following Bray & Karig (1985), several authors have computed the expected steady state fluid outflow due to compaction using simplifying assumptions (Langseth *et al.* 1990; Screaton *et al.* 1990; Le Pichon *et al.* 1990 *c*). They have shown that the expected outflow velocities do not exceed a fraction of a millimetre per year and that the total expected flux out of the toe is a few square metres per year (cubic metres per year per metre of length along strike of the wedge). The following discussion amplifies these two conclusions.

Le Pichon *et al.* (1990 *c*) have shown that a single porosity depth function from the basin to the wedge fits reasonably well existing data. We use this relation for the porosity φ as given by them

$$\varphi = \varphi_0 \exp(-az), \tag{2.1}$$

where $\varphi_0 = 0.7$, $a = 0.67$ km^{-1} and z is the depth in kilometres.

The average porosity of a vertical column is

$$\bar{\varphi} = (\varphi_0/az)[1 - \exp(-az)] \tag{2.2}$$

and the equivalent height of water contained in a vertical column is

$$H_{\rm w} = (\varphi_0/a)[1 - \exp(-az)]. \tag{2.3}$$

The maximum $H_{\rm w}$ is φ_0/a which is close to 1 km. For a thickness of sediment of 1 km, the height of water $H_{\rm w}$ is about 500 m and for 3 km, it is 900 m.

Increasing the thickness of the sedimentary column above 3 km would not increase significantly the amount of interstitial water. For typical entering thicknesses of sediment (1–4 km), the equivalent height of water $H_{\rm w}$ changes by a factor of less than 2 and is about 700 m $\pm 30\%$.

The actual volume of water entering the wedge per unit time is

$$V_{\rm w} = v_{\rm a} H_{\rm w}, \tag{2.4}$$

where $v_{\rm a}$, the velocity of accretion of sediment to the wedge is equal to the velocity of subduction $v_{\rm s}$ plus the growth velocity of the wedge. For a mature wedge, $v_{\rm a} \approx v_{\rm s}$ (Le Pichon *et al.* 1990 *c*) and thus

$$V_{\rm w} = v_{\rm s} H_{\rm w} \approx 700 v_{\rm s}. \tag{2.5}$$

Hence the volume of water entering the wedge per unit time does not vary much with the thickness of sediment z provided the entering sediment thickness is 1 km or larger. On the other hand, it is linearly dependent on the velocity of subduction.

The volume of water per unit time released between the toe of thickness z and the wedge column of thickness z' is

$$\Delta V_{\rm w} = v_{\rm s} z(\bar{\varphi}(z) - \bar{\varphi}(z'))/(1 - \bar{\varphi}(z')). \tag{2.6}$$

This formula is obtained by assuming conservation of the volume of grains as the columns thickens. Note that the horizontal velocity of grains with respect to the

Table 1

(ΔV_w is computed for a thickening of 1 km.)

| | z | | v_s | H_w | V_w | $\Delta V_\mathrm{w}/v_\mathrm{s}$ | ΔV_w | | U_z |
	km	$tg\alpha$	mm a^{-1}	km	m^2 a^{-1}	km	m^2 a^{-1}	$U_z/tg\alpha v_\mathrm{s}$	mm a^{-1}
Northern Barbados	1.0[a]	0.06[a]	20[a]	0.53	10	0.20	4	0.31	0.4
Eastern Nankai	3.0[b]	0.12[b]	20[b]	0.9	18	0.23	5	0.30	0.7
Northern Cascadia	3.0[c]	0.12[c]	45[c]	0.9	40	0.23	10	0.30	1.6
Central Oregon	3.0[d]	0.12[d]	30[c]	0.9	27	0.23	6	0.30	1.1

[a] After Le Pichon *et al.* (1990*c*). [b] After Le Pichon *et al.* (1990*a*). [c] After Davis *et al.* (1990). [d] After Moore *et al.* (1990).

backstop decreases away from the toe. The vertical velocity of escape of water through the sea-floor $U_{z'}$, obtained by differentiation, is (Le Pichon *et al.* 1990*c*)

$$U_{z'} = z'^{-1} \frac{(\bar\varphi(z') - \varphi(z'))}{[1 - \bar\varphi(z')]^2} z[1 - \bar\varphi(z)] \tan \alpha v_\mathrm{s}, \tag{2.7}$$

where α is the wedge taper angle. The maximum velocity of escape is obtained at the toe for $z' = z$ and decreases as the wedge thickens.

$$U_{z'} = \frac{(\bar\varphi(z) - \varphi(z))}{(1 - \bar\varphi(z))^1} \tan \alpha v_\mathrm{s}. \tag{2.8}$$

The maximum local steady-state outflow velocity does not continuously increase with thickness z of the entering column. It is maximum for a value of about 1.5 km and changes little ($\pm 10\%$), staying close to 0.3, between $z = 1$ and $z = 4$ km. This is because $U_z/v_\mathrm{s} \tan \alpha$ rapidly increases from 0 for $z = 0$ to 0.333 for $z = 1.66$ km but then slowly decreases to 0 at infinity.

To summarize, the local outflow velocity should not be affected by the thickness of the entering sediment column and should only depend on the product of the taper by the subduction velocity.

Table 1 compares different parameters for the areas we discuss in this paper. For Northern Cascadia and Central Oregon, we have adopted the subduction velocity given by Davis *et al.* (1990), taking into account the obliquity of subduction to the wedge in Central Oregon. The volume of water lost after 1 km of thickening, ΔV_w, varies between 5 and 10 m^2 a^{-1} and is controlled by the subduction velocity; $U_z/v_\mathrm{s} \tan \alpha$ has the same value of 0.3 for all the margins whereas U_z, the maximum escape velocity at the toe, changes from a minimum of 0.3 mm a^{-1} in Northern Barbados to 1.6 mm a^{-1} in Northern Cascadia. The average velocity over the first kilometre of thickening is minimum for Northern Barbados (0.23 mm a^{-1}) and maximum for Northern Cascadia (1.2 mm a^{-1}).

Thus, with local steady-state compaction, expected velocities are indeed very small. As a consequence, the heat flow should not be significantly affected (Langseth *et al.* 1990; Screaton *et al.* 1990; Le Pichon *et al.* 1990*c*). Note that even if all the water of the latest thrust package is channelized through the decollement or along the main thrust fault, the expected flow is not more than 5 or 10 m^2 a^{-1}.

Are there other possible sources of fluid within the sediment or below the sedimentary column within the oceanic crust? As discussed by Von Huene & Lee (1983) and more recently by Moore (1989), fluid structurally bound to clay minerals should be released upon burial and heating. Also, the hydrous oceanic crust liberates fluids upon heating (Peacock 1987, this symposium). Below, we briefly show that the order of magnitude expected is less or equal to the values computed above, that is less than a few square metres per year. More important, most of the release of water from smectite occurs at temperatures larger than 60–100 °C and from the oceanic crust at temperatures larger than 300 °C and consequently well to the inward side of the accretionary wedge.

Let us first consider the fluid contained within the oceanic crust. Peacock (1987) proposes a conservative volatile content of 2 % in weight for the 8 km thick oceanic crust. Using a density of 3000 kg m^{-3}, this is equivalent to $480v_\mathrm{s}$ m^2 a^{-1} or 10–20 m^2 a^{-1} for the wedges considered in table 1. As the estimate of 2 % in weight appears to be conservative these values could possibly be doubled to 20–40 m^2 a^{-1}. These values are comparable with the total amount of interstitial water released from the wedge in steady state.

Next, we consider the smectite–illite transformation. To get a rough estimate of the amount of water produced by the smectite–illite transformation at temperatures larger than 60 °C, we assume that three volumes of smectite give two volumes of illite and one of water. According to Tribble (1990), in the Barbados area, the proportion of smectite in the solid phase is 30 %. Then, using table 1, we can obtain the additional volume of water produced by transformation of smectite, which would be 1 m^2 a^{-1}. Assuming the same proportion of smectite in Oregon, which is certainly an overestimation, we obtain 6 m^2 a^{-1}. This value is significant, but is smaller than the original volume of interstitial water. However, at the depth at which this transformation occurs ($T > 60$ °C), the low porosity there ensures that the resulting fluid will be much less saline than sea water.

3. Water budget of the Eastern Nankai accretionary wedge

The outer Eastern Nankai accretionary wedge has been investigated along a 20 km section with *in situ* measurements, observations and samplings coupled with deep tow surveys and surface geophysical observations (Kobayashi *et al.* 1990; Le Pichon *et al.* 1990*a*). Fluid outlets have been identified from the presence of clam colonies, vestimentifera, calcareous concretions, chimneys and white patches. Studies of the chemistry of sea water and interstitial water give qualitative indications on the type of fluid venting whereas instantaneous and long term temperature gradient measurements within the sediments give indirect estimates of the velocity of fluid venting (Foucher *et al.* 1990*a*; Henry *et al.* 1990). An important discovery of this study is that, for a given clam colony type, the range of fluid advection velocities below it is relatively small. Thus an estimation of the surface area occupied by the different types of clams colonies can be converted into a water outflow estimate (Foucher *et al.* 1990*a*). Unfortunately, this method could only be used over a zone of fluid venting discovered near the toe of the 4000–3500 m deep outmost thrust unit. Although another important zone of fluid venting was identified at a depth of about 2000 m near the summit of the second 0.6 Ma old thrust unit, the bottom water temperature fluctuations are too large to estimate fluid advection from temperature

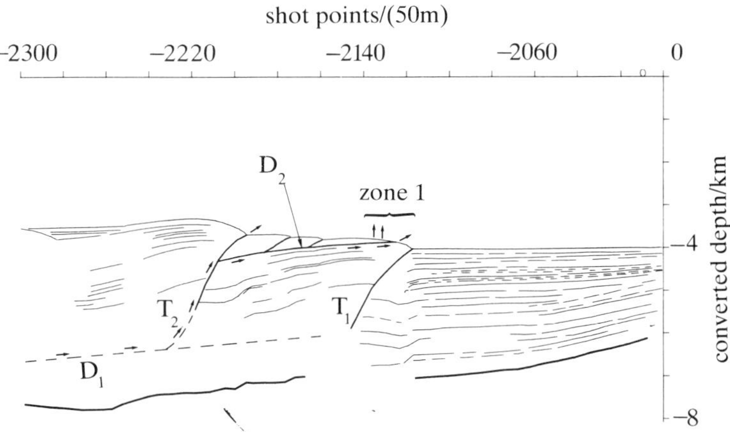

Figure 1. Line drawing from a seismic cross section of the toe of the Nankai wedge (ORI, NKT 5), at the location of the Kaiko–Nankai submersible dives. Arrows show the probable path followed by fluid from the main decollement D_1 to the vents. Fluid migrates along thrust T_2 and then along the shallow decollement D_2 which brings them to Zone 1. Evidence of fluid expulsion has been found at the outcrop of T_2 but most of the fluid is diverted to Zone 1. There, part of the fluid is expelled at the outcrop of D_2, but most of the colonies spread out on the plateau and are fed by a fluid which is diluted with sea water (modified from Chamot-Rooke *et al.* 1990).

gradients beneath the sea floor, which reflect these fluctuations. Thus the budget obtained only concerns the fluid vented out from the toe of the wedge and ignores the significant amount of fluid vented out from the top of the second unit.

The outer unit has a taper angle of about 7°; it is characterized by two successive thrust related anticlines linked to a 2000 m deep decollement (figure 1). The largest and best investigated zone of fluid venting is close to the apex of the outer fold at a depth of about 3800 m. Fluid venting is dense over a width of 500 m inward from the base of a 20 m high cliff. Fluid venting is laterally continuous over all the 2000 m length investigated.

The cliff cuts through horizontal turbidites and mudstones affected by numerous joints. Its base has been interpreted by Chamot-Rooke *et al.* (1990) to coincide with the emergence of an upslope shallow flat thrust linked to the decollement through the second thrust (figure 1). Note that the outcrop of the frontal thrust, which is concealed below a tectonically fresh talus, is not associated with obvious fluid venting.

Numerous white patches, which probably are bacterial mats, are present at the base of the 20 m cliff, together with chimneys and carbonate concretions whereas only scattered clams are observed. The white patches are the locations of the highest concentrations of methane and H_2S measured, up to 9×10^4 nl kg^{-1} of CH_4 compared with 5 nl kg^{-1} for normal sea water (Boulègue *et al.* 1990). Chlorinity there is decreased by 4–7 % and the fluid is highly oversaturated in carbonates. Boulègue *et al.* (1990) interpret the composition of these fluids as indicating that gas hydrates have been dissolved during the upward migration.

In contrast, fluid venting manifests itself mostly through the presence of numerous dense colonies of Calyptogena on the edge of the plateau above the cliff. There is no detectable chlorinity anomaly; the CH_4 concentrations are significantly less than in the white patches but still two to three orders of magnitude over the background level and the fluids are less oversaturated in carbonates. Boulègue *et al.* (1990)

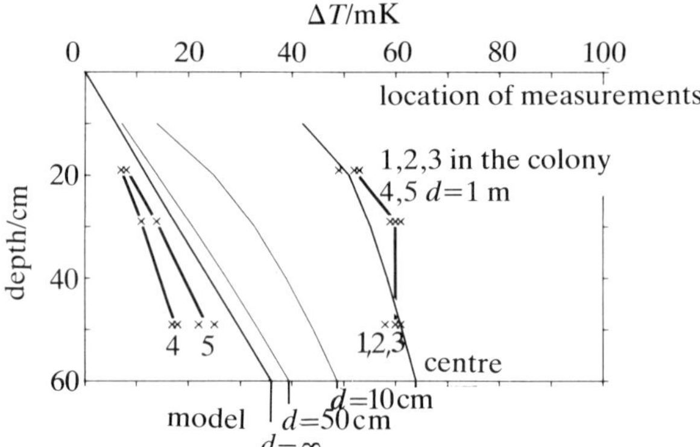

Figure 2. Temperature measurements and model, within a type A clam colony (Kaiko–Nankai cruise, dive 12, Sci. Obs.: K. Kobayashi). Temperature elevation above surface temperature is given as a function of depth within the colony and for different values of distance d from the rim of the colony.

conclude that the fluid coming out in the area of great concentrations of clam colonies is more diluted by sea water than the fluid coming out along the base of the cliff. The absence of large clams colonies at the base of the cliff may be due to excessive concentrations of sulphur there.

Venting sites were classified on a biological basis (Sibuet *et al.* 1990) and one dive was devoted to estimation of the surface area covered by each type of colony. The submersible followed straight parallel lines 100 m apart, exploring a 625 m by 500 m zone along the upper part of the slope and on the plateau. The surfaces occupied by colonies were measured on camera images. The area seen on the video represents 4% of the area surveyed. We thus obtain estimates of the proportion of the surface covered by each of the different types of colonies in the explored 4%. We further assume that the same proportions exist in the unexplored 96% of the area surveyed.

Most of the estimated fluid flux (85%) come from the so-called A-type colonies where the population density is about 1000 clams per m^2. During this dive, 122 A-colonies, covering a total surface area of 60 m^2, were observed and consequently the statistics made should be significant.

Thirty-six measurements of the temperature gradient over the first 50 cm of sediment were made within this venting zone. The temperature gradients are linear outside the colonies but show a rapid downward decrease below the A-colonies (Foucher *et al.* 1990*a*) (figure 2). The temperature gradient below one of these A-colonies was continuously recorded for two months. It is used to obtain a value of sediment thermal diffusivity compatible with a thermal conductivity of about 1 W m^{-1} K^{-1} (Henry *et al.* 1990). The corresponding thermal diffusivity for fluid advection is 2.5×10^{-7} m^2 s^{-1} (7.9 m^2 a^{-1}). It is used to model the upward fluid flows below colonies which reproduce the observed thermal gradients. Both unidimensional and cylindrical modelling were made. Values for four A-colonies vary between 70 and 150 m a^{-1} (Henry *et al.* 1990) (figure 2) and an average value of 100 m a^{-1} was adopted for the statistics. Henry *et al.* (1989) had used a similar method to estimate the fluid flow over a colony further east and had also found a Darcy velocity of about 100 m a^{-1}.

Multiplying the surface area of the different types of colonies by the corresponding average fluid advection velocities, one obtains a total mean flux of about 200 m^2 a^{-1}

in the toe area $((100 \text{ m a}^{-1} \times 60 \text{ m}^2)/(0.04 \times 625 \text{ m}))$. This flux ignores the limited amount of venting observed 5 km further inward along the outcrop of the second thrust. It is also limited by the fact that this thermal estimate cannot detect flows of less than 10 m a^{-1}. Finally, as stated above, it ignores the important zone of fluid venting discovered 10 km further inward near the summit of the second thrust unit. Consequently, the proposed flux is certainly significantly lower than the actual flux. Yet it is about 40 times larger than the amount obtained assuming local steady-state dewatering.

4. Central Oregon and Northern Cascadia fluid flux estimates

Direct measurements of fluid venting from clam colonies on the toe of the Central Oregon accretionary wedge are reported by Carson *et al.* (1990). The fluid appears to come from relatively shallow depth as the methane it contains is biogenic (Kulm & Suess 1990). No decreased salinity has been measured. The obtained Darcy velocities are about 65 m a^{-1} over two vents. Four such vents each having a surface area of about 9 m^2 were discovered within an area of the toe that is about 5 by 2 km. The authors have not indicated how much of these 10 km^2 were visually explored during their dives but the density of submersible tracks shown suggests that 4% of the 1.5 km wide most densely covered area was explored. If 4% is a correct proportion, the average flux from this 1.5 km wide portion of the toe is about $40 \text{ m}^2 \text{ a}^{-1}$ whereas the expected local steady-state dewatering would be about $6 \text{ m}^2 \text{ a}^{-1}$ (see table 1). This estimate ignores numerous other less important manifestations of fluid flow such as isolated concentrations of clams, tubeworms and scattered clams (Moore *et al.* 1990).

Davis *et al.* (1990) have made 110 closely spaced heat flow measurements over a section of the Northern Cascadia accretionary sedimentary prism off Vancouver Island and compared them with estimates of heat flow based on the depth of a bottom-simulating reflection (BSR) that is interpreted to mark the thermally controlled base of a methane hydrate layer. They find a systematic discrepancy, the BSR estimated heat flow being about 30% less than the measured surface heat flow. They explain the discrepancy through vertical upward fluid advection at an average rate of about 2.5 cm a^{-1} over a section about 15 km long centred on the toe of the wedge. The total equivalent flux is about $400 \text{ m}^2 \text{ a}^{-1}$ and should be compared to the expected local steady-state dewatering which would be about $15 \text{ m}^2 \text{ a}^{-1}$ for a section of 15 km length (see table 1). This last conclusion differs from that of Davis *et al.* (1990) who apparently erred in their estimates of flow due to dewatering.

5. Barbados fluid flux estimates

Along the Barbados $15° 30'$ N section where several holes have been drilled (Moore *et al.* 1988), heat flow measurements made during drilling reveal high heat flow that ranges from 96 to 192 mW m^{-2} in the upper 30–40 m with nonlinear temperature profiles. Below about 100 m, the temperature profiles are linear and the heat flow increases toeward from 40 to 90 mW m^{-2} (Fisher & Hounslow 1990). The upper high temperature gradient anomalies have been interpreted by Fisher & Hounslow (1990) as due to rapid transient fluid flow of water along faults. Sea-floor venting in faulted areas is confirmed by local surface heat flow anomalies at a scale of a few hundreds of metres which requires flows of a few tens of square metres per year (Foucher *et al.*

1990b; Lallemant *et al.* 1990). The deeper linear gradients were interpreted as due to flow along the decollement at velocities of about 10 m a^{-1} (Fisher & Hounslow 1990; Foucher *et al.* 1990 b). The resulting seaward fluid flow along the decollement would be about 400 m^2 a^{-1} (Foucher *et al.* 1990b). Fluid, sampled both in the decollement and near active faults, has a salinity reduced by 10–30% with respect to normal sea water (Gieskes *et al.* 1990). Near 12° 20′ N, small-scale high heat flow anomalies also coincide with sea-floor faulted areas and can be explained by surface venting of a few tens of square metres per year (Foucher *et al.* 1990). Two heat flow transects near 14° 20′ N and 14° 35′ N did not reveal any broadscale anomalies of high heat flow, but again showed some locally high values possibly associated with faults (Langseth *et al.* 1990).

Additional remarkable evidence for large-scale fluid venting was reported by Le Pichon *et al.* (1990b). They describe a 0.55 km^2 oval shaped mud volcano near 13° 50′ N, 12 km seaward of the Barbados accretionary complex deformation front, through which about 10^6 m^3 of fluid are advected upward every year, with a maximum upward Darcy velocity at the axis of 17 m a^{-1}. This estimate is based both on thermal and chemical considerations. The interstitial water sampled in the mud-volcano is half as saline as normal sea water. The edge of the mud volcano is more than 30000 and less than 100000 years old. The mud volcano is part of a large field of mud volcanoes there and implies a very large amount of lateral drainage at depth as, in steady state, the water, if it were produced by compaction, would correspond to the production of a 100 km length along strike of wedge.

The Barbados evidence thus indicates both the existence of large-scale seaward flow at depth along the decollement of deep originated less saline water, up to two orders of magnitude above local steady-state compaction estimates, and more local, possibly transient, surface anomalies implying advection of a few tens of square metres per year.

6. Discussion

We have shown that submersible investigations indicate the existence of fluid seeps on a scale of a few square metres. Indirect measurements based on subsurface temperature gradients (Foucher *et al.* 1990a; Henry *et al.* 1990) as well as direct measurements (Carson *et al.* 1990), indicate fluid flow velocities of the order of 100 m a^{-1}. However, thermal measurements in the first 50 cm of sediment cannot be used to estimate upward fluid velocities of less than 10 m a^{-1}. On the other hand, investigations made from the surface ships with heat flow probes indicate wider (100 m to a few kilometres) but much smaller heat flow anomalies with an increase over the background heat-flow of 30–100%, corresponding to surface flow velocities of a few centimetres per year to 10 m a^{-1} at most. However, pogo-stick-type thermal measurements can only be used to measure distributed and slow intergranular flow, not only because the probability to hit a vent is quite low (about 1% in a known active zone) but also because thermistors are usually too deep to detect the shallow curvature of the temperature profiles measured within vents.

One could argue that localized venting of large amounts of fluids should increase significantly the conductive heat flow outside the vents, which can be measured both from sea surface as well as during submersible investigations. However, in the deepest zone explored during the Kaiko–Nankai cruise (referred to as Zone 1), no significant increase of the background heat flow has been observed in the area of active venting. We show below that this result implies that deep originated fluids flow laterally into the venting area and are then diluted with sea water before they reach the clam colonies.

(a) Necessity of lateral fluid input

The total heat flux F_s though the surface of the active area on the plateau in Zone 1 can be expressed as the sum of two terms, one corresponding to the gradient outside the colonies, the other to the heat flux through clam colonies

$$F_s = \Lambda S f_0 + (\rho c)_f \theta Q_c, \tag{6.1}$$

where Λ is the thermal conductivity (about 1 W m^{-1} K^{-1}), S is the surface area of the active area on the plateau (300 m^2 per m of subduction zone), f_0 is the thermal gradient outside colonies (60 mK m^{-1}), $(\rho c)_f$ is fluid volumetric heat capacity (4×10^6 J m^{-3}), Q_c is the fluid flux through clam colonies (about 200 m^2 a^{-1}) and θ is the characteristic temperature elevation inside clam colonies. θ is of the order of 0.1 °C at most. In the following discussion, we also use thermal diffusivity $D = \Lambda/(\rho c)_f$ (7.9 m^2 a^{-1}). Note that with these parameters, heat expelled by fluid advection in clam colonies is seven times smaller than conductive heat flux outside clam colonies integrated over the active area.

If all the fluid flowing through clam colonies is seeping vertically from deep below the colonies, a model with uniform fluid advection at Darcian velocity Q_c/S in a uniform half-space can be used. Then, temperature at any depth z is:

$$T(z) = T_\infty (1 - \exp{(-z/h)}), \tag{6.2}$$

where
$$h = SD/Q_c = 12 \text{ m} \tag{6.3}$$

and as $F_s = (\rho c)_f T_\infty Q_c$
$$T_\infty = \theta + h f_0 = 0.81 \text{ °C.} \tag{6.4}$$

Surface temperature is here taken as zero. The temperature of the fluid source compatible with this model is T_∞. This corresponds to a very shallow depth (13.5 m), which implies that deep fluids are cooled, either by conduction or dilution. As a first step, we will determine the amount of cooling that can be expected from conduction to the sediment surface outside the area of venting if fluids come laterally.

(b) Fluid cooling along a sloping aquifer of low dip angle

A shallow decollement (or low-angle thrust) outcrops near the top of the slope, 20 m below the top of the frontal thrust anticline (Chamot-Rooke *et al.* 1990). This decollement connects with the second major thrust (see figure 1). Its dip angle can be constrained from seismic reflection data and is about 4°. The taper angle α of the overlying wedge is fairly constant at about 7°. Assuming steady state and neglecting horizontal heat transfer, temperature along the fluid path from the deep decollement to the feeding thrust fault and then to the shallow decollement which feeds Zone 1 vents (see figure 1) can be calculated. We assume for the shallow decollement below the main cluster of clam colonies a depth of 30 m. The temperature of the incoming fluids T_f is then computed (see Appendix A).

For a fluid flux Q_r along the shallow decollement smaller than 40 m^2 a^{-1}, the gradient increase due to fluid flow along this decollement is moderate and, below the surface zone of outflow, does not depend on the temperature of the fluid source. If fluids flowing out of clam colonies were undiluted ($Q_r = Q_c = 200$ m^2 a^{-1}), the approximation used no longer applies and very high heat flows of more than 300 mW m^{-2} are predicted by the exact solution (see Appendix A). It follows that the fluid flux Q_r along the shallow decollement cannot be of the same order of magnitude than the fluid flux observed in clam colonies. This implies that fluids from the deep source are diluted with sea water before they reach the colonies.

(c) Dilution models

For a given fluid dilution factor $d = Q_c/Q_r$, both the sloping aquifer model and the unidimensional advection model give ranges of values for the temperature at the depth of the shallow decollement. It is thus possible to determine the probable amount of dilution by looking for the compatibility domain of these two models. Mixing with sea water occurs within the shallow decollement at depth H. Thus, upward flow from decollement to the surface is equal to Q_c/S and temperature at depth 0 to H is given by equation (6.2). As explained in Appendix B a relationship between temperature in the decollement below the colonies and the temperature of incoming fluids can be obtained from the heat balance equation at the shallow decollement level.

But the structure of the temperature field in and around clam colonies has been interpreted as an indication that recharge with sea water occurs at a shallow depth (1 or 2 m), implying local convection around the colonies (Henry *et al.* 1990). We consequently consider another case when dilution occurs by convection near the surface. We assume that a uniform advection model applies in the domain between the shallow decollement and the base of the convection cells (see Appendix B). Then, the temperature predicted in the shallow decollement by this model must be equal to the temperature T_f' of incoming fluids.

Comparing these two models, convection near the surface requires higher dilution than mixing in the decollement, the actual dilution factor being most probably between 6 and 15. Temperature in the decollement should be very low if sea water is flowing into it, but should not be anomalous if dilution occurs near the surface. On the other hand conductive heat flow is expected to be significantly higher than normal north of the active zone if mixing occurs in the decollement as the predicted temperature for incoming fluid is higher.

(d) Numerical estimates of dilution

When applying either model, the major source of uncertainty is the value of f, the thermal gradient below the shallow decollement, which is supposedly equal to the 'regional' gradient.

There are few measurements which can be used as references for heat flow. One measurement made in the trench between Tenryu Canyon and Zenisu ridge gives 67 mW m^{-2} (Nahihara *et al.* 1989). As conductivity is in the range 0.7–1.3 W m^{-1} K^{-1} but more probably around 1 W m^{-1} K^{-1}, this value is quite compatible with the measured gradient of 60 mK m^{-1} in the toe area. Measurements were recently made closer to the Kaiko–Nankai area (Kinoshita & Kazumi 1990) on the lower tectonic unit to the northeast. At the northeastern extremity, the thermal gradient is 60 mK m^{-1}, while closer to the explored area but near the base of the upper tectonic unit, the thermal gradient is low, around 40 mK m^{-1}. In the absence of fluid advection, heat flow tends to decrease landward because of tectonic thickening of sediments in the wedge, and because of the slab burial effect (Toksoz *et al.* 1975; Le Pichon *et al.* 1990c; Molnar & England 1990). As Zone 1 is much closer to the deformation front, a surface gradient value of 40 mK m^{-1} probably determines an absolute minimum for heat flow coming from below the shallow decollement.

For a value of the background gradient of 40 mK m^{-1}, dilution should be between 6.5 and 9; for 50 mK m^{-1}, between 10 and 15; for 60 mK m^{-1}, between 20 and more than 30, but this last determination is also very dependent on other parameters such as thermal conductivity, aquifer slope and aquifer depth below the active zone.

Salinity contrast, rather than temperature gradient, is probably driving convection (Henry *et al.* 1990). As noted earlier, fluids originated from deep within an accretionary complex tend to have a low salinity, but the actual salinity of incoming fluids is not known in the Kaiko–Nankai area. A salinity difference of a few per mil between sea water and source is enough to start convection if conduit permeability is high, but the major constraint is that flow in the conduit, at a velocity of 100 m a^{-1}, has then to be driven by buoyancy of the diluted fluids. Assuming that fluids have, after dilution, a density a few grams per litre less than sea water, conduit permeabilities of the order of 10^{-10} m^2 are required (Henry *et al.* 1990). Dilution factors of more than 10 are thus unlikely.

(e) *Applicability to other wedges*

The model proposed for Kaiko–Nankai area might be applied to the 12° N section of the Barbados accretionary wedge, as the context is somewhat similar. There, fluids presumably flow along a major thrust into a shallow decollement, and are expelled through faults in the overlying thrust outcrop wedge of deformed sediment (Foucher *et al.* 1990*b*). The corresponding heat flow increase is moderate (40%), indicative of a deep fluid flux of the order of 20 m^2 a^{-1}. However, we do not know if vents are present on these thrust outcrop wedges and consequently whether there is some convection.

In the Oregon diving sites, the venting fluids are probably diluted, as their salinity is normal and as the large amount of fluid flux is not compatible with local shallow sources. Then convection would occur between the surface and permeable sand strata acting as conduits for lateral migration. As in the Kaiko–Nankai area, major vents with associated clam colonies were found at the top of the anticline, but other types of vents were also found in association with permeable sand strata on the eroded slope just below the active zone.

However, the very high heat flow at depth in the toe area of the Barbados Ocean Drilling Project (ODP) profile indicates that a large amount (a few 100 m^2 a^{-1}) of deep originated low salinity fluid flows there along the decollement (Foucher *et al.* 1990*b*; Langseth *et al.* 1990). Recent transient fluid flow has been proposed to explain high surface heat flows (Fisher & Hounslow 1990) but this does not account for the overall increase of the gradient at depth. Furthermore, the huge amount of fluid expelled through the mud volcano area at 13° 50′ N requires lateral flow at a 100 km scale. This leaves us with the problem of explaining the origin of tens of square metres per year of low salinity water at depth and their subsequent large-scale transient transport.

The widespread fluid flow (of the order of 400 m^2 a^{-1}) out of the Northern Cascadia wedge (Davis *et al.* 1990) cannot be explained either way, but one may ask what confidence can be given to the temperature estimated at the base of the hydrate layer as the exact composition of the hydrates is not known.

Phil. Trans. R. Soc. Lond. A (1991)

14-2

7. Conclusion

We have shown that on the Eastern Nankai area very large fluid outflows measured on the toe of the accretionary wedge include a large component of locally derived sea water, probably resulting from haline convection. However, the amount of low salinity water necessary to drive this convection is still of the same order of magnitude as the water produced by compaction of the whole sedimentary column. In the Barbados area previous work has given further evidence for large-scale flow of low salinity water along the decollement, this flow being an order of magnitude larger than possible flow from compaction of the whole sedimentary column.

Where does the large amount of low salinity water come from ? The smectite–illite transformation and the hydrate layer are obvious sources but cannot provide amounts larger than a few square metres per year to at most $10 \ \mathrm{m^2 \ a^{-1}}$. Another possible source is the oceanic crust if it includes a sizeable thickness of serpentinized peridotite. For example 2 km of 50% serpentinized peridotite would contain more water than the sedimentary column. This source would not be sufficient if the amount of low salinity water required is indeed larger than a few tens of square metres per year in steady state. Then, one would have to invoke long distance transport of fresh water from the adjacent subaerially exposed basins and (or) exotic recharge mechanisms such as the seismic pumping mechanism (Sibson *et al.* 1975).

Appendix A. Model of fluid flow along an aquifer

Steady state is assumed and diffusion along the horizontal (x) axis is neglected. The geometry is as shown on figure 3. The aquifer is thin, so that the fluid temperature $T_f(x)$ is uniform across it. The depth of the aquifer below the surface is $H(x)$. The fluid flux in the aquifer is Q. In steady state, heat flow below the aquifer is unaffected by the fluid flow because all the additional heat must be conducted upward to the surface. Thus we assume that the heat flow below the aquifer is uniform and constant, equal to the heat flow at the surface in the absence of fluid flow. As we also assume a uniform thermal conductivity (Λ), the temperature gradient below the aquifer (f) is a constant. Because horizontal heat diffusion is neglected, the temperature profile between the aquifer and the surface is linear and the gradient above the aquifer is T_f/H. The heat lost by water flowing along the aquifer is then related to the discontinuity of gradient across the aquifer by the following equation

$$(\rho c)_w \, Q \, \mathrm{d}T_f/\mathrm{d}x = \Lambda[T_f/H(x) - f]. \tag{A 1}$$

Solutions of this equation can be determined easily if the dip (α) of the aquifer is constant. With $H(x) = \tan \alpha x$, (A 1) becomes

$$T_f' - \nu T/x = -\nu \tan \alpha f, \tag{A 2}$$

where

$$\nu = \Lambda/(\rho c)_w \, Q \tan \alpha. \tag{A 3}$$

For $\nu \neq 1$, its solutions are

$$T_f(z) = Cz^\nu + (\nu/(\nu - 1))fz, \tag{A 4}$$

where C is a function of ν and of the temperature at the deep end of the aquifer.

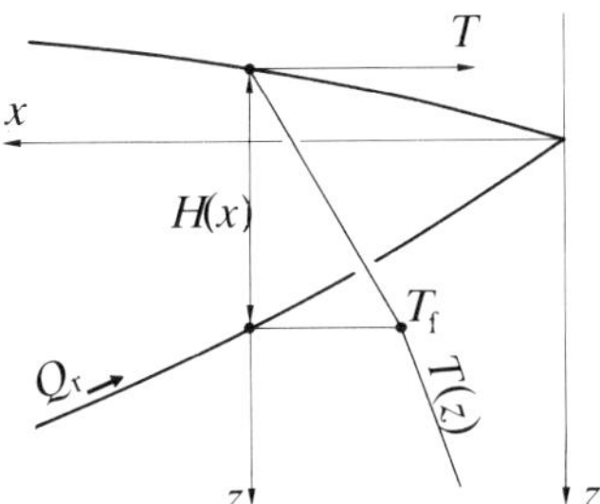

Figure 3. Schematic representation (with vertical exaggeration) of the low-angle aquifer model. Bold curves are the surface and the aquifer. Steady state is assumed and horizontal conduction is neglected. The fluid flux along the aquifer is Q_r. As fluid flows towards the outlet at $(0,0)$ it loses heat, causing a discontinuity of the temperature gradient across the aquifer, which lies at depth $H(x)$. T_f is the temperature of the fluid in the aquifer.

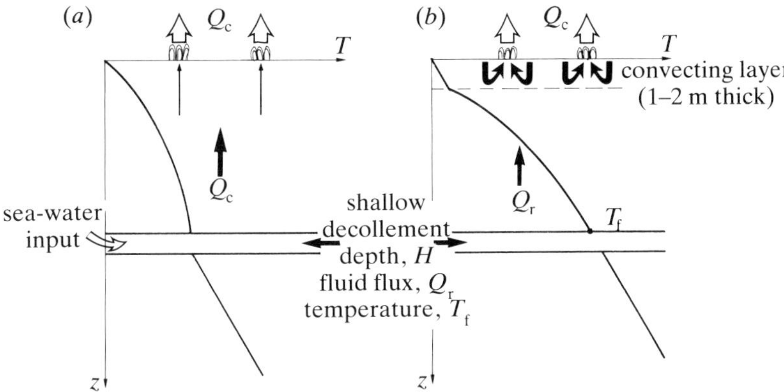

Figure 4. Models for the thermal structure below the colonies. Fluid comes along the shallow decollement at depth $H = 30$ m and seeps upward towards the clam colonies. For the thermal model it is assumed that upward fluid flow is uniform. Flux of incoming fluid is Q_r, but fluid flux Q_c through the clam colonies is much larger due to dilution with sea water. (a) Dilution with sea water occurs in the shallow decollement, causing a temperature drop. The fluid flux between the shallow decollement and the surface is Q_c. (b) Dilution occurs near the surface by convection around the colonies. The temperature in the shallow decollement is that of incoming fluid (T_f) and the fluid flux between the shallow decollement and the convecting layer is Q_r. Note that in both cases a gradient discontinuity occurs at the level where sea water is injected.

For $\nu < 1$, which corresponds to high fluid flux, the gradient tends to infinity where the aquifer reaches the surface. This artefact is a consequence of the assumption of no horizontal heat conduction, which is no longer valid. When $\nu \ll 1$, fluid temperature in the aquifer stays constant, equal to the source temperature until it comes very close to the outlet.

For $\nu > 1$, which corresponds to low fluid flux, the gradient tends to a finite limit f_1 at the outlet. This value is independent of the temperature of the fluid source

$$f_1 = \nu/(\nu-1)f. \qquad (A\,5)$$

In the cases considered in the text, $\nu > 1.5$. Then, the limit f_1 is a good approximation of the temperature gradient in a wide zone beginning at the outlet.

Phil. Trans. R. Soc. Lond. A (1991)

Appendix B. Dilution models

These models are based on the two following assumptions. Fluids come into the area of active venting laterally, along a shallow slightly dipping aquifer. A unidimensional advection model can be used to describe the temperature profile between this aquifer and the surface in the area of venting (See figure 4).

The temperature of fluids coming along the aquifer is known from the model explained in Appendix A. It can be computed from equation (A 5),

$$T_{\mathrm{f}} = \nu/(\nu-1)\,Hf, \tag{B 1}$$

where H (30 m) is the depth of the shallow decollement below the zone of active venting.

Very generally, the temperature profile resulting from uniform upward advection of fluid at velocity U in a half-space of thermal uniform thermal diffusivity D ($D = \Lambda/(\rho c)_{\mathrm{w}}$) is

$$T(z) = T_{\mathrm{surface}} + T_{\infty}[1-\exp{(-Uz/D)}]. \tag{B 2}$$

For simplicity, we take $T_{\mathrm{surface}} = 0$.

The total heat flow, including advection and conduction is uniform in the half-space, and equals $(\rho c)_{\mathrm{f}}\,UT_{\mathrm{limit}}$. In particular, it must be equal to the heat flow measured at the surface. Let us define $T_{\mathrm{a}} = F_{\mathrm{s}}/(\rho c)_{\mathrm{f}}\,Q_{\mathrm{c}}$ where F_{s} is the total heat flux through the surface in Zone 1, as defined in (6.1). We have

$$(\rho c)_{\mathrm{f}}\,Q_{\mathrm{c}}\,T_{\mathrm{a}} = F_{\mathrm{s}} = (\rho c)_{\mathrm{f}}\,SUT_{\infty}. \tag{B 3}$$

Then two end-member cases are considered, in one case dilution occurs near the surface, thus $SU = Q_{\mathrm{r}}$ and, from (B 3), $T_{\infty} = T_{\mathrm{a}}Q_{\mathrm{c}}/Q_{\mathrm{r}}$. In the other case dilution occurs in the shallow decollement, thus $SU = Q_{\mathrm{c}}$ and, from (A 3), $T_{\infty} = T_{\mathrm{a}}$. As in (6.3), let us define $h = DS/Q_{\mathrm{c}}$ and $d = Q_{\mathrm{c}}/Q_{\mathrm{r}}$. The temperature as a function of depth is, for the model with shallow convection,

$$T(z) = dT_{\mathrm{a}}[1-\exp{(-z/dh)}] \tag{B 4}$$

and for the model with mixing in the shallow decollement,

$$T(z) = T_{\mathrm{a}}[1-\exp{(-z/h)}]. \tag{B 5}$$

The next step is to relate the temperature $T(H)$ in the shallow decollement predicted by the uniform advection model with the temperature T_{f} of incoming fluids. If mixing with sea water occurs near the surface, one obviously has

$$T(H) = T_{\mathrm{f}}. \tag{B 6}$$

If mixing occurs in the shallow decollement, the heat balance across the decollement must be considered. The conductive flux from below is ΛfS, the heat input due to fluid flow along the shallow decollement into the zone of venting is $(\rho c)_{\mathrm{w}}\,Q_{\mathrm{r}}\,T_{\mathrm{f}}$ and the heat flux above this aquifer is equal to the flux through the surface F_{s}. Thus

$$T_{\infty} = T_{\mathrm{f}}/d + DSf/Q_{\mathrm{c}}. \tag{B 7}$$

We finally obtain for each case a set of equations which determines the dilution factor d from known parameters.

Note that, rigorously, the heat balance at the decollement should also be applied to the model with dilution near the surface. In this case, it implies continuity of the temperature gradient across the aquifer. This additional condition theoretically determines the value of F. However, this condition is not satisfied exactly, as the actual fluid flow pattern is not strictly one dimensional. In particular, part of the fluids continue to flow along the aquifer and release some heat. Thus we tolerate gradient discontinuities of the order of $10\ \mathrm{mK}\ \mathrm{m}^{-1}$.

References

Boulègue, J., Aquilina, L., Mariotti, A., Gamo, T. & Sakai, H. 1990 Chemistry of fluids in the Kaiko–Nankai area (Nankai accretionary prism). *Int. Conf. Fluids in Subduction Zones (Abst.)*, Paris, 5–6 November 1990.

Bray, C. J. & Karig, D. E. 1985 Porosity of sediments in accretionary prisms and some implications for dewatering processes. *J. geophys. Res.* **90**, 768–778.

Carson, B., Suess, E. & Strasser, J. C. 1990 Fluid flow and mass flux determinations at vent sites on the Cascadia margin accretionary prism. *J. geophys. Res.* **95**, 8891–8897.

Chamot-Rooke, N., Lallemant, S., Henry, P., Le Pichon, X., Cadet, J. P. & Lallemand, S. 1990 Tectonic context of fluid venting at the toe of the Nankai Accretionary Prism (Kaiko–Nankai diving cruise). *Int. Conf. Fluids in Subduction Zones (Abstr.)*, Paris, 5–6 November 1990.

Davis, E. E., Hyndmann, R. D. & Willinger, H. 1990 Rates of fluid expulsion across the Northern Cascadia Accretionary Prism: constraints from new heat flow and multichannel seismic reflection data. *J. geophys. Res.* **95**, 8869–8889.

Fisher, A. T. & Hounslow, M. W. 1990 Transient fluid flow through the toe of the Barbados accretionary complex: constraints from ocean drilling program Leg 110 heat flow studies and simple models. *J. geophys. Res.* **95**, 8845–8859.

Foucher, J. P., Henry, P., Sibuet, M., Kobayashi, K. & Le Pichon, X. 1990*a* Heat flow and fluid budget at the toe of the Nankai accretionary prism near 138° E. *Int. Conf. Fluids in Subduction Zones (Abstr.)*, Paris, 5–6 November 1990.

Foucher, J. P., Le Pichon, X., Lallemant, S., Hobart, M. A., Henry, P., Benedetti, M., Westbrook, G. K. & Langseth, M. G. 1990*b* Heat flow, tectonics and fluid circulation at the toe of the Barbados Ridge accretionary prism. *J. geophys. Res.* **95**, 8859–8868.

Gieskes, J. M., Vrolijk, P. & Blanc, G. 1990 Hydrogeochemistry of the Barbados accretionary wedge transect: Ocean Drilling Project Leg 110. *J. geophys. Res.* **95**, 8809–8818.

Henry, P., Lallemant, S., Le Pichon, X. & Lallemand, S. 1989 Fluid venting along Japanese trenches: tectonic context and thermal modeling. *Tectonophys.* **160**, 277–291.

Henry, P., Foucher, J. P., Le Pichon, X., Lallemant, S. & Chamot-Rooke, N. 1990 Thermal modelling of clam colonies: fluid flow velocity estimates from Kaiko–Nankai thermal data. *Intn. Conf. on Fluids in Subduction Zones*, Paris, 5–6 November 1990, Abstract.

Kinoshita, M. & Kazumi, Y. 1988 Heat flow – basic data. In *Preliminary report of the Hakuho–Maru cruise KH 86-5 (ODP site survey)* (ed. A. Taira), pp. 51–76. Tokyo: ORI.

Kobayashi, K. *et al.* 1990 General description of the Kaiko–Nankai Project. *Intn. Conf. on Fluids in Subduction Zones*, Paris, 5–6 November 1990, Abstract.

Kulm, L. D. & Suess, E. 1990 Relationship between carbonates deposits and fluid venting: Oregon accretionary prism. *J. geophys. Res.* **95**, 8899–8916.

Lallemant, S., Henry, P., Le Pichon, X. & Foucher, J. P. 1990 Detailed structure and possible fluid paths at the toe of the Barbados accretionary wedge (ODP Leg 110 area). *Geology* **18**, 854–857.

Langseth, M. G., Westbrook, G. K. & Hobart, M. 1990 Contrasting geothermal regimes of the Barbados Ridge accretionary complex. *J. geophys. Res.* **95**, 8829–8844.

Le Pichon, X. *et al.* 1990*a* Fluid venting in easternmost Nankai accretionary prism based on Kaiko–Nankai cruise results. *Intn. Conf. on Fluids in Subduction Zones*, Paris, 5–6 November 1990, Abstract.

Le Pichon, X., Foucher, J. P., Boulègue, J., Henry, P., Lallemant, S., Benedetti, F., Avedik, F. & Mariotti, A. 1990*b* Mud volcano field seaward of the Barbados accretionary complex: a submersible survey. *J. geophys. Res.* **95**, 8931–8945.

Le Pichon, X., Henry, P. & Lallemant, S. 1990*c* Water flow in the Barbados accretionary complex. *J. geophys. Res.* **95**, 8945–9868.

Molnar, P. & England, P. 1990 Temperatures, heat flux, and frictional stress near major thrust faults. *J. geophys. Res.* **95**, 4833–4856.

Moore, J. C. *et al.* 1988 Tectonics and hydrogeology of the northern Barbados Ridge, results from the Ocean Drilling Program Leg 110. *Geol. Soc. Am. Bull.* **100**, 1578–1593.

Moore, J. C. 1989 Tectonics and hydrogeology of accretionary prisms: role of the decollement zone. *J. struct. Geol.* **11**, 95–106.

Moore, J. C., Orange, D. & Kulm, L. D. 1990 Interrelationship of fluid venting and structural evolution: Alvin observations from the frontal accretionary prism: Oregon. *J. geophys. Res.* **95**, 8795–8808.

Nagihara, S., Kinoshita, H. & Yamano, M. 1989 On the high heat flow in the Nankai Trough area – A simulation study on a heat rebound process. *Tectonophys.* **161**, 33–41.

Peacock, S. M. 1987 Thermal effects of metamorphic fluids in subduction zones. *Geology* **15**, 1057–1060.

Screaton, E. J., Wuthrich, D. R. & Dreiss, S. J. 1990 Permeabilities, fluid pressures and flow rates in the Barbados ridge complex. *J. geophys. Res.* **95**, 8997–9007.

Sibson, R. H., Moore, J. McM. & Rankin, A. H. 1975 Seismic pumping – a hydrothermal fluid transport mechanism. *J. geol. Soc. Lond.* **131**, 653–659.

Sibuet, M., Fiala, A., Foucher, J. P. & Ohta, S. 1990 Spatial distribution of clams colonies at the toe of the Nankai accretionary prism near 138° E. *Int. Conf. on Fluids in Subduction Zones*, Paris, 5–6 November 1990, Abstract.

Toksöz, M. N., Minear, J. W. & Julian, B. R. 1971 Temperature fields and geophysical effects of a downgoing slab. *J. geophys. Res.* **76**, 1113–1138.

Tribble, J. S. 1990 Clay diagenesis in the Barbados accretionary complex: potential impact on hydrology and subduction mechanics. In *Proc. ODP, Sci. Results* (ed. J. C. Moore *et al.*) **110**, 97–110. College station, Texas: Ocean Drilling Program.

Von Huene, R. & Lee, H. 1983 The possible significance of pore fluid pressure in subductions zones. In *Studies of continental margin geology. Mem. Am. Ass. Petrol. geol.* **34**, 781–791.

Fluid expulsion from the Cascadia accretionary prism: evidence from porosity distribution, direct measurements, and GLORIA imagery

By Bobb Carson[1], Mark L. Holmes[2], Kea Umstattd[3],
Jeffrey C. Strasser[1] and H. Paul Johnson[4]

[1] *Department of Geological Sciences, Lehigh University, Bethlehem, Pennsylvania 18015, U.S.A.*
[2] *U.S. Geological Survey, School of Oceanography, University of Washington, Seattle, Washington 98195, U.S.A.*
[3] *Department of Geology, Carleton College, Northfield, Minnesota 55057, U.S.A.*
[4] *School of Oceanography, University of Washington, Seattle, Washington 98195, U.S.A.*

[Plate 1]

Fluid expulsion from the Cascadia accretionary prism off Oregon results from porosity reduction by compaction, and by cementation as methane-rich pore waters precipitate diagenetic carbonate deposits near the sediment-water interface. Porosity changes suggest that dewatering begins 5–6 km west of the base of the slope, in a proto-deformation zone. GLORIA imagery of surficial carbonate deposits confirms that fluid is actively expelled from this zone; there is no such evidence further west in Cascadia Basin. Within the uncertainties of the data, porosities do not decrease landward beneath the prism. This pattern is consistent with imbricate thrust faulting on the slope which provides the vertical load to induce compactive dewatering, and may physically import as much as 50% of the total fluid volume in the section. A simple vertical compaction model suggests that significant pore water volumes have been expelled from the lower slope, but at flux rates $(10^{-11}–10^{-12} \ \mathrm{m^3 \ m^{-2} \ s^{-1}})$ which are orders of magnitude less than those measured at individual vent sites $(10^{-6} \ \mathrm{m^3 \ m^{-2} \ s^{-1}})$. Faulting clearly controls some fluid expulsion, but GLORIA data suggest that repeated local discharge, cementation, and abandonment lead to dispersed accumulations of diagenetic carbonate.

1. Introduction

Fluid venting associated with subduction-induced sediment accretion occurs at numerous sites across the post-Oligocene portion of the Cascadia prism (Kulm & Suess 1990). Individual vents near the toe of the prism are of particular interest because they occur within a decipherable structural setting which suggests fluid transport paths. These paths may follow depositional unconformities (Lewis & Cochrane 1990), bedding, or fault zones (Moore *et al.* 1990). However, submersible and seismic reflection surveys conducted to date provide little information on the areal distribution of fluid discharge or the porosity distribution within the prism. This paper presents results from analysis of side-scan sonar (GLORIA) images and

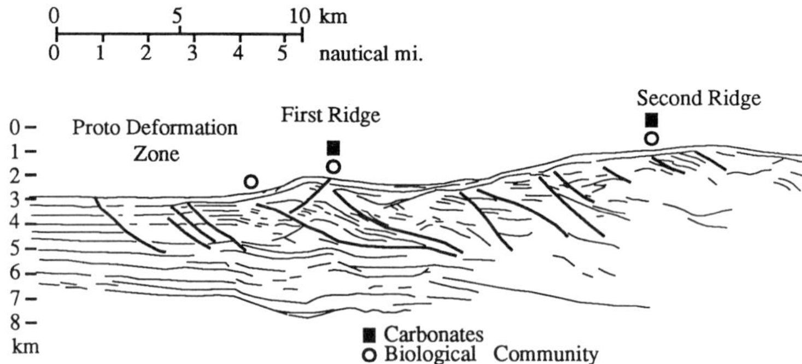

Figure 1. Interpreted depth section (after Snavely & Miller 1986) of the outer continental margin and Cascadia Basin off central Oregon. Boxes indicate approximate positions of carbonate deposits identified from submersible; circles indicate biological communities (after Moore *et al.* 1990). Faults are indicated by heavy lines.

relates them to seismic reflection and refraction data and fluid discharge measurements. The combined data-set constrains estimates of the magnitude and distribution of fluid flux on the Oregon margin.

The Oregon lower continental slope at 44° 40′ N originates as a seaward-verging, Late Pleistocene, ramp anticline (figure 1) which forms the lowermost (first) slope ridge. The frontal ramp joins the main decollement at a depth of *ca.* 2.2 km. West of the first ridge, a series of incipient thrusts that strike subparallel to the base of the slope define a proto-deformation zone, which extends 5–6 km into the Cascadia Basin. East of the first ridge is a slope basin filled with 0.5 km of Late Pleistocene sediments lying unconformably upon the thrust sheet. The second topographic ridge appears to be composed of several, older (Pleistocene–Pliocene) imbricate thrust sheets, but coherent reflections are sparse.

2. Evidence of fluid expulsion

(a) *Direct measurements*

Fluid expulsion has been measured at three sites on the Oregon margin (1428, 1900, and 2277, figure 2*a*). On the first ridge, flows of 1–6×10^{-6} m^3 m^{-2} s^{-1} are characteristic of small vents (9–16 m^2; Carson *et al.* 1990). At site 2277 vigorous expulsion, marked by discharge of methane bubbles and preliminary estimates of flow 3–5 times greater than those measured on the first ridge (E. Suess, personal communication), is recorded near the surface trace of a thrust fault, and apparently involves substantially larger discharge zones.

(b) *Porosity reduction*

Porosity distributions for the Cascadia margin (figure 3) are derived from compressional wave velocities determined by seismic refraction (Lewis 1990), and from migration analysis of multichannel reflection lines which corroborate the refraction data (Cochrane *et al.* 1990). Higher velocities beneath the first ridge of the lower slope were derived by Klaeschen & von Huene (1990), but are not used here. Refraction results (Cochrane *et al.* 1988) are utilized beneath the second ridge as poor reflector coherence precludes reliable migration analysis. The derived porosity distribution beneath the second ridge must be considered tentative.

Phil. Trans. R. Soc. Lond. A (1991)

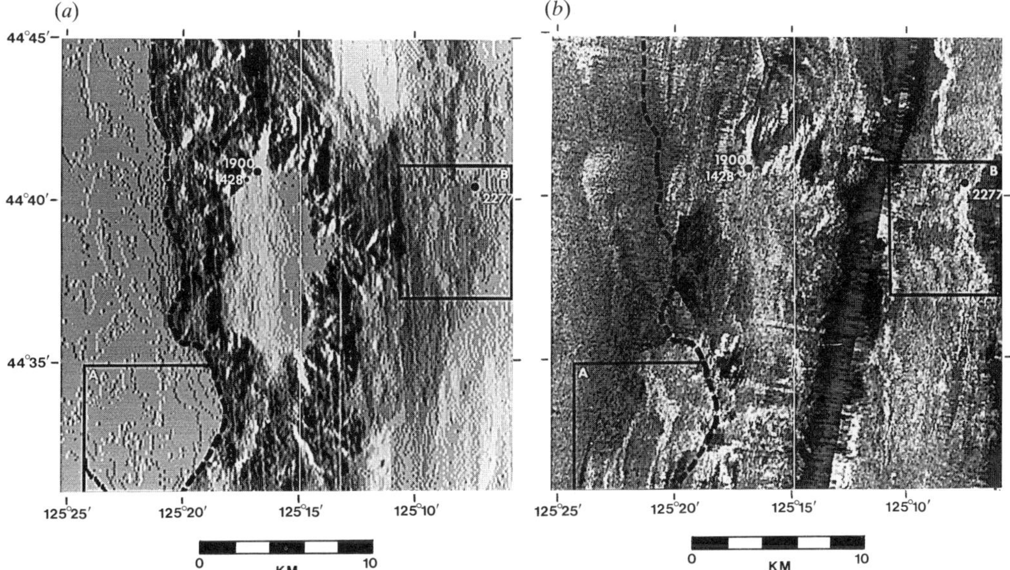

Figure 2. (*a*) Derivative map of SeaBeam bathymetry showing bottom slope of the lower margin and Cascadia Basin off Oregon. Subareas (A) and (B) exhibit minor variations in bottom slope. (*b*) GLORIA image of same areas as figure 2*b*. Base of the continental slope is indicated by the dashed line. Sites 1428, 1900, and 2277 are *Alvin* dive sites at which carbonates were sampled. GLORIA trackline (zone of no data) extends from 125° 10′ W at top of image to 125° 14′ W at the bottom. East of this track the bottom is insonified from the west. West of the track, the near bottom is insonified from the east. Cascadia Basin, on the far left side of the figure has been insonified from the west as part of the adjacent swath. The join between the two swaths can be seen at 125° 17′ W (top); it proceeds irregularly to 124° 24′ W (bottom).

The velocity–porosity conversion is based on more than 600 data points derived from well logs on the Washington margin, from shipboard logs of Deep Sea Drilling Program (DSDP) sites 174 and 175, and from analysis of returned samples (Strasser 1989). Regression on the velocity–porosity data is mediocre ($r = 0.82$), because carbonate cementation (see below) modifies the rigidity modulus as well as density in the compressional wave velocity equation (Lewis 1990). Although we suspect that carbonate cementation is relatively minor below several metres sub-bottom (based on carbonate occurrence at DSDP sites 174 and 175), any cementation would increase the seismic velocity and reduce the derived porosity in a way that differs from simple compactive change. Nevertheless, those effects are incorporated in our regression equation. The velocity–porosity relationship used here probably underestimates the actual porosity, as indicated by the low values calculated for Cascadia Basin deposits compared with measured porosities from DSDP site 174 (figure 3). However, the absolute values of porosity are of secondary importance because we are interested here in relative changes in porosity across the lower prism.

Porosities on the Oregon margin are apparently lower than porosities in distal portions (greater than 20 km west of margin toe) of Cascadia Basin (figure 3). The porosity loss across the deformation front, which averages about 4 % to a depth of 2.0 km, has been attributed to compaction associated with accretion (Carson 1977; Davis *et al.* 1990), and to exposure of previously buried section (Cochrane *et al.* 1988).

However, the regional landward reduction in sediment porosity inferred for many accretionary prisms (Bray & Karig 1985) is not apparent at the toe of the Oregon

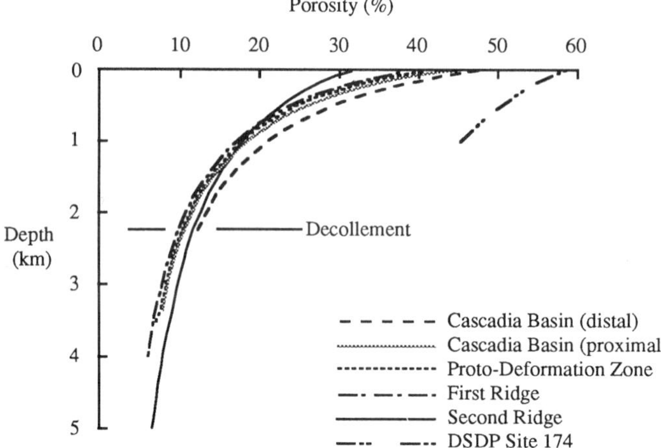

Figure 3. Porosity–depth profiles for Cascadia Basin and lower slope, based on seismic refraction results (first ridge, Lewis 1990; second ridge, Cochrane *et al.* 1988), interval migration velocities (Cascadia basin and proto-deformation zone, Cochrane *et al.* 1990), and the velocity–porosity distribution for the Cascadia margin (Strasser 1989). Porosity (n)–depth (m) regression lines are: (1) distal Cascadia Basin: $1/n = 2.14528 + 2.71205$E-3(m), $r = 0.94$, 19 d.f.; (2) proximal Cascadia Basin: $1/n = 2.31979 + 3.15217$E-3(m), $r = 0.92$, 27 d.f.; (3) proto-deformation zone: $1/n = 2.48078 + 3.20834$E-3(m), $r = 0.93$, 34 d.f.; (4) first ridge: $1/n = 2.46752 + 3.41516$E-3(m), $r = 0.99$, 26 d.f.; (5) second ridge: $1/n = 3.31991 + 2.29109$E-3(m), $r = 0.99$, 4 d.f.; (6) DSDP site 174: $1/n = 1.75218 + 4.57154$E-4(m), $r = 0.53$, 15 d.f.

slope (Strasser *et al.* 1989; Lewis 1990). Porosities in the proto-deformation zone, and the first ridge are indistinguishable, and beneath the second ridge are not significantly different (figure 3). The porosity distributions superficially suggest significant fluid expulsion from the proto-deformation zone, but fluid retention beneath the first ridge, and perhaps the second, since no systematic landward decrease in porosity is apparent.

(c) Diagenetic carbonate deposition

Where fluid discharge occurs on Cascadia and other convergent margins, authigenic carbonate deposits are formed by oxidation of dissolved methane (Kulm *et al.* 1986; Kastner *et al.* 1987; Le Pichon *et al.* 1987; Ritger *et al.* 1987; Kulm & Suess 1990). On the Cascadia margin some carbonates are exposed at the seafloor, but much of the carbonate deposition takes place by anaerobic methane oxidation within the sulphate-reducing zone (i.e. the upper 2–10 m; Suess & Massoth 1984) of the sediment column (Ritger *et al.* 1987; Han & Suess 1987). As these near-surface carbonates are positive indicators of pore fluid discharge, their extent and position on the accretionary prism are a record of the magnitude and structural/stratigraphic control of past expulsion.

Because carbonates have a significantly greater acoustic impedance (9–12×10^6 kg m^{-2} s^{-1}) than unconsolidated sediment (2×10^6 kg m^{-2} s^{-1}; Johnson & Helferty 1990), the diagenetic carbonate deposits can be imaged by side-scan sonar at the seafloor (reflection coefficient is 0.71–0.77) and beneath a veneer of mud, as the reflection coefficient across the water–mud interface is only 0.13, but from unconsolidated mud to carbonates is 0.64–0.71. Other factors affect the amplitude of GLORIA backscatter, but in areas where the slope and surface roughness are constant, carbonate deposits in unconsolidated sediments are readily detected. Consolidated

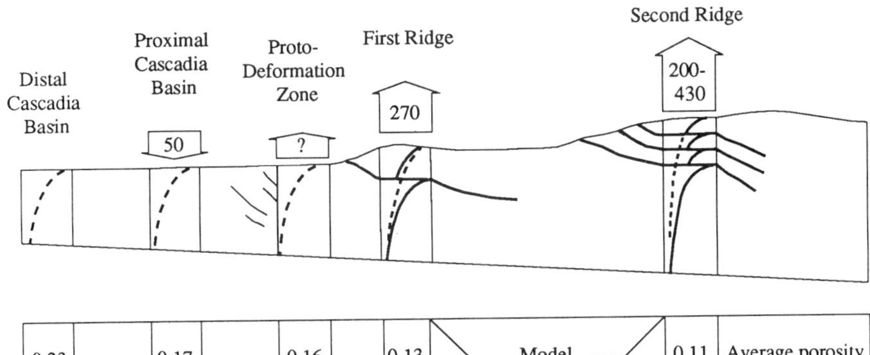

Distal Cascadia Basin	Proximal Cascadia Basin	Proto-Deformation Zone	First Ridge	Model	Second Ridge	
0.23	0.17	0.16	0.13	— Model —	0.11	Average porosity
2.1	3.2	3.4	4.3 / (1.5, 2.8)		4.3-4.8 / 6.0-6.5	Thickness (km)
480	530	530	570 / 840		940-1070 / 720-750	Porewater vol. ($m^3 m^{-2}$)
	-10	?	47		25-50	% present vol. expelled
			2.9×10^{-11} - 4.3×10^{11}		1.1×10^{-11} - 3.0×10^{-12}	Apparent rate of expul. ($m^3 m^2 s^{-1}$)

Figure 4. Schematic diagram of porosity, fluid inventory, and expulsion volumes for the Cascadia margin. Conceptual porosity–depth profiles associated with instantaneous thrust-faulting are shown by heavy lines; representations of observed profiles (figure 3) indicated by dashed lines. Calculated expulsion (or incorporation) volumes are shown in arrows, which imply nothing about mode of discharge. Pore fluid volumes for each location are tabulated beneath the diagram. See text for explanation of computations.

sandstones and shales are also strong reflectors (acoustic impedance is 7×10^6 kg m^{-2} s^{-1}; seafloor reflectance coefficient is 0.64) and occur in this area (Kulm *et al.* 1986). Although carbonates are superior reflectors relative to sandstones and shales, the acoustic impedance contrast (12.5%) is not sufficiently large that they can be reliably differentiated; identification must be made on the basis of geometry or independent (submersible) observation.

Sites of significant carbonate deposition (1900, 2277; figure 2*b*) observed from the submersible *Alvin* appear as regions of very strong backscatter in GLORIA imagery. Not all carbonate deposition is imaged, however; where carbonate precipitation is areally restricted (e.g. 1428, 9 m^2 as mapped from *Alvin*; figure 2*b*) or zones where precipitation occurs at subsurface depths greater than those penetrated by the acoustic signal (greater than 0.5–2 m) are not defined in the GLORIA images. Interpretation is complicated by the fact that bottom topography affects the angle of insonification, and high-amplitude backscatter (identical to returns from carbonate deposits) is produced at high reflectance angles. Nevertheless, we can assess the areal distribution of carbonate deposition in those regions where the bottom slope remains relatively constant (A, B; figure 2).

3. Fluid expulsion on the Oregon accretionary prism

(a) Cascadia Basin

Although average porosity decreases in Cascadia Basin from west to east, pore water volume increases as deposition thickens the sediment column (figure 4). At any given depth, however, the carbonate is lower near the margin (figure 3) which may indicate lateral textural variations or that the effects of tectonic stress extend seaward of the proto-deformation zone. The GLORIA data show no evidence of near-surface carbonate deposition in Cascadia Basin at distances of greater than 6 km west of the base of the continental slope.

(b) *Proto-deformation zone*

Blind thrusts extend to at least 2.2 km depth (figure 1) and show reversed polarity on multichannel seismic lines, suggesting higher, local porosity. If so, they may be active fluid conduits (J. C. Moore, personal communication). If the faults have delivered a significant fluid volume to the surface, the effect of that dewatering should be apparent in both the porosity and GLORIA data.

Porosity is lower at all comparable depths in the proto-deformation zone than in Cascadia Basin (figure 3). Integration and comparison of the porosity–depth profiles from proximal Cascadia Basin and the proto-deformation zone indicates indistinguishable pore fluid volumes, although the sediment section has thickened by 200 m (figure 4). Thickening results from some combination of deposition and small-displacement thrust faulting. The porosity data are insufficiently precise to indicate whether or not fluids are expelled in this region.

Although GLORIA imagery does not indicate strong backscattering uniformly across the proto-deformation zone, clear evidence of cementation is apparent between 44° 31′ N and 44° 35′ N (figures 2*b*, 5, plate 1). Flat-lying Cascadia Basin deposits (figure 2*a*) show high-amplitude backscatter in an anastamosing pattern which evolves into a slope-parallel fault trace to the south (44° 28′ N to 44° 12′ N). The geometry and backscattering levels preclude a clastic deposit and there is no apparent correlation with topography. Diffuse areas of strong reflectance to the north (44° 41′ N and 44° 44′ N, figure 2*b*) are more problematic; they may indicate carbonate deposition or small, sand-rich fan deposits.

(c) *First ridge*

Although porosity–depth distributions for the proto-deformation zone and first ridge are virtually identical (figure 3), the fluid budget is complicated by tectonic thickening (thrust-faulting). The ridge was built by imbricate thrusting of pre-existing proto-deformation deposits over proximal Cascadia Basin sediments. A conceptual model (figure 4, First Ridge; Strasser 1989) assumes instantaneous faulting of a 1500 m proto-deformation section over 2800 m of Cascadia Basin sediments. This tectonic movement substantially increases the pore fluid volume beneath the ridge (to 840 m^3 m^{-2}; the analysis is quite insensitive to the position of the fault), and with burial of the highly porous lower section, induces porosity reduction by vertical compaction. A rough estimate of the fluid discharged, obtained by subtracting the present fluid volume (570 m^3 m^{-2}) from the calculated, tectonically emplaced volume, is 270 m^3 m^{-2}; 47 % of the present fluid volume. The analysis clearly suggests substantial fluid loss.

GLORIA data across the first ridge and associated slope basin are difficult to interpret. Although very strong backscattering is observed at *Alvin* site 1900 (figure 2*b*), and around the perimeter of the slope basin, most of the highest-amplitude reflectance appears to be topographically controlled. The ridge is characterized by low reflectance along its crest and seaward flank. The slope basin exhibits generally strong backscattering that might reflect widespread, diffusive discharge and cementation. Even this pattern has a topographic component, however; the highest

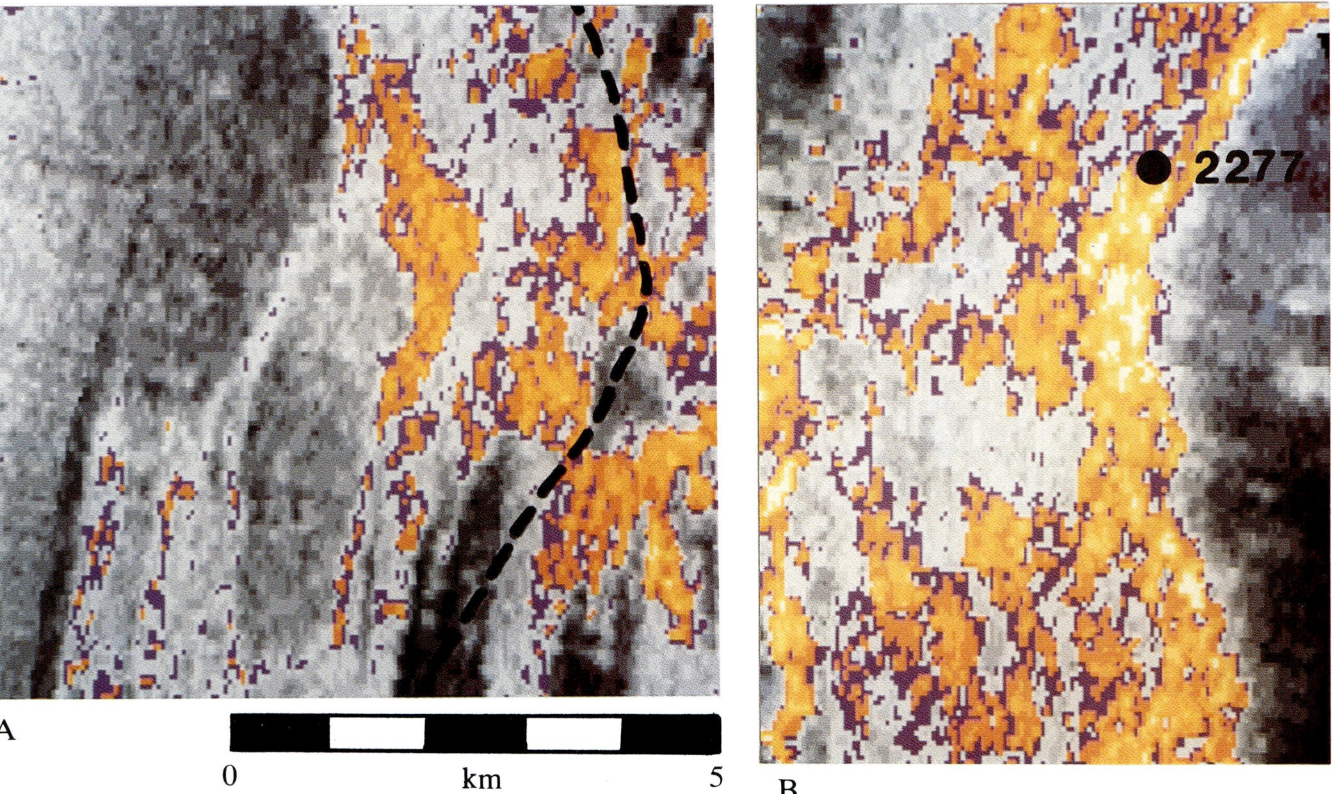

Figure 5. Enlargements of GLORIA imagery in subareas (A) and (B), figure 2b. Coloured areas comprise pixels in the upper 37% of the reflectance spectrum. Lower levels of backscattering are rendered in grey. Within the coloured zones, reflectance increases from purple to light yellow. Strongest reflectors lie south and northeast of Alvin site 2277, in the middle of a band of very strong backscattering. Dashed line in (A) is the base of the continental slope (figure 2a).

amplitudes are associated with the broad upwarping of the western half of the basin. Submersible surveys in the slope basin reveal no evidence of carbonate deposition.

If the first ridge was elevated between 200 000 and 300 000 years before present (BP) (Kulm & Fowler 1974), the calculated discharge volume (figure 4) would have been expelled at rates of 2.9–4.3×10^{-11} m^3 m^{-2} s^{-1}. These values are five orders of magnitude less than measured venting rates at sites 1428 and 1900 (Carson *et al.* 1990).

(d) Second ridge

Cementation on the second ridge is extensive (figure 2b, 5) but areally variable. Very strong reflectance in the GLORIA imagery occurs at site 2277, where large (greater than 3×10^4 m^2), surficial carbonate deposits were observed in several *Alvin* dives (J. C. Moore, personal communication). This site is associated with a fault (figure 1) that supports vigorous discharge. The GLORIA data, however, indicate that cementation is not restricted to this particular locality; strong backscattering characterizes the western flank of the ridge between 44° 37′ N and 44° 41′ N, at depths ranging from 1300–700 m. The reflectance pattern defines two particular zones that could be construed as fault traces: one is the very strong backscattering associated with site 2277; the other is a band of high reflectance which trends nearly north–south at 125° 8.5′ W, between 44° 41.5′ N and 44° 45.5′ N (figures 2b, 5). The latter follows a zone of increased bottom slope (figure 2a), but the strength of backscattering is out of proportion to the change in slope, and we infer that cementation and fluid expulsion have occurred here. The remainder of the slope (figure 5) is characterized by intermediate to high backscattering which occurs in a complex, distributed pattern that suggests almost random point discharges and cementation. Perhaps the pattern results from long-standing dewatering in which individual expulsion sites are activated and then abandoned with some regularity.

The porosity beneath the second ridge decreases from an anomalously low 32% at the surface (which may reflect the extensive cementation) to about 8% at a depth of 4 km (figure 3). The conceptual model of near-surface thrust-faulting to thicken the section (figure 4) assumes repetition of this porosity–depth relationship. Thickening due to faulting at depths greater than 2 km has an insignificant effect (less than 20 m^3 m^{-2}) on the calculated fluid inventory. However, the fluid volume is fairly sensitive to the number of imbricate thrust packets emplaced near the surface, and to the total thickness of the section. The variation in predicted porewater volume (940–1070 m^3 m^{-2}; figure 4) reflects emplacement of one versus three thrust packets over the same vertical extent (1700 m). Comparing these predicted volumes with the present estimated porewater volume (720–750 m^3 m^{-2}) suggests expulsion of 190–350 m^3 m^{-2}, or about 25–50% of the present fluid inventory. This is a maximum estimate as the model assumes that thickening has occurred exclusively near the surface. The extensive carbonate deposition observed from *Alvin* at site 2277 and the GLORIA data both indicate an extensive expulsion history.

Radiolarian biostratigraphy suggests that the second ridge was uplifted 1–2×10^6 years ago (L. D. Kulm, personal communication). If expulsion of the estimated 190–350 m^3 m^{-2} occurred over that period, average fluid discharge rates have been 1.1×10^{-11}–3.0×10^{-12} m^3 m^{-2} s^{-1}. As on the first ridge, these rates are orders of magnitude less than flow rates measured at vent site 2277 (E. Suess, personal communication).

4. Implications for fluid expulsion

GLORIA data confirm that fluids are delivered to the sediment–water interface in the proto-deformation zone (figure 5). Carbonate deposition clearly takes place at or near the sediment surface, but initially it may also occur in the shallow subsurface (less than 200 mBSF) as advecting methane-saturated fluids are anaerobically oxidized by sulphate reduction (Ritger *et al.* 1987; Han & Suess 1989). Because cementation inevitably leads to porosity reduction and decreased permeability near the top of the sediment column, continued flow is probably accommodated by repeated faulting and development of a fracture permeability (J. C. Moore *et al.*, this symposium). We suggest that this process and significant surface expulsion begin in the proto-deformation zone.

Surficial dewatering in the proto-deformation zone is terminated by overthrusting associated with advance of the prism toe. The advance presumably buries the surface and near-surface carbonates deposited in the proto-deformation zone and incorporates them in the tectonic stack, where they may be preserved or undergo dissolution and redistribution. Excess pore pressure is generated beneath the thrust sheet (Wang *et al.* 1990), perhaps driving flow upward through the hanging wall along pre-existing proto-thrust faults or new fractures developed during thrusting. Burial of cold, porous, near-surface sediments beneath the advancing thrust sheet creates a high-porosity reservoir, dominated by low-temperature fluids (Shi *et al.* 1988). To date, all fluids recovered from this area exhibit biogenic methane (E. Suess, personal communication), which implies burial temperatures less than 75 °C (Ritger *et al.* 1987).

Fluid expulsion on the first ridge is probably largely controlled by faulting (Moore *et al.* 1990), although diapiric structures may be locally important (Lewis & Cochrane 1990). By the time the ridge is elevated, advection has displaced the sulphate reducing zone to the surface (Han & Suess 1988) and dictates that cementation is almost wholly a surficial (less than 2–10 mBSF) phenomenon. Vent sites supported by fractures are frequently initiated and abandoned as cementation reduces surface permeability. Nevertheless, discharge is a local phenomenon; there is no obvious side-scan sonar evidence to support widespread expulsion and cementation.

The second ridge is characterized by extensive cementation. The greatest concentrations occur in two, well-defined zones that parallel the strike of the regional slope. One of these (Site 2277, figure 2*b*) is the surface trace of a thrust fault. Other strong backscattering on the second ridge is dispersed, however, and does not suggest fault control. It is not clear whether the complex cementation pattern (north as well as south and west of site 2277, figure 5) represents relict fault control complicated by post-depositional dissolution, or whether active venting is inhomogeneously dispersed across the slope. Davis *et al.* (1990) suggest areally pervasive vertical flow for the northern Cascadia margin. If the extensive backscattering on the second ridge indicates dispersed vertical flow, it is a marked departure from the character of discharge further seaward off Oregon, and may indicate development of a different dispersal mechanism, such as extensive small-scale fracturing.

Although uncertainties in the velocity–porosity relationship and in the importance of horizontal advection limit the usefulness of the thrust-sheet loading/vertical compaction model, it provides some insights. Clearly horizontal fluid transport by thrust sheet emplacement is an important component of the fluid budget. Not only does the thrust sheet provide the vertical load to induce compactive dewatering, but it may physically import as much as 50% of the total fluid volume in the section. Although the calculated discharge rates (10^{-11}–10^{-12} m^3 m^{-2} s^{-1}, figure 4) are suspect, they provide gross estimates of the flux that is supportable by compaction alone. Regardless of the error inherent in those estimates, they cannot be reconciled with measured rates of venting (1–6×10^{-6} m^3 m^{-2} s^{-1}; Carson *et al.* 1990).

The GLORIA data, which suggest that significant cementation associated with expulsion covers a relatively small portion of the margin, dictate that the disparity between the estimated rates of outflow derived from porosity change and measured expulsion rates at vents requires both channelling of flow by high permeability zones and non-steady-state discharge. Beyond this point, the problem is underconstrained; we need to determine the duration and magnitude of site-specific venting if we are to determine the volume of the fluid reservoir that supports it. Dating the carbonate deposits and geochemically relating their volume to the volume of depositing fluids would be an important contribution to resolving the fluid budget.

This work was funded by grants from the National Science Foundation (OCE-8613501) and JOI/U.S. Science Support Program (JSC-8-90). B. T. R. Lewis and G. Cochrane conducted the refraction survey and provided velocity analysis. G. Cochrane and R. von Huene generously shared velocity results from their MCS migration analyses. D. Twitchell and V. Paskevich provided invaluable assistance in processing the GLORIA and SeaBeam data. The typescript benefited significantly from reviews by J. C. Moore and an anonymous referee.

References

Bray, C. J. & Karig, D. E. 1985 Porosity of sediments in accretionary prisms, and some implications for dewatering processes. *J. geophys. Res.* **90**, 768–778.

Carson, B. 1977 Tectonically-induced deformation of deep-sea sediments off Washington and Oregon: mechanical consolidation. *Mar. Geol.* **24**, 289–307.

Carson, B., Suess, E. & Strasser, J. C. 1990 Fluid flow and mass flux determinations at vent sites on the Cascadia margin accretionary prism. *J. geophys. Res.* **95**, 8891–8897.

Cochrane, G. R., Lewis, B. T. R. & McClain, K. J. 1988 Structure and subduction processes along the Oregon–Washington margin. *Pure appl. Geophys.* **128**, 767–800.

Cochrane, G. R., Moore, J. C., Mackay, M. & Moore, G. F. 1990 Fluid flow in the Oregon accretionary prism from seismic data and vents. *Eos, Wash.* **71**, 1580.

Davis, E. E., Hyndman, R. D. & Villinger, H. 1990 Rates of fluid expulsion across the northern Cascadia accretionary prism: constraints from new heat flow and multichannel seismic reflection data. *J. geophys. Res.* **95**, 8869–8889.

Han, M. W. & Suess, E. 1989 Subduction-induced pore fluid venting and the formation of authigenic carbonates along the Oregon/Washington continental margin: implications for the global Ca cycle. *Paleogeogr. Paleoclimat. Paleoecol.* **71**, 119–136.

Johnson, H. P. & Helferty, M. 1990 The geological interpretation of side-scan sonar. II. Processing and analysis of images. *Rev. Geophys.* **28**, 357–380.

Kastner, M., Suess, E., Garrison, R. E. & Kvenvolden, K. 1987 Hydrology, geochemistry, and diagenesis along the convergent margin off Peru. *Eos, Wash.* **68**, 1499.

Klaeschen, D. & von Huene, R. 1990 Ocean Drilling Program site survey seismic record – Oregon. *Fluids in Accretionary Prisms* (abstr.), Paris.

Kulm, L. D. & Fowler, G. A. 1974 Oregon continental margin structure and stratigraphy: a test of the imbricate thrust model. In *The geology of continental margins* (ed. C. A. Burke & C. L. Drake), pp. 261–284. New York: Springer-Verlag.

Kulm, L. D., Suess, E., Moore, J. C., Carson, B., Lewis, B. T., Ritger, S., Kadko, D., Thornburg, T., Embley, R., Rugh, W., Massoth, G. J., Langseth, M., Cochrane, G. R. & Scamman, R. L. 1986 Oregon subduction zone: Venting, fauna, and carbonates. *Science, Wash.* **231**, 561–566.

Kulm, L. D. & Suess, E. 1990 Relationship between carbonate deposits and fluid venting: Oregon accretionary prism. *J. geophys. Res.* **95**, 8899–8915.

Le Pichon, X. *et al.* 1987 Nankai Trough and Zenisu Ridge: a deep-sea submersible survey. *Earth planet. Sci. Lett.* **83**, 285–299.

Lewis, B. T. R. 1990 Changes in P and S velocities caused by subduction related sediment accretion off Washington/Oregon, NATO Conference on Shear Waves in Marine Sediments. (Submitted.)

Lewis, B. T. R. & Cochrane, G. R. 1990 Relationship between the location of chemosynthetic benthic communities and geologic structure on the Cascadia subduction zone. *J. geophys. Res.* **95**, 8783–8794.

Moore, J. C., Orange, D. & Kulm, L. D. 1990 Interrelationship of fluid venting and structural evolution: *Alvin* Observations from the frontal accretionary prism, Oregon. *J. geophys. Res.* **95**, 8795–8808.

Ritger, S., Carson, B. & Suess, E. 1987 Methane-derived authigenic carbonates formed by subduction-induced pore water expulsion along the Oregon/Washington margin. *Geol. Soc. Am. Bull.* **98**, 147–156.

Shi, Y., Wang, C., Langseth, M. G., Hobart, M. & von Huene, R. 1988 Heat flow and thermal structures of the Washington/Oregon accretionary prism – a study of the lower slope. *Geophys. Res. Lett.* **15**, 1113–1116.

Snavely, P. D., Jr & Miller, J. 1986 The central Oregon continental margin, lines WO76-4 and WO76-5. In *Seismic images of modern convergent margin tectonic structure* (ed. R. von Huene). *AAPG Stud. Geol.* **26**, 24–29.

Strasser, J. C. 1989 Velocity-derived porosity distribution of the Oregon margin: implications for sediment dewater at the toe of the accretionary prism. M.S. thesis, Lehigh University, Department of Geological Sciences, Bethlehem, Pennsylvania.

Strasser, J. C., Carson, B. & Lewis, B. T. R. 1989 Velocity-derived porosity distribution of the lower Oregon margin accretionary prism: implications for dewatering. *Geol. Soc. Am., Abstr. Progr.* A **21**, 312.

Suess, E. & Massoth, G. J. 1984 Evidence for venting of pore waters from subducted sediments of the Oregon continental margin. *Eos, Wash.* **65**, 1089.

Wang, C.-Y., Shi, Y., Hwang, W.-T. & Chen, H. 1990 Hydrogeologic processes in the Oregon–Washington accretionary complex. *J. geophys. Res.* **95**, 9009–9023.

Colour plate printed in Great Britain by George Over Limited, London and Rugby

Numerical simulation of subduction zone pressure–temperature–time paths: constraints on fluid production and arc magmatism

By Simon M. Peacock

*Department of Geology, Arizona State University, Tempe,
Arizona 85287-1404, U.S.A.*

The location and sequence of metamorphic devolatilization and partial melting reactions in subduction zones may be constrained by integrating fluid and rock pressure–temperature–time (P–T–t) paths predicted by numerical heat-transfer models with phase diagrams constructed for metasedimentary, metabasaltic, and ultramafic bulk compositions. Numerical experiments conducted using a two-dimensional heat transfer model demonstrate that the primary controls on subduction zone P–T–t paths are: (1) the initial thermal structure; (2) the amount of previously subducted lithosphere; (3) the location of the rock in the subduction zone; and (4) the vigour of mantle wedge convection induced by the subducting slab. Typical vertical fluid fluxes out of the subducting slab range from less than 0.1 to 1 (kg fluid) $m^{-2} a^{-1}$ for a convergence rate of 3 cm a^{-1}. Partial melting of the subducting, amphibole-bearing oceanic crust is predicted to only occur during the early stages of subduction initiated in young (less than 50 Ma) oceanic lithosphere. In contrast, partial melting of the overlying mantle wedge occurs in many subduction zone experiments as a result of the infiltration of fluids derived from slab devolatilization reactions. Partial melting in the mantle wedge may occur by a two-stage process in which amphibole is first formed by H_2O infiltration and subsequently destroyed as the rock is dragged downward across the fluid-absent 'hornblende-out' partial melting reaction.

1. Introduction

Large-scale mixing of crustal and mantle materials occurs at subduction zones where oceanic lithosphere, capped by variably hydrated oceanic crust and sediments, descends beneath either oceanic or continental lithosphere. Over time, most of Earth's continental crust has been extracted from the mantle at subduction zones. The location of volcanoes above subducting slabs, as defined by inclined seismic planes, requires that partial melting occurs in subduction zones despite the cooling effect of subducting lithosphere. Much research has focused on constraining the source region of arc magmas by petrological and geochemical investigations of the products of arc magmatism (Gill 1981); such investigations may be considered analogous to inverse methods used in geophysics. In this paper, I use a forward method to investigate zone processes based on a two-dimensional numerical model of heat transfer and metamorphic reactions. Ultimately, the complementary nature of forward and inverse methods should result in a quantitative understanding of subduction zones and arc magmatism.

Numerical simulations of heat and mass transfer in subduction zones provide important constraints on processes occurring at depths greater than 50 km from which geologic samples are extremely rare. Numerous thermal models of subduction zones have been presented in the literature (Hasebe *et al.* 1970; Minear & Toksoz 1970; Oxburgh & Turcotte 1970; Toksoz *et al.* 1971; Andrews & Sleep 1974; Honda & Uyeda 1983; van der Beukel & Wortel 1986). Thermal models have been combined with petrologic models of the subducting slab and mantle wedge by Oxburgh & Turcotte (1976), Anderson *et al.* (1976, 1978, 1980), Delany & Helgeson (1978), Wyllie & Sekine (1982), and Wyllie (1988), among others.

In contrast to previous subduction zone models, the models presented in this paper specifically predict pressure–temperature–time (P–T–t) paths followed by rocks in the subducting slab and the overlying mantle wedge. Subduction zone P–T–t paths represent the P–T conditions encountered by rocks moving through an evolving (non-steady-state) thermal structure. Because of the material flow in subduction zones and the time-dependent thermal structure, subduction zone processes may be best understood by considering subducting slab and mantle wedge P–T–t paths. Subduction zone P–T–t paths can be combined with appropriate phase diagrams in order to constrain the sequence and location of metamorphic and melting reactions in the subducting slab. Of particular importance are reactions involving low-viscosity (aqueous) fluids that may trigger partial melting reactions in the overlying mantle wedge. Calculated mantle wedge P–T–t paths constrain the location of hydration, dehydration, and partial melting reactions in the sub-arc lithosphere and asthenosphere.

Recent estimates of subduction zone volatile budgets (Peacock 1990*a*) suggest that an order of magnitude more H_2O and CO_2 is subducted than can be accounted for in arc magmas. The release and ultimate fate of these volatiles may have a profound effect on the thermal, petrological, and rheological evolution of subduction zone. The results of the numerical simulations presented in this paper permit us to consider possible P–T paths followed by released volatiles and therefore place constraints on fluid processes in subduction zones.

2. Description of the numerical model

Numerical simulations of heat and mass transfer in subduction zones were conducted by using a two-dimensional finite difference model (figure 1). This model is similar to the finite difference models described in Peacock (1987*b*, 1990*b*), except that (1) oceanic geotherms were defined by calculating the conductive cooling of a mantle adiabat and (2) an analytical solution describing induced convection was incorporated into the mantle wedge. Briefly, the two-dimensional model simulates heat transfer in a subduction zone with a 26.6° dip. A spatial resolution of 1 km is used in the region of the subduction thrust to carefully monitor metamorphic reactions; elsewhere the spatial resolution is 5 km. The advective/conductive heat transfer equation was solved in two dimensions by using an explicit finite difference method. A more detailed description of the numerical method may be found in Peacock (1987*b*, 1990*b*).

Oceanic geotherms as a function of age were constructed by numerically calculating the conductive cooling of a zero-age adiabat (Jeanloz & Morris 1986). The zero-age adiabat was defined by a surface temperature of 1275 °C, a 3 °C km^{-1} adiabatic gradient in the upper 35 km (appropriate for mantle with 5–30% partial melt), and

Phil. Trans. R. Soc. Lond. A (1991)

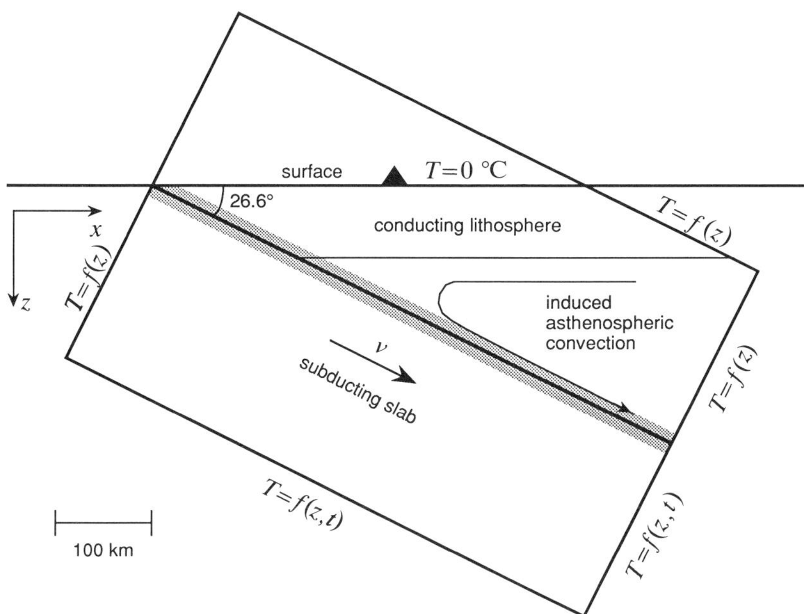

Figure 1. Geometry and boundary conditions used in the numerical model of subduction zone heat transfer. The two-dimensional 400×600 km grid is oriented at a $26.6°$ angle to the Earth's surface, such that the subduction thrust is parallel to one of the model axes. To carefully monitor reactions in the subducting oceanic crust, a 1 km finite-difference grid spacing is used within 10 km of the subduction thrust (stippled region); elsewhere the grid spacing is 5 km. Rocks in the subducting slab move downward, parallel to the subduction thrust at a constant velocity. Rocks in the overlying asthenosphere move along steady-state streamlines defined by Batchelor's (1967) analytical solution for two-dimensional fluid flow in a corner. The surface temperature is kept constant at 0 °C. Boundaries labelled $T = f(z)$ are kept at a constant temperature defined by the initial oceanic geotherm. Boundaries labelled $T = f(z,t)$ are reset during the advection time step and held constant during the conduction time step.

a $0.3\,°\mathrm{C\,km^{-1}}$ adiabatic gradient below 35 km (Stacey 1977; Basaltic Volcanism Study Project 1981; Jeanloz & Morris 1986). During cooling, the surface temperature was fixed at 0 °C and the thermal diffusivity at $10^{-6}\,\mathrm{m^2\,s^{-1}}$. The calculated oceanic geotherms agree well with oceanic heat flow measurements for $t < 80$ Ma; for older oceanic lithosphere, the conductive cooling model predicts slightly lower heat flow values than are observed.

Induced flow in the asthenospheric wedge overlying the subducting slab (figure 1) was simulated using an analytical solution for two-dimensional incompressible fluid flow in a corner (Batchelor 1967, pp. 224–227). The fluid is assumed to have a uniform viscosity throughout the wedge. A no-slip boundary condition is used at the base of the conducting lithosphere (defined as the 1300 °C isotherm) and a constant slip equal to the slab velocity is used at the top of the subducting slab. More complex induced flow models could be constructed, but this steady-state model represents an end-member model that may be compared with simulations run with no induced convection.

A series of numerical experiments were conducted in order to determine the primary variables that control rock P–T–t paths in subduction zones. Each numerical experiment simulated up to 100 Ma of thermal evolution. The following

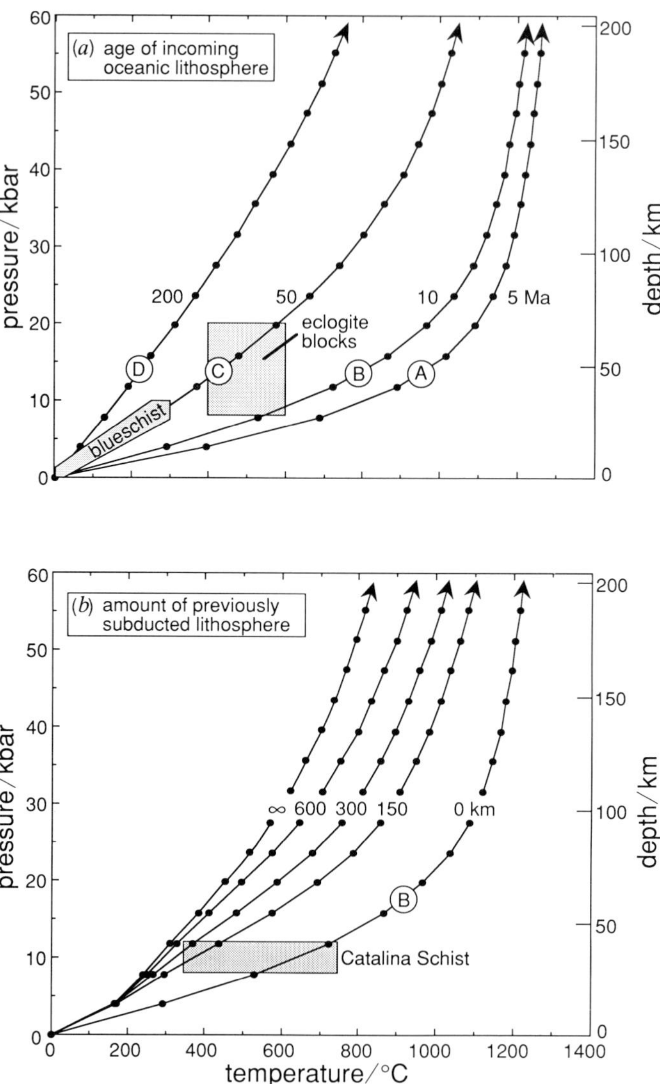

Figure 2. Subduction zone P–T–t paths predicted by the two-dimensional, heat-transfer model and selected pressure–temperature constraints based on investigations of ancient subduction zones. Solid circles represent intervals of 1 Ma; open circles represent intervals of 10 Ma. Selected P–T–t paths are labelled with letters for use in later figures. (a) Calculated P–T–t paths for the top of the subducted oceanic crust as a function of the age of the incoming oceanic lithosphere. Convergence rate is 3 cm a^{-1}. Prograde blueschist P–T paths (Ernst 1988) and P–T conditions recorded by eclogite blocks (Evans & Brown 1986) shown by stippled regions. (b) P–T–t paths followed by the top of the subducting oceanic crust as a function of the amount of previously subducted lithosphere (convergence rate × time). Convergence rate is 3 cm a^{-1}. The curve labelled '∞' represents the steady-state P–T–t path. The predicted cooling with time is consistent with the P–T conditions preserved in the Catalina Schist inverted metamorphic gradient (Platt 1975; Sorensen 1986, 1988). (c) P–T–t paths as a function of position (subducting oceanic crust, mantle wedge) in the subduction zone. Convergence rate is 3 cm a^{-1}. Downward conduction of heat into the subducting slab results in warmer P–T–t paths for the top of the subducting oceanic crust as compared with the base. Mantle wedge material, located 5 km from the top of the subducting slab, that is not dragged down with the subducting slab will follow isobaric cooling P–T–t paths. P–T–t path G represents the approximate path followed by mantle material that is dragged down by the

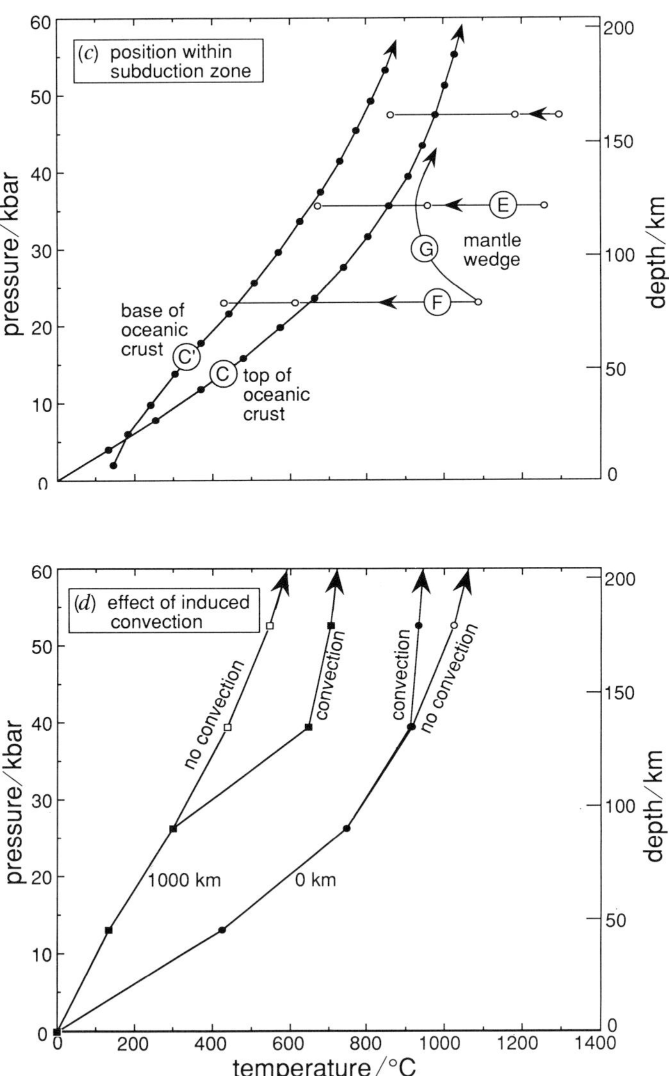

subducting slab. (*d*) *P–T–t* paths followed by the top of the subducting slab as a function of the vigour of induced convection and the amount of previously subducted material. Convergence rate is 10 cm a^{-1}. Initially, induced mantle wedge convection results in slightly cooler slab *P–T–t* paths. As subduction progresses, slab *P–T–t* paths remain warm because of induced mantle wedge convection.

parameters were varied during individual numerical experiments: (1) the initial age of the oceanic lithosphere; (2) the convergence rate; (3) the nature of metamorphic dehydration reactions; (4) the amount of shear heating; and (5) the presence of induced mantle convection. Each numerical experiment produces a data set containing the thermal structure of the subduction zone as a function of time from which *P–T–t* paths were calculated for selected rocks in the subducting slab and the mantle wedge.

3. Results

The results of the numerical experiments demonstrate that the primary factors controlling rock P–T–t paths are: (1) the initial thermal structure; (2) the amount of previously subducted lithosphere (convergence rate × time); (3) the position of the rock in the subduction zone; and (4) and the vigour of induced mantle convection.

The initial thermal structure is the dominant factor in determining the P–T–t path followed by the earliest subducted oceanic crust (figure 2a). Subduction zones that form in young, relatively hot, oceanic lithosphere result in warmer slab P–T–t paths as compared with subduction zones that form in older oceanic lithosphere. Subduction zones initiated in 5–10 Ma oceanic lithosphere result in slab P–T–t paths that exceed 1100 °C at 30 kbar†. In contrast, subduction zones formed in 200 Ma oceanic lithosphere result in slab P–T–t paths passing through $ca.$ 500 °C at 30 kbar. Blueschist P–T paths and the P–T conditions of eclogites are consistent with P–T–t paths calculated for $ca.$ 50 Ma oceanic lithosphere (figure 2a).

The amount of previously subducted lithosphere (convergence rate × time) is a major factor in controlling subduction zone P–T–t paths (figure 2b). The advection of heat by the subducting slab dominates the thermal structure of subduction zones. As more lithosphere is subducted, subducting oceanic crust P–T–t paths follow cooler trajectories. For a subduction zone formed in 10 Ma oceanic crust, the initial subducting oceanic crust P–T–t path intersects $ca.$ 1100 °C at 30 kbar; P–T–t paths for oceanic crust subducted after 600 km of previous subduction intersects $ca.$ 700 °C at 30 kbar. The predicted cooling with time is consistent with inverted metamorphic gradients preserved in palaeosubduction zones such as the Catalina Schist and Pelona Schist of southern California (Peacock 1987b). Qualitatively, P–T–t paths calculated for mature subduction zones resemble P–T–t paths calculated for subduction zones formed in older oceanic lithosphere. Cool, approximately steady-state conditions are achieved after 50–100 Ma of subduction.

Rocks in the subducting slab and the overlying mantle wedge follow strikingly different P–T–t paths (figure 2c). Subducting oceanic crust follows P–T–t paths characterized by increasing pressure and temperature. Conduction of heat downward from the overlying mantle wedge into the top of the subducting slab results in warmer P–T–t paths for the top of the oceanic crust as compared to the base (figure 2c). In the absence of mantle wedge convection induced by the subducting slab, mantle wedge rocks follow isobaric cooling P–T–t paths (figure 2c). In subduction zones with induced mantle convection, rocks in the mantle wedge near the subducting slab first cool approximately isobarically, and subsequently undergo compression with minor heating as they are dragged downward by the slab (figure 2c).

In addition to strongly influencing mantle wedge P–T–t paths, induced convection also affects subducting slab P–T–t paths (figure 2d). Slab P–T–t paths calculated with and without induced mantle wedge convection diverge at the base of the overriding lithosphere (100 km in figure 2d). Initially, induced convection results in slightly cooler slab P–T–t paths because the dragging down of cool mantle wedge material next to the slab results in reduced thermal gradients in the subduction thrust region and therefore reduced downward conduction. With continued subduction, induced convection results in substantially warmer slab P–T–t paths, as

---

† 1 bar = 10^5 Pa.

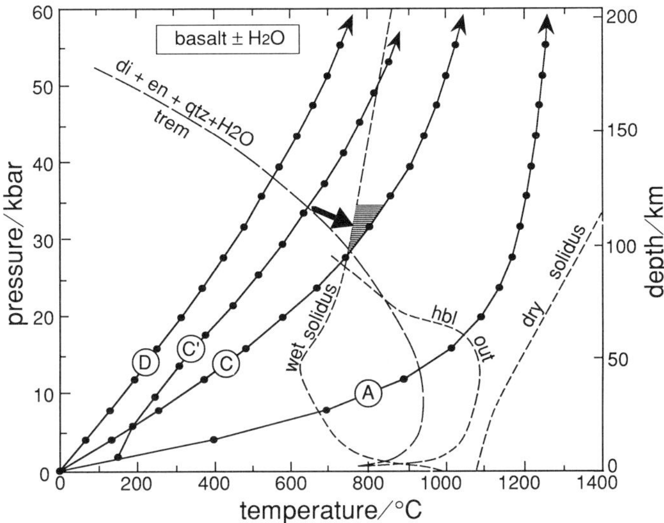

Figure 3. *P–T–t* paths (solid lines) and experimental melting relations (short dashed lines) for the hydrous basalt system. The addition of excess H_2O drastically decreases the melting temperature of basalt (dry solidus: Green 1982; wet solidus: Lambert & Wyllie 1972). The 'hbl out' curve represents fluid-absent partial melting of hornblende-bearing metabasalt (Green 1982). Long-dashed line represents calculated tremolite dehydration reaction using the thermodynamic data of Berman (1988). Subducted oceanic crust following *P–T–t* paths cooler than path C will dehydrate before reaching melting reactions, suggesting that arc magmas in cool subduction zones are not derived from partial melting of the oceanic crust. Warmer *P–T–t* paths predicted for subduction zones developed in young oceanic lithosphere result in substantial partial melting associated with the breakdown of hornblende at pressures less than *ca.* 28 kbar. If the top (path C) and base (path C') of the subducting oceanic crust straddle the wet solidus at *ca.* 30 kbar, then fluids released by dehydration reactions (arrow) could trigger partial melting in overlying oceanic crust (striped region).

compared with experiments with no convection, because the overlying mantle wedge is replenished with hot asthenosphere. The overall effect of induced mantle wedge convection on slab *P–T–t* paths is to retard the rapid cooling effect of subducting lithosphere shown in figure 2*b*.

Additional processes, such as shear heating (frictional heating, viscous dissipation), metamorphic reactions, and metamorphic fluid flow affect subduction zone *P–T–t* paths, but only to a minor extent. The thermal effect of shear heating is proportional to the shear stress and convergence rate of the subduction zone. As compared with numerical experiments with no shear heating, numerical experiments that incorporate shear heating (maximum shear stress is 250 bar, $v = 3$ cm a^{-1}) result in subducting slab *P–T–t* paths that are only *ca.* 50 °C warmer. Numerical experiments that incorporate prograde metamorphic reactions that consume 50 kJ kg^{-1} result in subducting slab *P–T–t* paths that are only 5 °C cooler (Peacock 1990*b*). The thermal effect of metamorphic fluid flow in subduction zones has been discussed in earlier papers (Peacock 1987*c*, 1990*b*). The results of these previous experiments suggest a maximum effect of fluid flow of *ca.* 100 °C on mantle wedge *P–T–t* paths if all fluids released by devolatilization reactions in the subducting oceanic crust flow upward into widely spaced (more than 10 km) channels in the mantle wedge. Other flow

348   *S. M. Peacock*

geometries, such as pervasive upward fluid flow or slab-parallel fluid flow, result in a negligible effect on mantle wedge and subducting slab P–T–t paths.

4. Discussion

(a) Dehydration and partial melting of the subducting slab

Calculated subduction slab P–T–t paths can be combined with a phase diagram for hydrous basalt (figure 3) to predict the sequence of metamorphic and igneous reactions encountered by the subducting oceanic crust. The location of slab dehydration reactions and zones of high fluid flux depends critically on the subsolidus basaltic phase diagram. Unfortunately, few subsolidus phase equilibria experiments of hydrous basaltic compositions have been performed at pressures greater than 10 kbar. The anhydrous nature of most eclogites suggests that much of the H_2O in subducted oceanic crust is driven off at relatively shallow depths (less than 50 km).

In a previous paper (Peacock 1990b), I presented the results of numerical experiments that incorporated several different models of dehydration in subducting oceanic crust. In these calculations the 7.5 km thick oceanic crust was assumed to contain 2 wt% bound H_2O (a probable maximum amount) distributed homogeneously throughout the crust. Fluids were released by continuous and discontinuous end-members of three different dehydration models: pressure-sensitive, temperature-sensitive, and amphibole (negative dP/dT) dehydration. For a convergence rate of 3 cm a^{-1}, typical fluid fluxes out of the subducting slab range from *ca.* 0.1 (kg fluid) m^{-2} a^{-1} for the continuous reaction models to more than 1 (kg fluid) m^{-2} a^{-1} for the discontinuous pressure and amphibole dehydration models. These fluid fluxes should be considered maxima because of the assumption that 2 wt% H_2O is released.

Under a relatively restricted set of conditions, fluids released by slab dehydration reactions could trigger partial melting in overlying oceanic crust (figure 3). During subduction, fluids released from the base of the subducting oceanic crust will travel up temperature if they flow up toward the subduction shear zone. Upward fluid flow will trigger fluid-present partial melting reactions in the subducting slab if the temperature at the top of the subducting oceanic crust exceeds the wet basaltic solidus. This scenario occurs only if P–T–t paths for the top and base of the subducting oceanic crust straddle a temperature-sensitive dehydration reaction and the wet basaltic solidus as depicted by paths C and C′ in figure 3.

More commonly, fluids released by metamorphic reactions in the subducting oceanic crust that travel up the subduction shear zone would not be expected to trigger melting reactions. During subduction, H_2O-rich fluids released by dehydration reactions in the oceanic crust may drive decarbonation reactions in overlying pelagic sediments. Fluids that migrate up the subduction shear zone could cause hydration and carbonation reactions in oceanic crust at shallower levels in the subduction zone, or such fluids could be expelled through the accretionary prism.

In relatively warm subduction zones, formed in young oceanic lithosphere (less than 50 Ma), subducting oceanic crust intersects the wet basaltic solidus at pressures less than 28 kbar. The amount of partial melting that will occur above the wet solidus will be proportional to the amount of free H_2O present in the rock. The porosity of subducted oceanic crust at depths of *ca.* 100 km is unknown, but presumably small. If we assume a porosity of 0.3% filled with pure H_2O (equivalent to 0.1 wt % free H_2O in the rock) and that the melt phase contains 5 wt % H_2O, then approximately 2 % partial melt would form by fluid-present melting at temperatures above the wet solidus. As the oceanic crust is subducted to deeper levels, *P–T–t* paths intersect a major partial melting reaction caused by the breakdown of hornblende; substantial amounts of magma could form by this fluid-absent partial melting reaction in warm subduction zones.

P–T–t paths for relatively cool subduction zones do not intersect the wet basaltic solidus until pressures greater than 50 kbar (figure 3) suggesting that partial melting of subducting oceanic crust is an unlikely magma source in mature subduction zones or subduction zones that form in old oceanic lithosphere. Under such conditions the subducting oceanic crust should dehydrate before encountering any partial melting reactions. If the calculated position of the tremolite dehydration reaction (figure 3) represents a good approximation to the *P–T* location of the amphibole dehydration in metabasalt, then this reaction would generate substantial amounts of H_2O at depths of 100–150 km.

(b) *Hydration and partial melting of the mantle wedge*

Studies of ancient and modern subduction zones demonstrate that some of the fluids released from the subducting slab migrate upward into the overlying mantle wedge. Evidence for fluid infiltration of the mantle wedge includes the presence of serpentinite diapirs in the Marianas forearc (Fryer *et al.* 1985) and pervasively hydrated and metasomatized ultramafic hanging walls of the Trinity thrust system (Peacock 1987*a*) and the Catalina Schist terrane (Bebout & Barton 1989).

The consequences of fluid infiltration of the overlying mantle wedge may be investigated by combining mantle wedge *P–T–t* paths with the phase diagram for peridotite bulk compositions (figure 4). The effect of fluid infiltration depends critically on the temperature of the portion of the mantle wedge being infiltrated. If H_2O infiltrates the mantle wedge at temperatures below the 'hornblende-out' and 'phlogopite-out' partial melting reactions, then infiltrating H_2O will react with peridotite to form phlogopite (plus amphibole at $P < 28$ kbar). The amount of phlogopite $\pm$ amphibole that can form from H_2O infiltration depends on the amount of K, Na, Ca, and Al present in the peridotite plus the amount of these elements carried in by the fluid. Once the rocks capacity to form these minerals is exceeded, further H_2O infiltration at temperatures above the wet solidus will cause fluid-present partial melting to occur. Based on the phase diagram presented in figure 4, after phlogopite $\pm$ amphibole formation, H_2O should coexist as a free fluid with peridotite at temperatures less than the wet peridotite solidus and greater than *ca.* 700 °C. In this narrow temperature range, H_2O should pass through the mantle wedge without causing hydration or partial melting reactions. The infiltration of H_2O into the mantle wedge at temperatures below *ca.* 700 °C should form additional hydrous phases such as talc and chlorite.

In the absence of induced convection, mantle wedge rocks follow isobaric cooling trajectories (figure 4*a*). In general, fluid infiltration of the mantle wedge could cause partial melting during the early stages of subduction if temperatures exceed the wet peridotite solidus. As subduction progresses and the mantle wedge cools, most fluid infiltration will occur at subsolidus conditions and should cause hydration rather than partial melting.

In the presence of induced convection, rocks in the mantle wedge follow more complex *P–T–t* paths (figure 4*b*, *c*). In subduction zones where induced asthen-

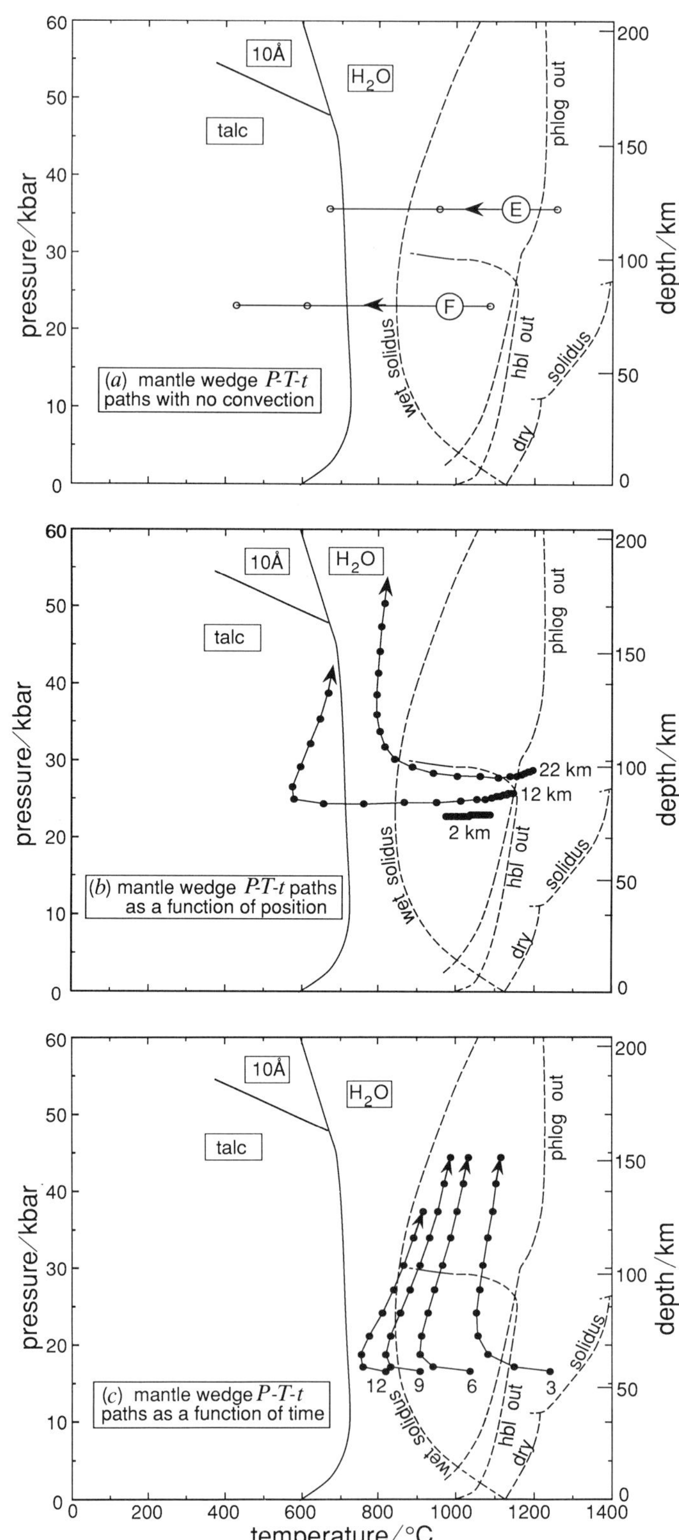

Figure 4. For description see opposite.

ospheric wedge flow occurs at relatively shallow depths (less than 100 km), peridotite containing hornblende formed by H_2O infiltration can be dragged across the 'hornblende-out' partial melting reaction (figure 4c). In subduction zones, large amounts of partial melt may be generated by this reaction (Tatsumi 1989).

The pressure-dependent portion of the 'hornblende-out' partial melting reaction in peridotite bulk compositions may be the source of arc magmas in subduction zones that have convecting asthenosphere above the subducting slab at shallow depths (less than 100 km) (figure 4c). In the numerical experiments described in this paper, the lithosphere/asthenosphere boundary is fixed by the 1300 °C isotherm in the initial oceanic geotherm. Thus, shallow convecting asthenosphere is only present in the experiments conducted for young (less than 10 Ma) oceanic lithosphere. In experiments simulating subduction zones formed in older oceanic lithosphere, mantle wedge *P–T–t* paths lie at pressures greater than the 'hornblende-out' partial melting reaction. If lower temperature material convects (e.g. the lithosphere/asthenosphere boundary is fixed at, say, 1100 °C), then the 'hornblende-out' partial melting reaction could be an important source of arc magmas in many subduction zones.

In modern subduction zones, asthenosphere appears to be present beneath most, if not all, magmatic arcs (Gill 1981). For example, beneath the northeast Japan volcanic arc, a low-velocity–low-Q region (asthenosphere) exists directly beneath the 35 km thick crust (Yoshii 1979). Because the Japan subduction zone formed in old (*ca.* 150 Ma) oceanic lithosphere, one might expect the overriding Japan plate to have a thick lithosphere. The apparent absence of lithospheric mantle beneath Japan may be explained by (1) tectonic erosion during subduction initiation; (2) thermal erosion caused by induced asthenospheric flow; or (3) crust/lithosphere decoupling during back-arc extension.

Future numerical experiments will focus on simulating modern subduction zones where the lithospheric structure of the subducting and overriding plates differ, developing more realistic petrogenetic grids, and evaluating different rheological models for the mantle.

Figure 4. Experimental partial-melting (dashed lines) and subsolidus phase relations (solid lines) applicable to the peridotite + H_2O system. As in the basalt system, the presence of excess H_2O decreases the melting temperature of peridotite (dry solidus: Takahashi & Kushiro 1983; wet solidus: Mysen & Boettcher 1975). Amphibole and phlogopite are stable on the solidus up to *ca.* 30 kbar and greater than 50 kbar, respectively. The 'hbl out' and 'phlog out' curves represent fluid-absent partial melting of hornblende-bearing and phlogopite-bearing peridotite, respectively (hbl out: Green 1973; phlog out: Wendlendt & Eggler 1980). Minerals in boxes represent the stable hydrous phase that coexists with forsterite + enstatite in the simplified MgO–SiO_2–H_2O peridotite system (Kitahara *et al.* 1966; Yamamoto & Akimoto 1977). Additional hydrous phases, such as phlogopite and amphibole, will be stable in the more complex peridotite chemical system. Thermodynamic calculations suggest that serpentine may coexist with forsterite and enstatite at temperatures less than 600 °C and pressures greater than *ca.* 20 kbar. Convergence rate is 3 cm a^{-1}. (*a*) In the absence of induced mantle wedge convection (*P–T–t* paths E and F), the mantle wedge cools isobarically. (*b*) *P–T–t* paths followed by convecting mantle wedge. Rocks are initially located 2, 12 and 22 km beneath 75 km thick lithosphere, and 150 km from the asthenospheric wedge tip. (*c*) *P–T–t* paths followed by convecting mantle wedge as a function of time in millions of years. Rocks start near asthenospheric wedge tip and are dragged down with the slab. Initial oceanic geotherm is 10 Ma.

5. Summary

Computer simulation of heat transfer provides a powerful means by which to investigate the location and sequence of metamorphic and igneous reactions that occur at depth in subduction zones. The results of the experiments described above suggest that partial melting of the subducting oceanic crust should only occur during the early stages of subduction initiated in young (less than 50 Ma) oceanic lithosphere. For many subduction zones, partial melting appears to be the result of H_2O infiltration into the convecting mantle wedge. Partial melting of the mantle wedge may occur (1) by H_2O infiltration at temperatures above the wet peridotite solidus after phlogopite $\pm$ amphibole formation or (2) by a two-stage process in which amphibole is first formed and subsequently destroyed as the rock is dragged downward across the 'hornblende-out' partial melting reaction.

I thank the organizers of this meeting for inviting me to present the results of my research. I thank Professor Bickle and Professor Barnicoat for their insightful reviews. This research was supported by National Science Foundation grant EAR-87-20343.

References

Anderson, R. N., Uyeda, S. & Miyashiro, A. 1976 Geophysical and geochemical constraints at converging plate boundaries. I. Dehydration in the downgoing slab. *Geophys. Jl R. astr. Soc.* **44**, 333–357.

Anderson, R. N., DeLong, S. E. & Schwarz, W. M. 1978 Thermal model for subduction with dehydration in the downgoing slab. *J. Geol.* **86**, 731–739.

Anderson, R. N., DeLong, S. E. & Schwarz, W. M. 1980 Dehydration, asthenospheric convection and seismicity in subduction zones. *J. Geol.* **88**, 445–451.

Andrews, D. J. & Sleep, N. H. 1974 Numerical modeling of tectonic flow behind island arcs. *Geophys. Jl R. astr. Soc.* **38**, 237–251.

Basaltic Volcanism Study Project 1981 *Basaltic volcanism on the terrestrial planets.* New York: Pergamon Press.

Batchelor, G. K. 1967 *An introduction to fluid dynamics.* Cambridge University Press.

Bebout, G. E. & Barton, M. D. 1989 Fluid flow and metasomatism in a subduction zone hydrothermal system: Catalina Schist terrane, California. *Geology* **17**, 976–980.

Berman, R. G. 1988 Internally-consistent thermodynamic data for minerals in the system. *J. Petrol.* **29**, 445–522.

Delany, J. M. & Helgeson, H. C. 1978 Calculation of the thermodynamic consequences of dehydration in subducting oceanic crust to 100 kb and $> 800\ ^\circ$C. *Am. J. Sci.* **278**, 638–686.

Ernst, W. G. 1988 Tectonic history of subduction zones inferred from retrograde blueschist *P–T* paths. *Geology* **16**, 1081–1084.

Evans, B. W. & Brown, E. H. (eds) 1986 *Blueschists and eclogites.* Geol. Soc. Am. Memoir 164.

Fryer, P., Ambos, E. L. & Hussong, D. M. 1985 Origin and emplacement of Mariana forearc seamounts. *Geology* **13**, 774–777.

Green, D. H. 1973 Experimental melting studies on a model upper mantle composition at high pressure under water-saturated and water-undersaturated conditions. *Earth planet. Sci. Lett.* **19**, 37–53.

Green, T. H. 1982 Anatexis of mafic crust and high pressure crystallization of andesite. In *Andesites* (ed. R. S. Thorpe), pp. 465–487. London: Wiley.

Gill, J. 1981 *Orogenic andesites and plate tectonics.* New York: Springer-Verlag.

Hasebe, K., Fujii, N. & Uyeda, S. 1970 Thermal processes under island arcs. *Tectonophys.* **10**, 335–355.

Honda, S. & Uyeda, S. 1983 Thermal processes in subduction zones – a review and preliminary approach on the origin of arc volcanism. In *Arc volcanism: physics and tectonics* (ed. D. Shimozuru & I. Yokoyama), pp. 117–140. Tokyo: Terra Scientific.

Jeanloz, R. & Morris, S. 1986 Temperature distribution in the crust and mantle. *A. Rev. Earth planet. Sci.* **14**, 377–415.

Kitahara, S., Takenouchi, S. & Kennedy, G. C. 1966 Phase relations in the system $MgO–SiO_2–H_2O$ at high temperatures and pressures. *Am. J. Sci.* **264**, 223–233.

Lambert, I. B. & Wyllie, P. J. 1972 Melting of gabbro (quartz eclogite) with excess water to 35 kilobars, with geological applications. *J. Geol.* **80**, 695–708.

Minear, J. W. & Toksoz, N. M. 1970 Thermal regime of a downgoing slab and new global tectonics. *J. geophys. Res.* **75**, 1397–1419.

Mysen, B. O. & Boettcher, A. L. 1975 Melting of a hydrous mantle: I. Phase relations of natural peridotite at high pressures and temperatures with controlled activities of water, carbon dioxide, and hydrogen. *J. Petrol.* **16**, 520–548.

Oxburgh, E. R. & Turcotte, D. L. 1970 Thermal structure of island arcs. *Geol. Soc. Am. Bull.* **81**, 1665–1688.

Oxburgh, E. R. & Turcotte, D. L. 1976 The physico-chemical behavior of the descending lithosphere. *Tectonophys.* **32**, 107–128.

Peacock, S. M. 1987*a* Serpentinization and infiltration metasomatism of the Trinity peridotite, Klamath province, northern California: implications for subduction zones. *Contrib. Mineral. Petrol.* **95**, 55–70.

Peacock, S. M. 1987*b* Creation and preservation of subduction-related inverted metamorphic gradients. *J. geophys. Res.* **92**, 12763–12781.

Peacock, S. M. 1987*c* Thermal effects of metamorphic fluids in subduction zones. *Geology* **15**, 1057–1060.

Peacock, S. M. 1990*a* Fluid processes in subduction zones. *Science, Wash.* **248**, 329–337.

Peacock, S. M. 1990*b* Numerical simulation of metamorphic pressure–temperature–time paths and fluid production in subducting slabs. *Tectonics* **9**, 1197–1211.

Platt, J. P. 1975 Metamorphic and deformational processes in the Franciscan Complex, California: Some insights from the Catalina Schist terrane. *Geol. Soc. Am. Bull.* **86**, 1337–1347.

Sorensen, S. S. 1986 Petrologic and geochemical comparison of the blueschist and greenschist units of the Catalina Schist terrane, southern California. *Geol. Soc. Am. Mem.* **164**, 59–75.

Sorensen, S. S. 1988 Petrology of amphibolite-facies mafic and ultramafic rocks from the Catalina Schist, southern California: metasomatism and migmatization in a subduction zone metamorphic setting. *J. metam. Geol.* **6**, 405–435.

Stacey, F. D. 1977 *Physics of the Earth*, 2nd edn. New York: Wiley.

Takahashi, E. & Kushiro, I. 1983 Melting of a dry peridotite at high pressures and basalt magma genesis. *Am. Mineral.* **68**, 859.

Tatsumi, Y. 1989 Migration of fluid phases and genesis of basalt magmas in subduction zones. *J. geophys. Res.* **94**, 4697–4707.

Toksoz, M. N., Minear, J. W. & Julian, B. R. 1971 Temperature field and geophysical effects of a downgoing slab. *J. geophys. Res.* **76**, 1113–1138.

van der Beukel, J. & Wortel, R. 1986 Thermal modelling of arc-trench regions. *Geol. Mijnbouw* **65**, 133–143.

Wendlandt, R. F. & Eggler, D. H. 1980 The origins of potassic magmas: 1. Melting relations in the systems $KAlSiO_4–Mg_2SiO_4–SiO_2$ and $KAlSiO_4–MgO–SiO_2–CO_2$ to 30 kilobars. *Am. J. Sci.* **280**, 385–420.

Wyllie, P. J. 1988 Magma genesis, plate tectonics, and chemical differentiation of the earth. *Rev. Geophys.* **26**, 370–404.

Wyllie, P. J. & Sekine, T. 1982 The formation of mantle phlogopite in subduction zone hybridization. *Contrib. Mineral. Petrol.* **79**, 375–380.

Yamamoto, K. & Akimoto, S. I. 1977 The system $MgO–SiO_2–H_2O$ at high pressures and temperatures – stability field for hydroxyl–chondrodite, hydroxyl–clinohumite and 10 Å-phase. *Am. J. Sci.* **277**, 288–312.

Yoshii, T. 1979 A detailed cross-section of the deep seismic zone beneath northeastern Honshu, Japan. *Tectonophys.* **55**, 349–360.

[127]

A physical model for the volume and composition of melt produced by hydrous fluxing above subduction zones

By J. Huw Davies[1] and M. J. Bickle[2]

[1]*Institute of Theoretical Geophysics, Department of Earth Sciences, University of Cambridge, Downing Street, Cambridge CB2 3EQ, U.K.*

[2]*Department of Earth Sciences, University of Cambridge, Downing Street, Cambridge CB2 3EQ, U.K.*

Thermal models of subduction zones, restrict the melt source region to a domain at sufficiently high temperature with water present (either as a free phase or in hydrous minerals). Water, released into the mantle by slab dehydration, traverses the wedge horizontally by a combination of (i) vertical movement as a fluid phase and (ii) fixed in amphiboles carried by the induced mantle flow; only in mantle hotter than amphibole stability can melts escape upwards. We develop a one-dimensional model for the source region fluxed with water. The induced mantle flow advects heat laterally to balance the latent heat of melting, in a column where the liquidus of the melt is depressed by its water content. Melt flux, fraction, temperature and water content are calculated assuming steady state. Melt compositions are predicted from the melt fraction distribution as a function of depth, constrained by the experimental data of Green. On investigating a range of plausible models, we find that the average degrees of melting predicted vary from *ca.* 2 to 8%. The predicted primary magmas are mafic high magnesium basalts with water contents ranging from 1.6 to 6 wt%, and temperatures from 1160 to 1290 °C. Models with shallower depths of segregation have higher degrees of melting and lower water contents. The volumes predicted by the physical model are a strong function of the water flux assumed to enter the source region. Previous estimates of arc growth would suggest either low water fluxes or that not all the melt reaches the arc crust.

---

1. Introduction

The petrogenesis of magmas generated above subduction zones is as controversial today as it was when the petrological implications of subduction were first realized (Green & Ringwood 1968). The problem in part stems from the multiplicity of potential magma sources and in part from the additional petrological complexities of melting in a hydrous system. There are a number of lines of evidence for the involvement of water in the petrogenesis of arc magmas. Volcanicity is often explosive, lavas are hydrous and may contain OH-bearing phenocryst phases, high-pressure modules may contain amphibole, while eruption temperatures are often between 1100 and 1200 °C, lower than their dry compositional equivalents (Gill 1981, p. 65). A number of isotopic tracers in arc magmas, notably ^{10}Be (Tera *et al.* 1986), ^{207}Pb and ^{87}Sr and possibly some of the more incompatible trace elements (Pearce 1982) are derived from the subducted hydrated oceanic crust and sediments.

Thermal models of subduction zones show that temperatures in the downgoing lithosphere are too low to allow extensive melting of the subducted crust (Davies & Stevenson 1991; Honda 1985). However, experiments on amphibole stability indicate that downgoing oceanic lithosphere will dehydrate by a near-isobaric amphibole to eclogite transition at *ca.* 2.5 GPa (Tatsumi 1986). Figure 1 illustrates how water is transported away from the subduction zone. Water released from the subducted oceanic crust would amphibolitize the overlying mantle and then be carried down by the mantle flow in the wedge to a similar amphibole breakdown reaction in peridotite at *ca.* 3.0 GPa (Tatsumi 1986). From there the free water rises vertically through the already amphibolitized mantle until it reaches fresh mantle. There it reacts again forming amphibole. It is then carried down again by the induced mantle flow. The process repeats at *ca.* 3.0 GPa where the amphibole breaks down again. The net effect is the lateral transport of water across the mantle wedge, towards the wet solidus (Davies & Stevenson 1991). Addition of water to mantle above its wet solidus would cause melting. In this paper we model melt production by such hydrous fluxing, the aim being to quantify the rate of melt production, the degree of melting, melt water contents, temperatures and their major element compositions.

2. Thermal model

Davies & Stevenson (1991) have reviewed the numerous thermal models of subduction zones published over the past two decades. The differences between models mainly reflect either the incorporation of mechanisms which have since been shown to be of much less significance, especially strain heating at the slab surface; or the neglect of others that are accepted to be significant, such as induced wedge flow. The resulting thermal models (Andrews & Sleep 1974; Honda 1985; Davies & Stevenson 1991) do not differ greatly, and can be used as a basis for discussing the source region of subduction zone magmatism. The prediction of volumes and compositions of magmas from these thermal models will provide further tests and constraints on these models.

We shall consider that the water reaches the source region by the mechanism outlined in the introduction (Davies & Stevenson 1991), illustrated in figure 1. Note that we will get some melting at the water saturated solidus, but the volumes of melt produced are so low (since the amount of free water present is so low) that they will percolate upwards but are highly unlikely to segregate into cracks. Hence these melts remain in local chemical equilibrium and in effect they are similar to the hydrous silica-rich phase in that they transport water vertically, continuing the net horizontal movement of the water. The start of the source region is defined as the point where in the vertical path of the wet melt, it reaches mantle that is hotter than the thermal stability of amphibole. The width of the source region in this model is controlled by the position of the amphibole breakdown curve. It is assumed that over this width and height one gets sufficient melting to allow segregation into cracks. The segregation height could be at the maximum temperature of the source region, but given our poor understanding of magma segregation we have investigated the effect of varying this height.

It is unclear how big a role adiabatic decompression plays in generating subduction zone magmas relative to hydrous fluxing (Davies & Stevenson 1991). In this model we restrict ourselves totally to the hydrous fluxing end-member. Adiabatic decompression of dry mantle was addressed in McKenzie & Bickle (1988). Note

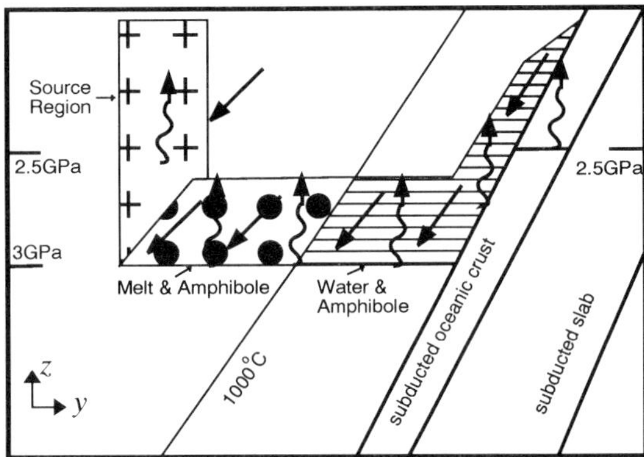

Figure 1. Schematic diagram of the proposed water transport mechanism and the orientation of the assumed source region where amphibole is no longer stable. The straight arrows represent the motion of matrix, while the wiggly arrows represent motion of either melt or water. Note there is a region where amphibole and melt is stable, and another where water and amphibole is stable. The source region and z axis are assumed to be vertical; the y axis horizontal, and the x axis is assumed to be horizontal into the paper, i.e. along strike of the subduction zone. The model considers v_s to be the horizontal component of the matrix velocity.

melting cools the mantle; this is true for melting due to hydrous fluxing as well as melting due to adiabatic decompression.

3. Wet melting

The main controls on the amount of melting are the rate of heat advected into the melt region by the induced flow in the mantle wedge, the water flux from the subducted plate and the melt thermodynamics; that is the relationship between melt temperature, water content and melt fraction. To a first approximation this can be modelled by a one-dimensional column of partial melt rising into hotter mantle. Hot unmelted dry mantle flows into the side of the column. This melts by reaction with the wet magma and any excess water flowing up the column. The latent heat of melting results in a temperature drop. Cooler partly melted (i.e. residual) mantle leaves the melt region on its downstream side.

We assume steady state, with constant water flux up the column. Melt, water dissolved in the melt, and any excess water are assumed to travel with the same velocity. The melt is in local thermal equilibrium with solid. We ignore both advection of heat by the melt and conduction of heat up the melt column. This can be shown *a posteriori* to be reasonable. Heat will be conducted horizontally into the column but this would correspondingly reduce the heat advected into the column. We ignore possible conductive gradients across the melting column.

The amount of melting is calculated from conservation of energy. We balance the latent heat of melting with the heat advected in by the solid flow. To do this we need to be able to calculate the (weight) fraction of peridotite which melts as a function of temperature and water content. Experiments on liquidus depression give an empirical linear relationship between mole fraction water and liquidus depression, given a mole equivalent mass of *ca.* 255 g mole^{-1} for melt, not dissimilar to the values derived by Burnham (1979). However, this relationship does not adequately describe melting of wet peridotite as melt composition changes across the melt region. For example an olivine tholeiite melt which has a dry liquidus consistent with the dry peridotite solidus has a water saturated liquidus at 3 GPa of *ca.* 1100 °C (Stern *et al.* 1975) which is inconsistent with the 3 GPa peridotite wet solidus at *ca.* 1000 °C (Green 1973). This compositional dependency is approximated by adopting a larger constant a_1 in the relation describing liquidus depression (see equation (1) below). The larger value for a_1 correctly describes the separation of dry and water saturated solidi. Melt temperature increases as a function of increasing melt fraction. This is accommodated by relating the melt temperature, to the dry melt temperature, at the same melt fraction and pressure, by

$$T_l = T_d - a_1 X_w^m, \tag{1}$$

where T_l is the wet melt temperature, a_1 is a constant, X_w^m is the mole fraction content of water and $T_d(X, P)$ is the temperature of a dry melt at temperature P and degree of melting X, calculated from the parametrization of McKenzie & Bickle (1988, eqns (18)–(20)). The liquidus depression data (Stern *et al.* 1975; Green 1972) indicate no detectable variation in a_1 with pressure between 2 and 3 GPa, the region of interest. Maximum solubility of water in melts is taken as 25 wt% at 3 GPa (Green 1973; Stern & Wyllie 1973). The pressure dependence of maximum water solubility is unimportant as the melts are only saturated over a small vertical interval.

The conservation of energy equation is

$$L \, dM/dt = v_s C_p \rho_s (T_i - T_l)/\Delta y, \tag{2}$$

where L is the latent heat of melting, v_s is the matrix velocity, C_p is the specific heat capacity at constant pressure, ρ_s is the mantle density, Δy is the source width, while T_i is the mantle temperature just before it enters the source region. Note since dM/dt is the mass rate of melting per cubic metre, $\Delta y \, dM/dt$ gives the melt production rate at height z per metre of arc length. The width of the melt region, Δy, does not affect the solution.

The equation for the mole fraction of water in the melt is

$$X_w^m = \frac{M_e}{18.02(1 - W_w^m)/W_w^m + M_e} = \frac{M_e}{18.02 W/F + M_e} \tag{3}$$

and

$$X_w^m = \frac{M_e}{18.02(1 - W_w^m(\text{sat}))/W_w^m(\text{sat}) + M_e}, \tag{4}$$

where X_w^m is the mole fraction of water in the melt, W_w^m is the weight fraction of water in the melt, M_e is the mole equivalent mass for the melt, W is the melt flux (kg m^{-2} s^{-1}), while F is the water flux (kg m^{-2} s^{-1}), if the melt is water-saturated, which it must be at infinitesimal degrees of melting in a hydrous system. Equations (1) and (3) together self-consistently describe the behaviour of the wet solidus and liquidus. The above parametrization is also continuous with the dry parametrization.

From the conservation of mass we can relate the melting rate to the melt flux gradient by

$$dM/dt = dW/dz. \tag{5}$$

Where z is the height above the base of the melting column. The bottom of the melting column, z_0 is at $z = 0$ km (at a depth of 100 km); while z_1, the top of the melting column, is at a height, $z = 40$ km (i.e. a depth of 60 km) for the basic model. Hence from substituting equations (5) and (1) into equation (2) we get

$$dW/dz = (v_s C_p \rho_s / L \Delta y)(T_i - T_d + a_1 X_m^w). \tag{6}$$

Table 1

	parameters		
symbol	name	value	units
C_p	specific heat capacity	10^3	$\mathrm{J\ kg^{-1}\ {}^\circ C^{-1}}$
L	latent heat	4.5×10^5	$\mathrm{J\ kg^{-1}}$
Δy	effective source region width	2×10^4	m
M_e	molar equivalent mass of melt	255.0	$\mathrm{g\ mole^{-1}}$
a_1	constant relating X_w^m to $(T_1 - T_d)$	642.0	$^\circ C$
F	water flux	0.6×10^{-8}	$\mathrm{kg\ m^{-2}\ s^{-1}}$
ρ_s	density of solid	3.34×10^3	$\mathrm{kg\ m^{-3}}$
v_s	velocity perpendicular to isotherms	10^{-9}	$\mathrm{m\ s^{-1}}$
z_0	bottom (at 100 km depth)	0.0	km
z_1	top (at 60 km depth)	40.0	km
T_i^{min}	T_i at z_0	1000	$^\circ C$
T_i^{max}	T_i at z_1	1250	$^\circ C$
W	melt flux	—	$\mathrm{kg\ m^{-2}\ s^{-1}}$
$\mathrm{d}M/\mathrm{d}t$	rate of melting	—	$\mathrm{kg\ m^{-3}\ s^{-1}}$

The degree of melting, X, is the ratio of the rate of melting to the solid flux,

$$X = (\mathrm{d}W/\mathrm{d}z)\,\Delta y/\rho_s v_s. \tag{7}$$

We vary T_i from T_i^{min} (usually 1000 °C) linearly to T_i^{max} at the top of the source region. The Davies & Stevenson (1991) thermal model suggests T_i^{max} of *ca.* 1250 °C. Since T_i varies with height in the source region this problem has to be solved numerically. Equation (6) must be solved by iteration, since T_d is a nonlinear function of X; this can be seen since to evaluate X we require T_d via $\mathrm{d}W/\mathrm{d}z$.

We assume that the melts rise slowly through the source region, remaining in hydrous as well as thermal equilibrium. By hydrous equilibrium we mean that the water content of the melt is the water content of the whole local system. We take no account of the change in the bulk composition with height as a result of advection, except to assume that water contents are appropriately diluted. Since the system is in steady state the relative contribution of melt produced at height z to the final melt segregating from the top of the source region is in proportion to its rate of production.

The parameters used for the basic model are in table 1. Many of the parameters are uncertain while others naturally vary. The temperatures T_i, melting column height, $z_1 - z_0$ and velocity structure v_s approximate models of Davies & Stevenson (1991). The water flux from subducted crust of $0.6 \times 10^{-8}\ \mathrm{kg\ m^{-2}\ s^{-1}}$ is equivalent to dehydration of 2 km of amphibolite containing 1 % H_2O subducting at 7 cm a^{-1}. This is less than some previous estimates (Peacock 1990); but much of the additional water held in minerals other than amphibole would be released at lower temperatures and pressures, and would escape to the surface. We have investigated the effect of variability of these parameters by evaluating the results for our basic model, changing only one parameter at a time.

4. Results

The effect of varying one parameter at a time, away from the basic model listed in table 1, is illustrated in figure 2. In varying the parameter a_1, we needed to increase

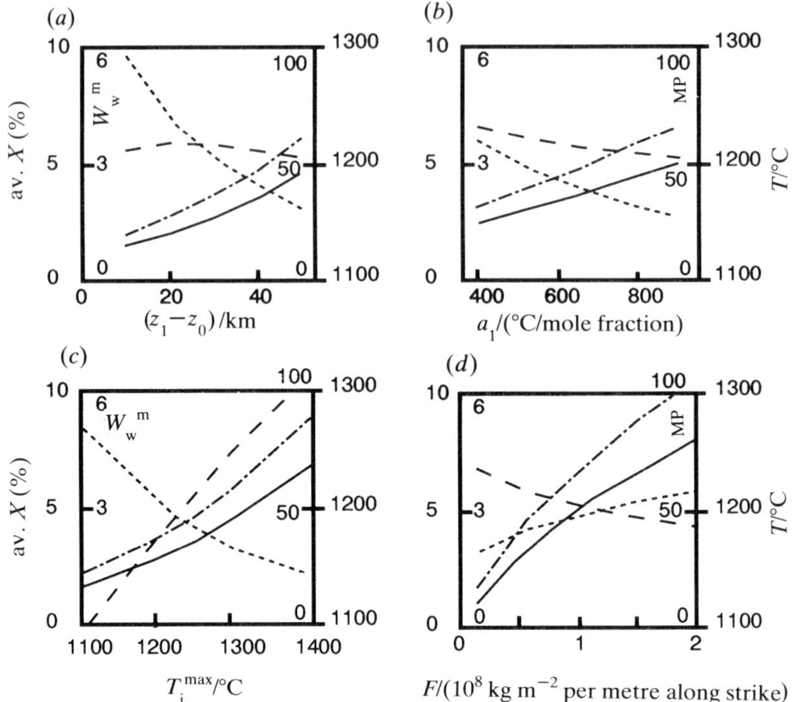

Figure 2. This shows how (i) the average degree of melting, whose scale is on the outside of the left hand side of the figures; (ii) the weight percent water in the primary melt just before segregation, $(W_w^m(z_1))$, whose scale is on the inside of the left hand side of the figures; (iii) the temperature of the primary melt just before segregation $(T_1(z_1))$, whose scale is on the outside of the right hand side of the figures; and (iv) rate of melting (km^3 Ma^{-1} km^{-1} along arc), whose scale is on the inside of the right-hand side of the figures; change as we vary (a) the height of the melting column i.e. $(z_1 - z_0)$, (b) the liquidus depression constant a_1, (c) the maximum temperature T_i^{max}, and (d) the water flux, F. All the other parameters are the standard parameters found in table 1. ———, Average degree of melting (%); ------, wt% water in melt; —·—·—·—, magma production (MP) (km^2 Ma^{-1}); —————, primary magma temperature (°C).

T_i^{min} when a_i was decreased to allow melting. In figure 2a, it is seen that decreasing the column height decreases the rate of melting. A decrease in the rate of melting is also observed in figure 2c with decreasing maximum temperature. Thicker lithospheres would lead to both shorter melting columns and cooler wedges. Hence with thicker lithospheres we would expect less melting, with the melts being wetter and cooler, and the result of lower degrees of melting. Notice that as we decrease the water flux (figure 2d), the rate of melting decreases, as do the water contents of the melts. Decreasing the velocity of the advected solid has a similar effect to increasing the water flux, with the exception that the rate of melting is increasing rather than decreasing with velocity. It was found that varying the latent heat of melting had little influence on the system! The predicted rates of melting are within the wide range of estimates of arc additions quoted by Ito *et al.* (1983), but are generally higher than the estimates of Reymer & Schubert (1984) (33 km^3 km^{-1} Ma^{-1}) and Brown & Mussett (1981) (17 km^3 km^{-1} Ma^{-1}). The lower volumes suggest low water fluxes and/or that all the melt produced does not reach the arc crust. The average degrees of melting (2–8%) are generally lower than mid-ocean ridge basalts (MORB) (greater than 10%).

The temperatures of melt at segregation (maximum temperature) are generally lower than those suggested by the equilibration studies of Tatsumi *et al.* (1983). This could be due to the fact that the lavas equilibrated were actually an integration of magmas from a range of pressures, and hence finding the multisaturation point at a single depth might be misleading. They are slightly lower than the liquidus of the high Mg basalt (ID16) of Nye & Reid (1986), 1280 °C. The range of temperatures predicted by the models are not unreasonable given the eruption temperatures of subduction zone lavas (Gill 1981, p. 64).

The water contents of the segregating magmas at 1.6–6 wt % are at the high end of the range (0–5 wt %) suggested by the numerous lines of evidence listed by Gill (1981, p. 116). The predominance of relatively dry magmas suggests either a low water flux, or very near surface degassing. The water contents are sufficiently low that without crystallization they will not exsolve their water until the melts reach the crust.

5. Predicted compositions

There are very few published experimental studies on the compositions of wet melts and there is much controversy (Green 1976). As discussed in the literature there are problems with quench phases, and also iron loss; these problems are illustrated by the lack of agreement among the various experiments (Green 1976; Kushiro 1990; Mysen & Boettcher 1975; Nehru & Wyllie 1975). We have used the three water-saturated experiments of Green (1976) to constrain predictions of compositions from our simple physical model of the source region. To extend the results of Green to parameters other than those of his experiments we have used the compositional parametrization derived by Watson & McKenzie (1991) (an enhancement of McKenzie & Bickle (1988)) for dry melting. We repeat the point made in McKenzie & Bickle (1988) that what we really want are Rayleigh type melting compositions, but we use batch melting compositions since it is unclear how to extract the requisite information on Rayleigh melting from the current experiments. We have evaluated a single shift in melt fraction and pressure that minimizes the sum of the absolute values of the relative misfits between the displaced dry parametrization and Green's three experiments. This is crude, in that we are hoping that wet melting has trends like dry melting. Given the lack of wet melting experiments with degrees of melting, we have no reasonable alternatives. We find that the best fit for the surface is with $dP_{\text{sat}} = -1.55$ GPa, and $dX_{\text{sat}} = 0.033$. We ignored Ti, Na and K, since the pyrolite used by Green was based on a Hawaiian basalt which has incompatible element concentrations inappropriate for the mantle. The actual fits to the experiments were reasonable, but given more data on hydrous melting it would be advisable to develop a separate parametrization for the saturated experiments, since the current fit suggests, that the compositional behaviour of wet melting is not identical to the compositional behaviour of dry melting. The average misfit was 5.5% while the largest misfit was 20%.

For water-undersaturated conditions we linearly decrease the shifts found for the water-saturated fits in proportion to their molar water contents towards zero for dry conditions; i.e. the compositions are given by

$$C_{\text{wet}}(X, P) = C_{\text{dry}}(X + dX_{\text{shift}}, P + dP_{\text{shift}}), \tag{8}$$

where
$$dX_{\text{shift}} = dX_{\text{sat}}(X_{\text{m}}^{\text{w}}/X_{\text{m sat}}^{\text{w}}) \tag{9}$$

and
$$dP_{\text{shift}} = dP_{\text{sat}}(X_{\text{m}}^{\text{w}}/X_{\text{m sat}}^{\text{w}}), \tag{10}$$

Table 2

(AOB, HAB, from Tatsumi *et al.* (1983), ID16 from Nye & Reid (1986), 25% and 10%, are from Plank & Langmuir (1988), while OTB, New Georgia, JOR-44, and AT-1 are all from compilation by Gust & Perfit (1987).)

	compositions								
	element (wt%)								
model	Si	Ti	Al	Fe	Mg	Ca	Na	K	the rest
basic model	47.6	1.5	15.5	9.3	11.9	11.2	2.5	0.3	0.4
AOB	49.1	1.0	15.5	9.4	11.6	9.7	2.5	1.1	—
HAB	49.4	0.9	15.7	9.8	12.1	9.4	2.3	0.3	—
OTB	49.7	0.7	15.0	10.6	13.0	9.0	1.6	0.3	—
ID16	48.9	0.7	16.0	9.0	11.4	10.9	2.2	0.5	—
New Georgia	49.8	0.5	12.2	10.1	12.7	10.9	2.0	0.25	—
JOR-44	52.6	0.8	16.6	7.5	9.4	8.5	3.5	0.1	—
AT-1	49.1	0.9	18.1	10.0	8.7	3.3	—	—	—
25%	49.8	0.6	13.3	9.4	13.6	12.0	1.3	—	—
10%	50.3	1.1	17.1	9.4	10.6	8.4	3.0	—	—

where $\mathrm{d}P_{\mathrm{sat}}$ and $\mathrm{d}X_{\mathrm{sat}}$ are the shifts that best fit Green's water saturated experiments to the dry parametrization of Watson & McKenzie (1991), and $X^{\mathrm{w}}_{\mathrm{m\,sat}}$ is the molar fraction of water in the melt at saturation.

We predict the compositions for primary melts at segregation using

$$C = \int_{z0}^{z1} XC\,\mathrm{d}z \Big/ \int_{z0}^{z1} X\,\mathrm{d}z, \tag{11}$$

i.e. the point and depth averaged composition (McKenzie & Bickle 1988). The equivalent anhydrous composition for the basic model (i.e. using the parameters listed in table 1) and a few estimated primary magmas are given in table 2.

The primary compositions discovered in the models are high magnesium low silica basalts. It is seen from table 2 that the predicted primary composition is in the middle of the range of estimates of island arc primary magmas. Remember that Ti, Na and K were not used to constrain the parametrization, hence these values have little meaning. It was found that even though the models covered a wide range of degrees of melting, little variation was found in the compositions, this was largely the result of the higher degree of melting trading off with lower $\mathrm{d}X_{\mathrm{shift}}$, due to lower water contents. The largest change in composition was found when the segregation height was made very shallow, i.e. a variation in pressure rather than in degree of melting. Given the crudeness of the parametrization, and the complexity of the experiments (see start of this section), the correspondence is heartening. It is clear though, that the experimental work needs to be extended to higher pressures (2–3 GPa) and lower degrees of melting (0–0.1); as well as the water-undersaturated régime.

6. Conclusion

We have developed a simple model for melting by hydrous fluxing in a subduction zone. If the varying estimates of arc additions and water contents are correct then our model suggests that the source region is fluxed by a range of water flux. Since estimated magma production rates are larger than arc additions rates this suggest either low water fluxes (1 wt % H_2O in upper 1 km of subducting crust only), or that not all the melt reaches the crust. Predicted degrees of melting are generally lower than for MORB. The poorly constrained compositions of the primary magmas predicted by the model are in the middle of previous estimates of primary magmas. The uncertainty in estimating the model's input temperature highlights the inadequacy of a one-dimensional model, and combined with the probable influence of horizontal variations in melting on compositions, point to the need to solve the problem self-consistently in two dimensions.

H. D. acknowledges the support of an NERC fellowship, and M. J. B. the support of the Geological Survey of Canada, and the Royal Society and NSERC for a study visit grant to Canada. We thank S. Sparks for detailed comments which greatly improved an earlier version of this paper. We also acknowledge P. Beattie for his comments on the present paper.

References

Andrews, D. J. & Sleep, N. H. 1974 Numerical modelling of tectonic flow behind island arcs. *Geophys. Jl R. astr. Soc.* **38**, 237–251.

Brown, G. C. & Mussett, A. E. 1981 *The inaccessible Earth*. London: George Allen and Unwin.

Burnham, C. W. 1979 The importance of volatile constituents. In *The evolution of the igneous rocks* (ed. H. S. Yoder Jr), pp. 439–482. Princeton University Press.

Davies, J. H. & Stevenson, D. J. 1991 Physical model of source region of subduction zone magmatism. *J. geophys. Res.* (Submitted.)

Gill, J. 1981 *Orogenic andesites and plate tectonics*. Berlin: Springer-Verlag.

Green, D. H. 1973 Contrasted melting relations in a pyrolite upper mantle under mid-oceanic ridge, stable crust and island arc environments. *Tectonophys.* **17**, 285–297.

Green, D. H. 1976 Experimental testing of "equilibrium" partial melting of peridotite under water-saturated, high pressure conditions. *Can. Mineral.* **14**, 255–268.

Green, T. H. 1972 Crystallization of calc-alkaline andesite under controlled high-pressure conditions. *Contrib. Mineral. Petrol.* **34**, 150–166.

Green, T. H. & Ringwood, A. E. 1968 Genesis of the calc-alkaline rock suite. *Contrib. Mineral. Petrol.* **18**, 105–162.

Gust, D. A. & Perfit, M. R. 1987 Phase relations of a high-Mg basalt from the Aleutian Island Arc: implications for primary island arc basalts and high-Al basalts. *Contrib. Mineral. Petrol.* **97**, 7–18.

Honda, S. 1985 Thermal structure beneath Tohoku, northeast Japan – a case study for understanding the detailed thermal structure of the subduction zone. *Tectonophys.* **112**, 69–102.

Ito, E., Harris, D. M. & Anderson Jr, A. T. 1983 Alteration and geologic cycling of chlorine and water. *Geochim. cosmochim. Acta* **47**, 1613–1624.

Kushiro, I. 1990 Partial melting of mantle wedge and evolution of island arc crust. *J. geophys. Res.* **95**, 15929–15939.

McKenzie, D. P. & Bickle, M. J. 1988 The volume and composition of melt generated by extension of the lithosphere. *J. Petrol.* **29**, 625–679.

Mysen, B. O. & Boettcher, A. L. 1975 Melting of a hydrous mantle. II. Crystals and liquids formed by anatexis of mantle peridotite at high pressures and high temperatures as a function of controlled activities of water, hydrogen and carbon dioxide. *J. Petrol.* **16**, 549–593.

Nehru, C. E. & Wyllie, P. J. 1975 Compositions of glasses from St. Paul's peridotite melted at 20 kilobars. *J. Geol.* **83**, 455–471.

Nye, C. J. & Reid, M. R. 1986 Geochemistry of primary and least fractionated lavas from Okmok volcano, central Aleutians: implications for arc magmagenesis. *J. geophys. Res.* **91**, 10271–10287.

Peacock, S. M. 1990 Numerical simulation of metamorphic pressure–temperature–time paths and fluid production in subducting slabs. *Tectonics* **9**, 1197–1211.

Pearce, J. A. 1982 Trace element characteristics of lavas from destructive plate boundaries. In *Andesite* (ed. R. S. Thorpe), pp. 525–548. New York: Wiley.

Plank, T. & Langmuir, C. H. 1988 An evaluation of the global variations in the major element chemistry of arc basalts. *Earth planet. Sci. Lett.* **90**, 349–370.

Reymer, A. & Schubert, G. 1984 Phanerozoic addition rates to the continental crust and crustal growth. *Tectonics* **3**, 63–77.

Stern, C. R., Huang, W. L. & Wyllie, P. J. 1975 Basalt-andesite-rhyolite-H_2O: crystallization intervals with excess H_2O and H_2O-undersaturated liquidus surfaces to 35 kilobars with implications for magma genesis. *Earth planet. Sci. Lett.* **28**, 189–196.

Tatsumi, Y. 1986 Formation of the volcanic front in subduction zones. *Geophys. Res. Lett.* **13**, 717–720.

Tatsumi, Y., Sakuyama, M., Fukuyama, H. & Kushiro, I. 1983 Generation of arc basalt magmas and thermal structure of the mantle wedge in subduction zones. *J. geophys. Res.* **88**, 5815–5825.

Tera, F., Brown, L., Morris, J., Sacks, I. S., Klein, J. & Middleton, R. 1986 Sediment incorporation in island-arc magmas: inferences from [10]Be. *Geochim. cosmochim. Acta* **50**, 535–550.

Watson, S. & McKenzie, D. 1991 Melt generation by plumes: a study of Hawaiian volcanism. *J. Petrol.* (In the press.)

Solubility of apatite, monazite, zircon, and rutile in supercritical aqueous fluids with implications for subduction zone geochemistry

By John C. Ayers and E. Bruce Watson

Rensselaer Polytechnic Institute, Troy, New York 12180–3590, U.S.A.

Solubilities of accessory minerals (apatite, monazite, zircon and rutile) in supercritical aqueous fluids have been measured to evaluate the role of these fluids in the mobilization of accessory mineral-hosted trace elements. We have characterized the effects on solubility of pH, X_{H_2O} (addition of CO_2), pressure ($P = 1.0$–3.0 GPa), temperature ($T = 800$–1200 °C), and dissolved silicate and NaCl concentration.

Fluorapatite solubility in pure H_2O is low, not more than 0.4 wt % at all conditions studied, but increases strongly with decreasing pH. Changes in P, T, X_{H_2O}, M_{NaCl} (the molality of NaCl), and dissolved silicate concentration have comparatively little effect on apatite solubility. Monazite is even less soluble in H_2O (not more than 0.2 wt %). Limited data suggests that monazite solubility increases with increasing P and T and with decreasing pH, but is insensitive to M_{NaCl}.

Zircon reacts with H_2O to form baddeleyite (ZrO_2) + silica-rich fluid. ZrO_2 solubility in H_2O and 1 M HCl is less than 0.2 wt %. Zircon, and therefore ZrO_2, solubility in *quartz-saturated* fluids $\pm$ HCl $\pm$ NaCl and in H_2O–CO_2 fluids is also very low. Rutile is more soluble than the other minerals examined, in the wt % range, and its solubility increases with increasing P and T.

Results indicate that high P–T aqueous fluids can dissolve significant amounts of Ti but very little Zr, and little phosphate unless the fluids are acidic. In most cases, apatite, monazite and zircon will remain present during episodes of aqueous fluid metasomatism and therefore will exert control, as 'residual phases', over element distribution. The higher solubility of rutile relative to other accessory minerals at high pressure may result in the depletion of high field strength elements relative to large ion lithophile elements observed in subduction zone volcanics.

1. Introduction and previous work

The association of volatile fluid metasomatism and loss or gain of accessory minerals suggests that volatile fluids dissolve, transport, and precipitate these minerals (Gieré 1990; Rubin *et al.* 1989; Sorenson & Grossman 1989). At issue here is whether supercritical aqueous fluids are capable of dissolving accessory minerals in sufficient quantity to be effective transport agents of the elements these minerals host. To test this we have measured the solubilities of the accessory minerals apatite, monazite, zircon, and rutile in various fluids. Before discussing the results we review the properties of each phase.

(a) Aqueous fluids

We know little about the chemical properties of aqueous fluids above 0.5 GPa. Metamorphic fluids often contain large amounts of dissolved salts, including NaCl,

KCl, $FeCl_2$ and $CaCl_2$ (Selverstone *et al.* 1990). Dissolved salts may be significantly dissociated at the high pressures of this study, and this serves to increase the ionic strength of the solution, which in turn may increase the solubility of ionic solids. Solubilities of silicate minerals in supercritical fluids are rarely more than several wt % at crustal conditions, but show a wide variation depending on fluid composition, ionic strength, mineral composition, and P and T (Holland & Malinin 1979). Aqueous fluids in equilibrium with most silicate rocks are rich in SiO_2, Al_2O_3, Na_2O and K_2O, while CaO, MgO, and FeO are not as soluble. Addition of CO_2 to the fluid greatly reduces the solubilities of silicate components (Eggler 1987).

(b) Apatite

Fluorapatite (FAp) is the most common end-member of the isomorphous series $Ca_5(PO_4)_3(OH,F,Cl)$. We therefore chose the nearly pure Durango FAp for study. The solubility reaction for FAp may be written:

$$Ca_5(PO_4)_3F_{(xt)} \rightleftharpoons 5\,Ca^{2+} + 3\,PO_4^{3-} + F^-. \tag{1.1}$$

Even in the simplest four-component system (Ca, P, O and F), each of the aqueous species produced can form many different complexes. The phosphate ion may undergo stepwise association with H^+ to form the species HPO_4^{2-}, $H_2PO_4^-$ and H_3PO_4. As a result, high activity of H^+ (low pH) should increase the solubility of apatite. Calcium may form the complexes $CaOH^+$, $CaPO_4^-$, $CaHPO_4$ and $CaH_2PO_4^+$, while F^- may associate to form HF. The value of K_{eq} for (1.1) at 25 °C as determined experimentally by McCann (1968) is $8.6\,(\pm 1.3)\times 10^{-61}$. Measured solubilities in p.p.m. FAp increase strongly with decreasing pH, from 3.0 at pH 7.0 to 253 at pH 4.0, but appear insensitive to changes in ionic strength (McCann 1968).

The aqueous solution chemistry of apatite is further complicated by the following exchange reactions:

$$Ca_5(PO_4)_3F_{(xt)} + H_2O \rightleftharpoons Ca_5(PO_4)_3OH_{(xt)} + HF, \tag{1.2}$$

$$Ca_5(PO_4)_3F_{(xt)} + HCl \rightleftharpoons Ca_5(PO_4)_3Cl_{(xt)} + HF. \tag{1.3}$$

(c) Monazite

Monazite is a rare-earth phosphate REE(PO_4), with Ce and other light REE generally being the most abundant REE. A likely solubility reaction involving neutral species is:

$$\text{REE}(PO_4)_{(xt)} + 3H_2O \rightleftharpoons \text{REE}(OH)_3 + H_3PO_4. \tag{1.4}$$

The effect pH will have depends on the relative values of the dissociation constants of REE$(OH)_3$ and H_3PO_4.

Limited data on solubility constants coupled with low concentrations of REE and P in seawater suggest that monazite is probably insoluble in aqueous fluids near ambient conditions. However, monazite in the epizone is considered to be of hydrothermal origin, suggesting it may be soluble under certain conditions (Overstreet 1967).

(d) Zircon

Zircon ($ZrSiO_4$) is resistant to weathering and is concentrated in heavy mineral deposits. This behaviour, combined with the immobility of Zr during diagenesis and low-grade metamorphism, suggests a low solubility of zirconium compounds.

At ambient conditions the neutral tetrahydroxide species predominates for dissolved ZrO_2 over a wide range of pH, resulting in simple dissolution reactions and solubilities that are independent of pH. The dissolution of zircon involves the congruent solution of zircon followed by the precipitation of baddeleyite ZrO_2 (Tole 1985):

$$ZrSiO_{4(xt)} + 4H_2O \rightleftharpoons Zr(OH)_4 + H_4SiO_4, \tag{1.5}$$

$$Zr(OH)_4 \rightleftharpoons ZrO_{2(xt)} + 2H_2O. \tag{1.6}$$

Charged Zr species (e.g. $Zr(OH)_3^+$, $Zr(OH)_2^{2+}$, $Zr(OH)^{3+}$ or Zr^{4+}) will become increasingly important at high pressure. It seems likely that the neutral or possibly the $+1$ charged species will predominate at the conditions of our experiments.

At ambient conditions zircon is very insoluble, K_{eq} for (1.5) being 2.48×10^{-17}, which for congruent solution and $\gamma_i = 1$ corresponds to 4.8×10^{-9} mole l^{-1} or 0.1 p.p.b. Si (Tole 1985). Maurice (1949) found that at 400 °C and 0.09 GPa for $ZrO_2:SiO_2 = 1$ the stable assemblage in acidic solutions is zircon; in near-neutral solutions it is zircon + baddeleyite; and in alkaline solutions it is baddeleyite + various zircono-silicates. Maurice (1949) states that zircon is very soluble in acidic solutions but presents little data to support this.

(e) Rutile

The insolubility of rutile (TiO_2) makes titanium very immobile during weathering and low-grade metamorphism. In water below 400 °C and at low pressures, $Ti(OH)_4$ is the dominant species in solution (Barsukova *et al.* 1979). The solubility reaction is:

$$TiO_{2(xt)} + 2H_2O \rightleftharpoons Ti(OH)_4. \tag{1.7}$$

In 1 M NaCl the concentration of $Ti(OH)_4$ is 3.3×10^{-8} mole kg^{-1} at 300 °C (Barsukova *et al.* 1979). In solutions containing F^-, the solubility of rutile is higher and becomes pH dependent due to formation of complexes such as $Ti(OH)_4F_2^{2-}$ and $Ti(OH)_2F_4^{2-}$. Rutile is therefore most soluble in aqueous solutions with high fluoride content and low pH (Barsukova *et al.* 1979).

2. Experimental procedures

To make efficient and precise measurements of low mineral solubilities we developed a technique involving measurement of the weight loss of a single crystal (Ayers & Watson 1989). Briefly, the technique involves placing a weighed crystal in a capsule with a known amount of fluid, running at P and T, then reweighing the fluid and crystal. To model natural solubility phenomena as closely as possible and facilitate interpretation, we used simple systems, with no more than three phases in a given experiment, and well-characterized natural crystals close to pure end-member compositions. Table 1 lists the experimental conditions and results.

All runs were done in a piston cylinder apparatus that maintains temperature at ± 5 °C and pressure $\pm 5\%$. Samples were run at nominal pressure. The ambient fO_2 of the assemblies is high, above Ni/NiO but in the fields of water stability and Fe-bearing olivine stability (Brenan & Watson 1991). This assures that H_2O ($\pm CO_2$) will be the dominant species in the fluid (Eggler 1987). Several different capsules and pressure assemblies were used in these experiments, the details of which will be discussed in a future communication.

As a test of the single crystal technique we measured the solubility of rutile at 2.0 GPa and 1100 °C in 24 h runs using it and the double capsule method of Schneider & Eggler (1986). The two methods give the same results. Runs of greater duration

Table 1. *Experimental results*

(Tech., experimental technique, single crystal (Xt) or double capsule (DC); caps is the type of capsule used (ex: Ti+Pt signifies a Ti outer capsule with a Pt liner). Err. is the 1-sigma weighing error in wt%. Solubilities in wt% (= (wt loss crystal/wt fluid) 100). BD, below detection. Fo, forsterite; Ab, albite; 1 M = 1 molal of species indicated dissolved in H_2O_5 for H_2O-CO_2 fluids, $X_{H_2O} = 0.5$.)

run	solid	fluid	P/GPa	T/(10^3 °C)	t/h	tech.	caps	pH	wt%	err.
R07	rutile	H_2O	2.0	1.1	23	Xt	Pt	—	0.96	0.04
R10	rutile	H_2O	2.0	1.1	42	Xt	Pt	—	1	0.03
R12	rutile	H_2O	2.0	1.1	24	DC	Pt	—	1	0.08
R13	rutile	H_2O	2.8	1.1	24	DC	Pt	—	5.9	0.03
R20[a]	rutile	H_2O	2.0	1.1	22.5	Xt	Ti+Pt	—	1.6	0.05
Ap02	FAp	H_2O	2.0	1.1	24	Xt	Ni	—	0.35	0.06
Ap06	FAp	H_2O	1.0	1.1	24	Xt	Ni	5	0.44	0.05
Ap08	FAp	H_2O	2.0	1.1	23	Xt	Ni+Pt	—	0.33	0.08
Ap10	FAp	H_2O	2.0	1.1	48	Xt	Nb+Pt	2.1	0.25	0.05
Ap11	FAp	H_2O	1.0	1.0	26	Xt	Ni+Pt	5	0.17	0.05
Ap12	FAp	1 M NaCl	1.0	1.0	24	Xt	Ti+Pt	—	0.14	0.05
Ap13	FAp+Fo	H_2O	1.0	1.0	24	Xt	Ti+Pt	—	0.17	0.05
Ap15	FAp	H_2O	1.0	1.2	24	Xt	Ti+Pt	1.8	0.13	0.05
Ap17[a]	FAp+Ab	H_2O	1.0	1.0	45	Xt	Ti+Pt	—	3.0	0.05
Ap18	FAp	H_2O	1.0	1.0	25.5	Xt	Ni	—	BD	0.05
Ap20	FAp+Ab	H_2O	1.0	1.0	24	Xt	Ti+Pt	8	0.067	0.05
Ap21	FAp	1 M HCl	1.0	1.0	66	Xt	Ti+Pt	0.52	2.1	0.05
Ap22	FAp	H_2O	1.0	0.8	46.5	Xt	Ti+Pt	7	0.1	0.05
Ap24	FAp	H_2O	3.0	1.0	24	Xt	Ti+AgPd	7	BD	0.05
Ap25	FAp	H_2O–CO_2	1.0	1.0	24	Xt	Ti+Pt	—	0.06	0.04
M01	monazite	H_2O	2.0	1.1	24	Xt	Ni	—	0.09	0.04
M02	monazite	H_2O	2.8	1.1	24	Xt	Ni	—	0.2	0.05
M03	monazite	H_2O	1.0	1.0	24	Xt	Ni+AgPd	—	0.14	0.05
M05	monazite	H_2O	1.0	0.8	24	Xt	Ti+AgPd	neut	BD	0.05
M07[b]	monazite	1 M HCl	1.0	1.0	24	Xt	Ti+Pt	acidic	0.36	?
M08	monazite	1 M NaCl	1.0	1.0	24	Xt	Ti+Pt	neut	0.12	0.06
Z02	zircon	H_2O	2.0	1.1	24	Xt	Ni	—	3.4	0.05
Z06	zircon	H_2O	2.0	1.1	26	Xt	Ni+AgPd	neut	6.9	0.06
Z08	zir+qtz	H_2O	1.0	1.0	24	Xt	Ni+AgPd	neut	0.17	0.05
Z09	zir+qtz	1 M HCl	1.0	1.0	24	Xt	Ni+AgPd	acidic	0.12	0.05
Z11	zir+qtz	1 M NaCl	1.0	1.0	24	Xt	Ni+AgPd	> = 10	low	—
Z13	zircon	H_2O	1.0	1.0	24	Xt	Ni+AgPd	neut	5.3	0.05
Z14	zircon	1 M HCl	1.0	1.0	24	Xt	Ti+Pt	acidic	5.0	0.05
Z15	zircon	H_2O–CO_2	1.0	1.0	24	Xt	Ti+Pt	—	BD	0.05

[a] Solubility high due to recrystallization in temperature gradient.

[b] Crystal broke during experiment; measured solubility may be artificially high.

listed in table 1 confirm that 24 h is sufficient to achieve equilibrium, as demonstrated in other studies of fluid/mineral systems (Brenan & Watson 1991; Schneider & Eggler 1986).

For some experiments the pH of quenched fluid was measured using litmus paper, and for three experiments (Ap 10, 15 and 21) using a pH microelectrode. Although quench pH is not equivalent to pH at run conditions, it is a measure of the relative activity of H^+ at P and T.

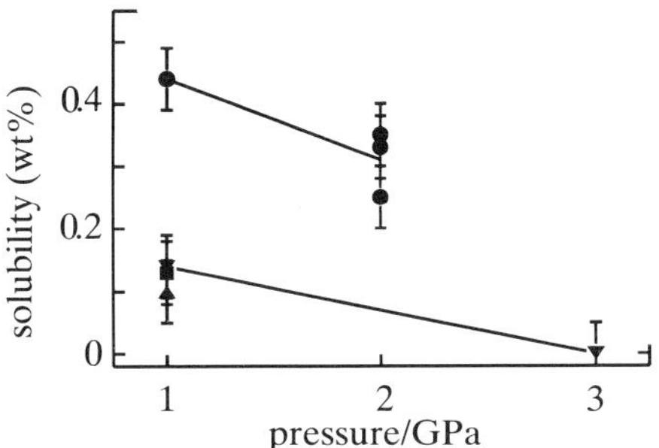

Figure 1. Apatite solubility in wt% as a function of pressure and temperature. Errors bars represent $\pm 1\sigma$ weighing errors. ▲, 800 °C; ▼, 1000 °C; ●, 1100 °C; ■, 1200 °C.

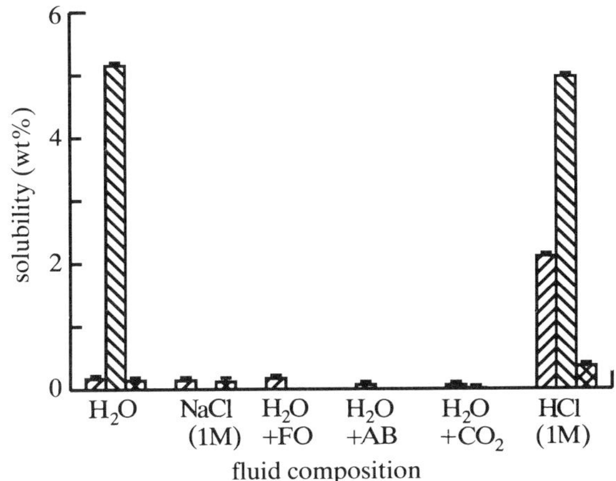

Figure 2. Solubilities of apatite, zircon, and monazite in wt% at 1.0 GPa and 1000 °C as a function of fluid composition. Error bars represent 1σ weighing errors. ▨, Apatite; ◪, zircon; ▩, monazite. Fo, forsterite; Ab, albite; 1 M = 1 molal of the indicated species.

3. Experimental results

(a) Apatite

To our surprise, apatite did not melt or recrystallize even at conditions as extreme as 1200 °C and 1.0 GPa. The solubility of apatite in H_2O decreases with increasing pressure (figure 1). Temperature dependence is less systematic, with solubility increasing over the range 800–1100 °C and then dropping to a lower value at 1200 °C.

The most important result is the generally low solubility compared with silicates, which have solubilities of several wt% at similar conditions (Schneider & Eggler 1986). The total range of measured solubility is 0.10 wt% at 1.0 GPa and 800 °C to 0.44 wt% at 1.0 GPa and 1100 °C. This range is very small compared with the range at ambient conditions as a function of pH, suggesting that P and T have a much smaller affect on solubility than pH. In fact, experiments for which quench pH was measured show solubilities very close to values at ambient conditions at the same pH. Low quench pH values suggest that exchange reactions like (1.2) and (1.3) occur.

The effect on solubility of changing fluid composition is variable as shown in figure 2, which includes results for apatite, monazite, and zircon at 1.0 GPa and 1000 °C. No change in the solubility of FAp on addition of NaCl to H_2O is evident. This

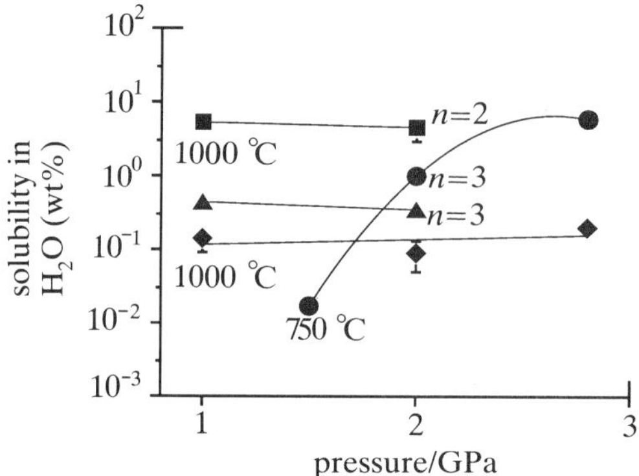

Figure 3. Solubilities in wt % of apatite, monazite, zircon and rutile in H_2O as a function of pressure at 1100 °C (except where noted). Error bars represent $\pm 1\sigma$ weighing errors except where n, the number of determinations, is greater than one. In this case error bars represent $\pm 1\sigma$ on multiple determinations. ●, Rutile; ▲, apatite; ◆, monazite; ■, zircon.

suggests there is neither complexing of Na^+ or Cl^- with dissolved apatite components nor a dependence of solubility on ionic strength, agreeing with the results of McCann (1968).

To test whether dissolved silicate material influences solubility, runs with added forsterite (Fo) and albite (Ab) were done. Results show no effect of dissolved silicate concentration on apatite solubility. The effect of CO_2 on apatite solubility was tested by loading oxalic acid to give a fluid composition $X_{H_2O} = 0.5$, $X_{CO_2} = 0.5$. Solubility dropped to about one third of the value for pure H_2O. Finally, the effect of pH was tested by adding 1 M HCl. As expected, solubility greatly increased to 2.1 wt %, showing that pH is the most important determinant of apatite solubility at high P and T just as it is at ambient conditions.

(b) Monazite

Monazite is even less soluble than apatite at the conditions studied. Solubility appears to increase with increasing P (figure 3), and at 1.0 GPa it increases with increasing T (table 1). Solubilities range from detection (not more than 0.05 wt %) at 1.0 GPa and 800 °C to 0.2 wt % at 2.8 GPa and 1100 °C.

Considering fluid composition (figure 2), the solubility of monazite in 1 M NaCl is the same as in H_2O at the same conditions. As with apatite, ionic strength does not affect the solubility. This may be a result of monazite dissolving as neutral species, or more unlikely, complete association of NaCl at these conditions. For 1 M HCl an increase in solubility is similar to the results obtained for apatite: decreasing the pH increases the solubility. Presumably low pH keeps phosphate in the protonated form, decreasing the activity of PO_4^{3-} and thereby increasing solubility of these minerals.

(c) Zircon

Zircon proved more difficult to study because of precipitation of baddeleyite (see (1.5) and (1.6)) on the surface of the crystal in runs containing pure H_2O. In some cases baddeleyite was removed by heating the crystal in concentrated HF, in which zircon is virtually insoluble. This allowed measurement of the weight loss of *zircon*, upon which the stated solubilities are based. Analysis of Zr in quench silicate suggests that the solubility of ZrO_2 in the fluid is low, between 700–1800 p.p.m. in equilibrium with zircon and quartz.

In pure H_2O the solubility of zircon at 1.0 GPa and 1000 °C is 5.3 wt%, and at 2.0 GPa and 1100 °C the average of two runs is 5.2%, suggesting P and T have little effect (figure 3). In 1 M HCl baddeleyite still formed, and the measured solubility is again close to 5 wt% (figure 2). Addition of CO_2 to the fluid, however, decreased the solubility to below detection. Like that of other silicates (Eggler 1987), zircon solubility decreases with dilution of H_2O, in this case by CO_2.

Experiments on the assemblage zircon–baddeleyite–H_2O did not give a clear idea of how much zircon could be dissolved by a silica-bearing fluid. We therefore tried experiments on the assemblage zircon–quartz–H_2O, but found retrieved crystals to have several micrometres thick layers of quench silicate mineral on their surfaces. To obtain a solubility we estimated the weight of the layer using two techniques: by calculation from its thickness and probable density, and by soaking the crystal in concentrated HF to remove the silica layer and reweighing. Averaging the results gave the weight loss of zircon and the calculated solubilities listed in table 1. Solubility of zircon in quartz-saturated fluids is very low, regardless of whether HCl or NaCl is present.

(d) Rutile

Recrystallization of rutile in the temperature gradient of the piston cylinder assembly required us to use a different technique. Rutile was run in horizontally positioned noble metal capsules surrounded by graphite; these runs yielded results showing a high degree of precision (table 1).

In runs in which recrystallization occurred, observation of run products and of the position of the capsule with respect to the hotspot suggested that rutile solubility increases with increasing temperature. Solubility also increases with increasing pressure, from 1 wt% at 2.0 GPa, 1100 °C to 5.9 wt% at 2.8 GPa, 1100 °C (figure 3). The solubility of rutile at 1.5 GPa and 750 °C is 0.017 wt% (Schneider & Eggler 1986), in agreement with the P–T trends observed in this study.

(e) Summary

Our measurements suggest that most accessory minerals are not very soluble in supercritical aqueous fluids of near-neutral pH. A comparison of solubilities in H_2O at 2.0 GPa and 1100 °C shows an increase in the order monazite < apatite < rutile < zircon (figure 3). The high apparent solubility of zircon results from its consumption by silica-undersaturated H_2O to form baddeleyite + silica-rich fluid. The actual solubility of ZrO_2 is low. Rutile is significantly soluble in water at very high P and T, and because its solubility strongly depends on P and T, it may be more or less soluble than the other minerals depending on conditions. The phosphates apatite and monazite are less soluble, and their solubilities less dependent on P and T. Their solubilities are, however, pH dependent, and in low-pH fluids apatite in particular can be very soluble. The fluid-compositional variables M_{NaCl}, dissolved silicate concentration, and X_{CO_2} do not increase the solubilities of apatite, zircon, or monazite (figure 2).

One important variable that we have yet to investigate is a_{F^-}. Studies at lower P–T (Barsukova *et al.* 1979) show that the ionic character of F^- makes it a good complexing anion for Ti, and maybe for Zr and REE (Gieré 1990). It is clear that not just any fluid can transport incompatible elements in significant quantities.

Phil. Trans. R. Soc. Lond. A (1991)

4. Petrological implications

(a) Comparison of accessory mineral solubilities in aqueous fluids and silicate melts

Accessory minerals are generally less soluble in aqueous fluids than in silicate melts at the same conditions (table 2). Silicate melts therefore have a greater potential for transport of accessory mineral components than aqueous fluids. However, the differences in the abilities of silicate melts and aqueous fluids to dissolve accessory minerals shown in table 2 probably represent maxima. Accessory minerals are less soluble in more silica-rich or differentiated silicate melt compositions. Also, lowering of pH and possibly the presence of complexing anions such as F^- will increase accessory mineral solubilities in aqueous fluids. Thus some aqueous fluids may be more effective solvents than some silicate melts.

(b) Estimation of fluid/rock ratios from solubility measurements

It is instructive to calculate the minimum fluid/rock ratio required to add a given amount of each phase to a rock. The mass balance equation is:

$$M^{\text{rock}}C^{\text{rock}}_{\text{AM}} = M^{\text{fl}}C^{\text{fl}}_{\text{AM}} + M^{\text{rock}}_{\text{i}}(C^{\text{rock}}_{\text{AM}})_{\text{i}}, \qquad (4.1)$$

where $M^{\text{rock}}_{\text{i}}$ is the initial mass of rock, and the concentration of accessory mineral (C_{AM}) in a phase is in wt%. In the simplest model a fluid saturated in the accessory mineral AM enters the rock, deposits its entire load of the mineral and exits the rock. In this case, the initial and final masses of rock are roughly equal (i.e. $M^{\text{rock}} \approx M^{\text{rock}}_{\text{i}}$). Assuming, then, that there is no pre-existing AM in the rock, we can calculate the fl/rock ratio *per wt% AM introduced* (i.e. $C^{\text{rock}}_{\text{AM}} = 1$) from:

$$M^{\text{fl}}/M^{\text{rock}} = 1/C^{\text{fl}}_{\text{AM}}. \qquad (4.2)$$

Using measured solubilities in H_2O at 2.0 GPa and 1100 °C (table 2) gives fl/rock values necessary to introduce 1 wt% of each accessory mineral as *ca.* 3 for Ap, *ca.* 1 rutile, *ca.* 11 for monazite, and *ca.* 6 for zircon (for quartz-saturated H_2O).

This simple model almost certainly greatly underestimates the required fluid/rock ratio, because a fluid will precipitate only a small part of its dissolved accessory mineral content unless its is completely consumed in hydration reactions (which is not possible for the fluid/rock ratios we are discussing). In addition, for processes occurring specifically in subduction zones, temperatures corresponding to 1.0–2.0 GPa are considerably lower than 1100 °C, so accessory mineral solubilities will be lower than the values used above. Fluid/rock ratios required to add a given amount of accessory mineral will be correspondingly higher.

One way a fluid can transport large quantities of components of insoluble accessory minerals is to carry only some of those components. For example, a fluid carrying PO_4^{3-} and F^- may move through a depleted harzburgite that contains negligible Ca. Upon entering an eclogite $a_{Ca^{2+}}$ will increase and apatite will precipitate. In this way a smaller volume of fluid is required to add a given amount of accessory mineral to a rock.

(c) Implications of rutile solubility for HFSE depletion in island-arc basalts

In the past, the depletion of high field strength elements (HFSE) relative to REE in island arc basalts (IAB) was often attributed to residual rutile in the source region of IAB. However, Ryerson & Watson (1987) showed that rutile is too soluble in basaltic

Table 2. *Mineral solubilities in* H_2O *and basalts in* wt%

mineral	H_2O	basalt	reference for basalt
apatite	0.4	6.0	Watson (1980)
monazite	0.1	1.8	Rapp *et al.* (1987)
rutile	1.0	5.0	Ryerson & Watson (1987)
zircon	0.2	0.6	Watson & Harrison (1983)

melt to be residual during partial melting. Fractionation of HFSE from REE by some process is required by the chemistry of IAB, and that fractionation may be caused by aqueous fluids. The source of fluid is the subducted slab.

Dehydration of amphibole in the slab occurs at roughly 35 kbar (3.5 GPa) and 800 °C (Peacock 1990). At these conditions rutile is much more soluble in H_2O than phases such as monazite and apatite that concentrate REE. As fluid ascends into the overlying mantle wedge it heats up to about 1100 °C (Peacock 1990), allowing it to dissolve even more rutile. Therefore fluid derived from dehydration of the subducted slab could flux the overlying mantle wedge, preferentially dissolving rutile and depleting the wedge in HFSE relative to REE and at the same time hydrating it to lower the solidus temperature. Subsequent partial melting of the source region would result in characteristically HFSE-depleted IAB. This scenario of fluid production in the slab by amphibole dehydration, fluxing of the overlying mantle wedge, and subsequent melting of peridotite in the wedge to form IAB is supported by thermal models and phase equilibria (Davies & Stevenson 1991; Peacock 1990). However, depending on the thermal model used it is possible that rutile may be less soluble than the REE-concentrating phases when fluid is released from the slab and metasomatizes the source region of IAB.

In closing, we note that there exist several problems with invoking rutile aqueous solubility to explain the HFSE anomaly. Most significant among these may be that IAB shows negative anomalies for all HFSE including Zr and Hf, and yet the latter two are controlled not by rutile but zircon, which we have shown to be relatively insoluble in aqueous fluids. Also, it is difficult to devise a model in which aqueous fluids can exist in a subduction zone and cause large-scale chemical transport over long periods of time. Partial melting will consume any fluids, so any modification of the source region must occur before basalt generation. If there is induced convection in the mantle wedge, fluids must continuously chemically modify fresh peridotite introduced to the source region (Davies & Stevenson 1991). In addition, aqueous fluids in equilibrium with silicate minerals show a marked preference for alkali metals and alkaline earths relative to REE (Brenan & Watson 1991), yet there is no fractionation of these elements in IAB.

Regardless of the role of rutile and aqueous fluids in subduction zones, our data show that Ti is significantly soluble in aqueous fluids at high P and T. The scale and significance of aqueous fluid transport of Ti is controlled by the thermal structure of the region, phase stabilities, and the fluid transport mechanism.

(d) Geologic evidence of aqueous fluid transport of accessory mineral components

Examples in the literature of accessory mineral mobility linked to volatile fluids are abundant. Based on their study of garnet amphibolites from a palaeosubduction zone, Sorenson & Grossman (1989) concluded that accessory mineral components were introduced by fluids. More direct evidence of fluid transport is given by

Selverstone *et al.* (1990) in their study of eclogites from a palaeosubduction zone in the Tauern Window of Austria. Fluid inclusions in these rocks contain H_2O, CO_2, and various salts, along with daughter minerals that include rutile, apatite, NaCl, and carbonate. The presence of the daughter minerals suggests that phosphorus and titanium (among other elements) are highly soluble in saline fluids at high pressure (2.0 GPa).

Addition of a whole suite of accessory minerals including zirconolite, rutile and apatite to a contact metamorphic marble has been documented by Gieré (1990). This study concludes that fluids derived from adjacent calcalkaline plutons metasomatized the country rock, and that fluorine and phosphorus in the fluid complexed with REE, HFSE, and actinides (Gieré 1990). In another setting, hydrothermal zircons in an intrusive igneous complex have been described by Rubin *et al.* (1989). Fluids derived from late-stage melts enriched in F^- are believed to transport zircon (Rubin *et al.* 1989).

In many cases of accessory mineral enrichment reported in the literature the enrichment is attributed to hydrothermal fluids, but the enriched rocks are in close proximity to a zone of partial melting (Gieré 1990; Rubin *et al.* 1989; Sorenson & Grossman 1989). The connection may result from exsolution of an aqueous fluid from a late-stage crystallizing magma high in F^- (Gieré 1990; Rubin *et al.* 1989). Fluorine is known to complex with Ti (Barsukova *et al.* 1979) and may complex with the other highly ionic ions such as Zr and REE. Such fluids also have low pH (Ayers & Eggler 1991), which should increase the solubility of many accessory minerals. Accessory minerals would precipitate as pH increases due to fluid-country rock interaction. One potential problem with this scenario is that F^- partitions preferentially into the melt relative to the fluid (see references in Ayers & Eggler 1991). Further study on the effect of F^- on the aqueous solubility of accessory minerals in required.

Funding for this study was provided by NSF Grant no. EAR-8904177. We thank James Brenan for his help.

References

Ayers, J. C. & Eggler, D. D. 1991 Saline aqueous fluids in equilibrium with silicate melt at mantle conditions: trace element partition coefficients and evidence for ionic behavior. *Geochim. cosmochim. Acta.* (Submitted.)

Ayers, J. C. & Watson, E. B. 1989 Solubility of accessory minerals in H_2O at upper mantle conditions. *EOS* **70**, 506.

Barsukova, M. L., Kuznetov, V. A., Dorofeyeva, V. A. & Khodakovskiy, L. I. 1979 Measurement of the solubility of rutile TiO_2 in fluoride solutions at elevated temperatures. *Geokhimiya* **7**, 1017–1027.

Brenan, J. B. & Watson, E. B. 1991 Partitioning of trace elements between olivine and aqueous fluids at high P–T conditions: implications for the effect of fluid composition on trace element transport. *Chem. Geol.* (Submitted.)

Davies, J. H. & Stevenson, D. J. 1991 Physical model of source region of subduction zone volcanics. *J. geophys. Res.* (Submitted.)

Eggler, D. H. 1987 Solubility of major and trace elements in mantle metasomatic fluids: experimental constraints. In *Mantle metasomatism* (ed. M. A. Menzies & C. J. Hawkesworth), pp. 21–41. London: Academic Press.

Gieré, R. 1990 Hydrothermal mobility of Ti, Zr and REE: examples from the Bergell and Adamello contact aureoles (Italy). *Terra Nova* **2**, 60–67.

Holland, H. D. & Malinin, S. D. 1979 The solubility and occurrence of non-ore minerals. In *Geochemistry of hydrothermal ore deposits*, 2nd edn (ed. H. L. Barnes), pp. 461–501. New York: Wiley.

Maurice, O. D. 1949 Transport and deposition of the non-sulphide vein minerals. V. Zirconium minerals. *Econ. Geol.* **44**, 721–731.

McCann, H. G. 1968 The solubility of fluorapatite and its relationship to that of calcium fluoride. *Archs. Oral Biol.* **13**, 987–1001.

Overstreet, W. C. 1967 The geologic occurrence of monazite. *U.S. Geol. Surv. Prof. Pap.* no. 530. USGS.

Peacock, S. M. 1990 Numerical simulation of metamorphic pressure–temperature–time paths and fluid production in subducting slabs. *Tectonics* **9**, 1197–1211.

Rapp, R. P., Ryerson, F. J. & Miller, C. F. 1987 Experimental evidence bearing on the stability of monazite during crustal anatexis. *Geophys. Res. Lett.* **14**, 307–310.

Rubin, J. N., Henry, C. D. & Price, J. G. 1989 Hydrothermal zircons and zircon overgrowths, Sierra Blanca Peaks, Texas. *Am. Min.* **89**, 865–869.

Ryerson, F. J. & Watson, E. B. 1987 Rutile saturation in magmas: implications for Ti–Nb–Ta depletion in island arc basalts. *Earth planet. Sci. Lett.* **86**, 225–239.

Schneider, M. E. & Eggler, D. H. 1986 Fluids in equilibrium with peridotite minerals: implications for mantle metasomatism. *Geochim. cosmochim. Acta* **50**, 711–724.

Selverstone, J., Franz, G. & Thomas, S. 1990 Fluids at high pressure: inferences from 20 kbar eclogites and associated veins in the Tauern Window, Austria (abstr.). *V. M. Goldschmidt Conference Program and Abstracts*, p. 81.

Sorenson, S. S. & Grossman, J. N. 1989 Enrichment of trace elements in garnet amphibolites from a paleo-subduction zone: Catalina Schist, southern California. *Geochim. cosmochim. Acta* **53**, 3155–3178.

Tole, M. P. 1985 The kinetics of dissolution of zircon ($ZrSiO_4$). *Geochim. cosmochim. Acta* **49**, 453–458.

Watson, E. B. 1980 Apatite and phosphorus in mantle source regions: an experimental study of apatite/melt equilibria at pressures to 25 kbar. *Earth planet. Sci. Lett.* **51**, 322–335.

Watson, E. B. & Harrison, T. M. 1983 Zircon saturation revisited: temperature and composition effects in a variety of crustal magma types. *Earth planet. Sci. Lett.* **64**, 295–304.

Fluid influence on the trace element compositions of subduction zone magmas

By A. D. Saunders, M. J. Norry and J. Tarney

Department of Geology, University of Leicester, Leicester LE1 7RH, U.K.

Subduction zones represent major sites of chemical fractionation within the Earth. Element pairs which behave coherently during normal mantle melting may become strongly decoupled from one another during the slab dehydration processes and during hydrous melting conditions in the slab and in the mantle wedge. This results in the large ion lithophile elements (e.g. K, Rb, Th, U, Ba) and the light rare earth elements being transferred from the slab to the mantle wedge, and being concentrated within the mantle wedge by hydrous fluids, stabilized in hydrous phases such as hornblende and phlogopite, from where they are eventually extracted as magmas and contribute to growth of the continental crust. High-field strength elements (e.g. Nb, Ta, Ti, P, Zr) are insoluble in hydrous fluids and relatively insoluble in hydrous melts, and remain in the subducted slab and the adjacent parts of the mantle which are dragged down and contribute to the source for ocean island basalts. The required element fractionations result from interaction between specific mineral phases (hornblende, phlogopite, rutile, sphene, etc.) and hydrous fluids. In present day subduction magmatism the mantle wedge contributes dominantly to the chemical budget, and there is a requirement for significant convection to maintain the element flux. In the Precambrian, melting of subducted ocean crust may have been easier, providing an enhanced slab contribution to continental growth.

1. Introduction

Subduction zones are the main tectonic setting where volatiles – particularly water – are returned to the mantle. These volatiles exert a major control on the melting relationships in the subducting slab and in mantle wedge of the over-riding plate. They also determine the nature of the fluxes that are responsible for transporting chemical elements from the slab into the wedge and, ultimately, into the island arc and calc-alkaline magmas which have contributed to continental growth throughout Earth history.

The chemical nature of, and pathways followed by, hydrous fluxes in the subduction zone environment are not well constrained. Many intraoceanic and continent-based arc magmas have high concentrations of large-ion lithophile (LIL) elements (Rb, Th, K and Ba) relative to high field strength (HFS) elements (Ti, Zr, Hf, Nb, Ta or P), and it has been suggested that breakdown of hydrous mineral phases in the subducted oceanic crust releases volatiles which preferentially transport the more water-soluble LIL-elements into the mantle wedge of the over-riding plate (Saunders *et al.* 1980; Pearce 1983; Gill 1981).

Slab dehydration may be only the first stage of a complex series of fractionation processes in subduction zones. As fluids enter the mantle wedge, hydration mechanisms serve to stabilize new minerals, such as hornblende or phlogopite, which themselves may become vulnerable to thermal breakdown as the thermal structure of the subduction zone evolves. Fluids migrating through the wedge will selectively scavenge incompatible elements from their more compatible partners. Such fluid migration and fractionation will lead to different chemical 'pathways' and different mantle wedge residence times for the various elements: for example, He, Rb, U, Be and Th may have particularly short residence times in the mantle wedge (Saunders *et al.* 1987; Sigmarsson *et al.* 1990). Moreover, the mantle wedge itself may be an important contributor of material to the ascending fluids or melts, and this could become a dominant supplier if induced convection effectively replenishes this wedge source.

At issue, then, is the relative importance of (i) the element flux derived from the slab (subducted ocean basalts plus sediments), which we shall term the *slab-derived flux*, and (ii) the element flux from the mantle wedge, which we shall term the *wedge-derived flux*. Recent studies have suggested that in oceanic arcs, which are underlain by young, depleted lithosphere and asthenosphere, the slab-derived flux may be dominant (Hole *et al.* 1984; Ewart & Hawkesworth 1987), whereas in continent-based arcs, a lithospheric or sub-lithospheric wedge-derived component is important (Saunders *et al.* 1980; Hawkesworth 1982; Pearce 1983; Hickey *et al.* 1986; Rogers & Hawkesworth 1989; Hawkesworth & Ellam 1989; Hickey-Vargas *et al.* 1989).

Continental flood basalts (and equivalent dyke swarms), erupted in tectonic settings far removed from contemporaneous subduction zones, often have trace element patterns similar to those of subduction zone magmas, possibly reflecting the magmas inheriting this signature from, or during ascent through, the continental lithosphere. This could imply that the suprasubduction zone trace element character occurs throughout extensive regions of the subcontinental lithosphere. If so, is this suprasubduction zone signature inherited by, or established at, subduction zones? If the former, how and where is the LIL-element enrichment achieved?

In this contribution, we assess the role of fluxes in the transport of trace elements in this environment. Crucial to any models of flux transport is the nature of the medium through which the fluids flow, and the mineral assemblages with which they equilibrate. It is the minerals which primarily control element fractionations, but their stability is critically dependent on the *P–T–X* conditions in different parts of the subduction zone. Thus we turn first to the postulated thermal structure of subduction zones.

2. Thermal and physical structure of subduction zones

The implications of whether or not subducted slabs melt beneath arc systems, or merely dehydrate and release an aqueous-rich fluid are considerable (Peacock 1990, 1991), as the compositions of the slab-derived component will vary accordingly. For example, melting of a high-pressure slab assemblage such as eclogite is likely to release a silicic melt substantially depleted in the heavy REE and Sc, elements strongly partitioned into garnet. On the other hand, dehydration during the breakdown of chlorite, amphibole or serpentine minerals will release a hydrous fluid phase capable of transporting substantial amounts of Si, K, light REE and LIL-elements from the slab (Boettcher & Wyllie 1968; Nakamura & Kushiro 1974; Tatsumi *et al.* 1986).

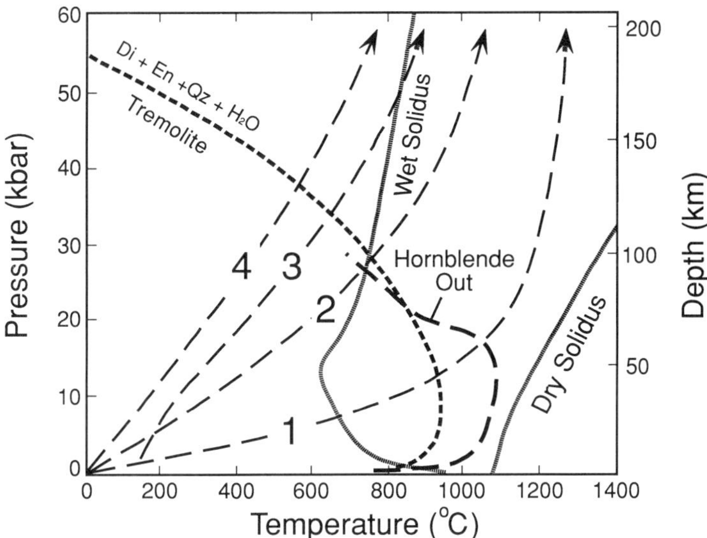

Figure 1. Pressure-temperature diagram illustrating the possible thermal conditions operative in subduction zone systems, modified from Peacock (1990). Lines 1–4 are representative thermal gradients along the top of the subducting oceanic slab in old, cold systems (line 4), through to subduction zones where there is induced mantle convection in the hanging wall. Phase relations in wet and dry basalt from Green (1982) and Lambert & Wyllie (1972); in wet and dry peridotite from Takahashi & Kushiro (1983), Mysen & Boettcher (1975). Hornblende and phlogopite stability curves are from Green (1973, 1982) and Wendlendt & Eggler (1980).

Peacock (1990, 1991) has shown that subducted oceanic crust is likely to intersect the wet basalt solidus only if the thermal gradient is high, as during the subduction of very young, hot oceanic crust, or during the initial stages of subduction (Tarney *et al.* 1981) when the adjacent mantle temperatures are still high; or during Archaean times when the geotherm was steeper (Tarney & Weaver 1987; Martin 1987) (figure 1). Trace element data for Archaean magmas are consistent with melting of a hydrous hornblende- or garnet-bearing mafic source (Tarney & Weaver 1987), but most modern arc volcanics do not have REE profiles consistent with extensive garnet fractionation. However, in areas of recent ridge subduction, where young (zero-age) oceanic crust has been subducted, such as Baja California, southern Chile and the Aleutian Islands (Rogers & Saunders 1989; Defant & Drummond 1990), the associated arc volcanics are unusually depleted in heavy REE, which might infer melting of eclogite under unusual thermal conditions. However, even under these conditions it is debatable whether the wet melting curve of eclogite is reached, and there are alternative explanations involving hornblende and garnet in the mantle wedge (Saunders *et al.* 1987; Rogers & Saunders 1989).

Slab-induced convection in the overhanging mantle wedge will influence the thermal structure of the subduction system, affect the stability of hydrous minerals, and influence the composition of the arc magmas. If the arc system is based on old and relatively thick continental lithosphere, and the angle of subduction is shallow, asthenosphere may be excluded from the subarc region. In this extreme situation, arc magmatism may not be possible. In arcs based on relatively young oceanic lithosphere, induced convection beneath the thin lithosphere could produce a deep asthenospheric wedge adjacent to the slab. These two 'end-member' wedge states will provide potentially quite different mantle source material for the arc magmas,

Phil. Trans. R. Soc. Lond. A (1991)

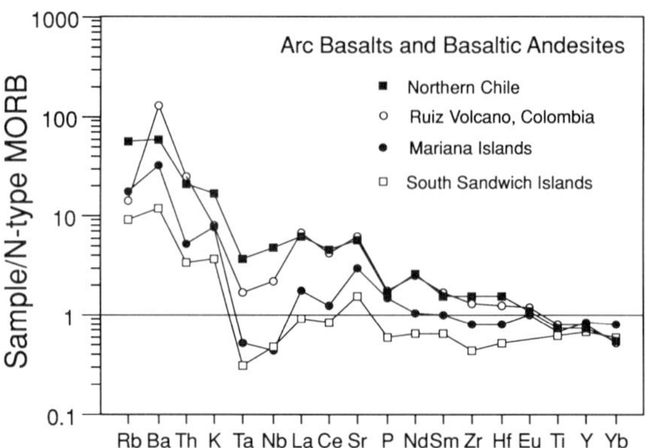

Figure 2. Representative analyses of basalts and basaltic andesites from oceanic arcs (South Sandwich Islands and Mariana Islands) and continent-based arcs. Nb and Ta are positioned according to their relative incompatibility in four-phase mantle peridotite. Note that the magnitude of the Nb and Ta anomaly decreases in the continent-based arcs, and that the abundance of most elements, including the HFS elements, increases in these lavas. Data sources: Northern Chile-sample A791 from Cerro Overo (Thorpe *et al.* 1984); Colombia-sample MB 260 from Ruiz Volcano (Marriner & Millward 1984); Mariana Islands-sample AS1 from Ascuncion Island (Hole *et al.* 1984); South Sandwich Islands, SSM4.1 from Montague Island (P. E. Baker and I. Luff, unpublished data). Normalizing values from Sun & McDonough (1989).

and result in different residence times for slab-derived material: for example, asthenospheric source material can be continually replenished, and removed, by mantle flow; whereas lithospheric mantle provides a more permanent storage for slab-derived components. The rheological state of the lower lithosphere boundary is critical. Importantly, Falloon & Green (1990) have shown that, in carbonated subcontinental lithosphere, carbonatitic melts could coexist with residual phlogopite and garnet at depths greater than 100 km. So it is possible that induced mantle flow could occur in old sub-continental lithosphere, and this would have the effect of circulating new mantle material, that had suffered earlier metasomatism, into the source regions of continental arcs.

3. Trace element characteristics of subduction-related magmas

Trace element comparisons are best made using multielement mantle- or MORB-normalized diagrams or 'spiderdiagrams' (Wood *et al.* 1979*a*, *b*; Tarney *et al.* 1979, 1980; Sun 1980; Thompson *et al.* 1983, 1984). For reference purposes, we plot a range of basalts and andesites from oceanic island arcs and continental-margin arcs in figure 2. Included in this compilation are data for post-subduction magmas from Mexico. The following features are apparent: all of the patterns exhibit a pronounced, but variable, trough or anomaly at Nb (and Ta); the HFS elements to the right of the diagrams have much lower abundances than the LIL elements to the left; there are considerable variations in inter LIL-element ratios (e.g. Rb/Ba, K/Ba; or Rb/Th); and continental magmas have higher LIL- and HFS-element abundances than their oceanic counterparts.

In these diagrams the elements are arranged in order of increasing incompatibility (right to left) for a normal upper mantle spinel peridotite mineral assemblage (Sun 1980; Tarney *et al.* 1980, 1981; Thompson *et al.* 1983, 1984; Sun & McDonough 1989).

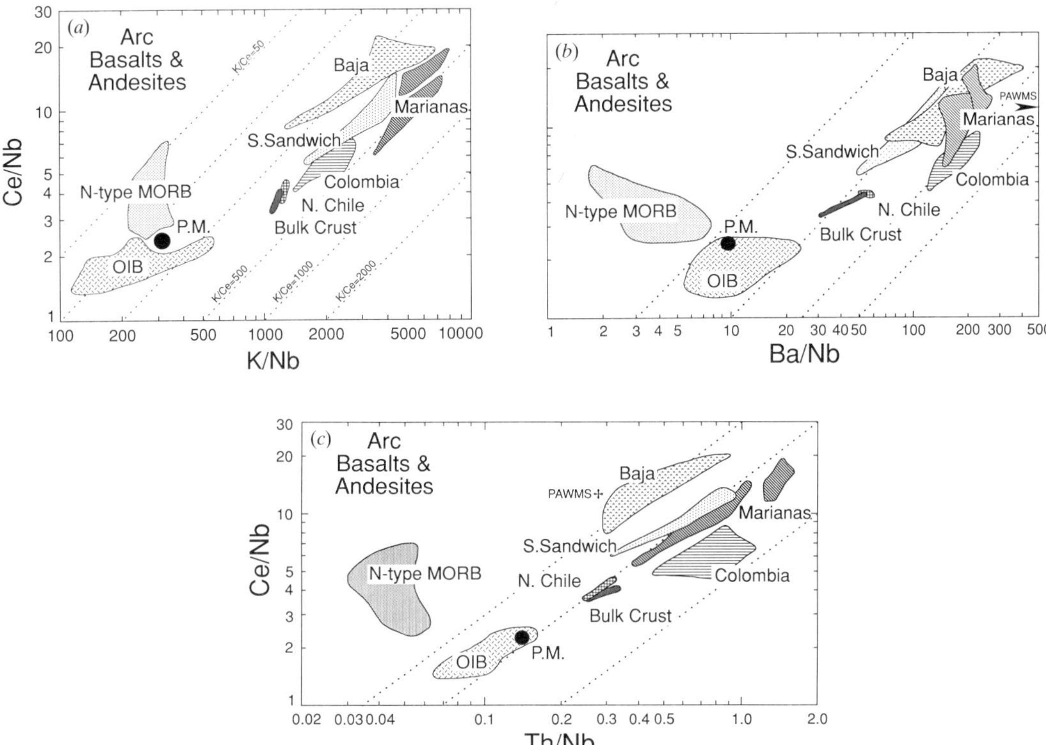

Figure 3. Ce/Nb ratio versus K/Nb, Ba/Nb and Th/Nb ratio, for subduction-related basalts and andesites, N-type mid-ocean ridge basalts (N-type MORB) and ocean island basalts (OIB). Fields for MORB and OIB, and primordial mantle (PM) values, from Saunders *et al.* (1988). Data sources for subduction-related volcanics given in figure 2, plus Baja California from Saunders *et al.* (1987) and Rogers & Saunders (1989); southern Chile data from Hickey-Vargas *et al.* (1989). 'Bulk crust' refers to average continental crustal estimates from Taylor & McLennan (1985) and Weaver & Tarney (1984); PAWMS (Pacific authigenic weighted mean sediment) from Hole *et al.* (1984). For most samples, and to improve precision at low abundances in arc basalts, Nb is calculated from Ta using a Nb/Ta ratio of 16.

Entry of another mineral into the fractionating assemblage with specific affinity for one or more elements can effectively change its position on the incompatibility scale, resulting in a prominent positive or negative anomaly on the spiderdiagram. Such is the case for Nb in most subduction-related magmas; that the high distribution coefficient indicated is a consequence of deep-level processes is shown by the fact that, say, K and Nb behave coherently in shallow-level magmas. We are left with the fundamental observation (Saunders *et al.* 1980) that Nb, along with other HFS elements like Ta and Ti, is strongly depleted relative to the LIL elements, in arc parental magmas. This feature can be illustrated in larger numbers of samples using ratio–ratio diagrams. By using Nb as a denominator, and adjacent LIL elements (e.g. La, Ce, K, Ba or Th) as numerators, we can produce a useful and immediate representation of the 'magnitude' of the anomaly (figure 3) in comparison with representative data for MORB and ocean island basalts. Note that all of these elements are considered to have low partition coefficients in olivine, pyroxene, plagioclase, spinel and garnet. Consequently, we would not expect individual magmatic suites to show much dispersion on the ratio–ratio plots, and this is, on the whole, observed.

The exceptions mainly reflect analytical uncertainty at the low abundances of Nb and Ta in primitive arc tholeiites.

There is only limited variation in Ba/Ce, K/Ce or Th/Ce ratios in the various arc lavas, which is perhaps surprising considering the range of values which are found in possible source rocks. For example, subducted pelagic sediments have low Th/Ce ratios (*ca.* 0.002) and high Ba/Ce ratios (*ca.* 140), in addition to high LIL/Nb ratios. Note also that all recent arc rocks, for which data are available, have LIL/Nb ratios greater than that of average continental crust compositions estimated by Weaver & Tarney (1984) and Taylor & McLennan (1985).

4. Mineral phases which may fractionate Ti, Nb, Ta, Zr, Hf and P

Mineral phases which are likely to sequester significant amounts of HFS elements, either in the slab or the mantle wedge, are clearly important in the petrogenesis of subduction zone magmas. Phases which contain significant Nb and Ta are the titaniferous minerals ilmenite, sphene, rutile and perovskite. Some amphiboles can contain significant Nb, Ta, Ti and Zr. Zircon and apatite contain stoichiometric Zr and P respectively. At high pressures, Zr, Ti and P may have enhanced solubilities in garnet and pyroxene. Pyrochlore contains stoichiometric Nb, but this mineral is normally restricted to rocks with very high Nb contents (e.g. carbonatites) and its behaviour and stability in a subduction zone environment are unknown.

Amphibole megacrysts have measured distribution coefficient (D) values for Nb and Ta ranging from 0.2 to 0.6 (Irving & Frey 1984). Values increase for intermediate and acid rocks, reaching 4.0 (Pearce & Norry 1979). Green & Pearson (1986) investigated the distribution of Nb and Ta between various magmas and magnetite, ilmenite, rutile and sphene. For Nb, the reported D value for magnetite was 0.7, for ilmenite values are in the range 2–5, sphene in the range 3–8, whereas rutile shows the highest values at 26–30. Reported D values for Ta are correspondingly higher, especially for sphene and rutile. As with hornblendes, D values rise markedly with increasing silica content of the associated liquid. Although there are no published D values for perovskite, concentrations of Nb in this mineral in igneous rocks reach 5000 p.p.m. (Wedepohl 1970) implying that D values must be high for most magma compositions. However, because perovskite is only stable in equilibrium with magmatic liquids with very low silica activities (Carmichael *et al.* 1970) it is unlikely that perovskite would able to equilibrate with orthopyroxene under subsolidus conditions in the subarc mantle, and would more likely react to form another Ti-bearing phase.

Pargasitic amphibole could be stable in the downgoing slab or in the mantle wedge, provided that the water pressure did not exceed about 30 kbar (3 GPa) (Allen *et al.* 1975). K-rich amphiboles are stable to rather greater depths (Sudo & Tatsumi 1990). Survival of amphibole in the mantle wedge could confer negative Nb and Ta anomalies on the equilibrium liquids. However, the D values for Nb and Ta in amphibole are relatively small, and substantially less than Zr, Ti, P (?) and the HREE, which would result in a strong decoupling between these elements and Nb, which is not observed in subduction-related magmas. Moreover, post-subduction magmatic suites, whose unusual compositions (particularly their high K/Rb ratios (greater than 3000)) are attributed to breakdown of amphibole (Saunders *et al.* 1987; Rogers & Saunders 1989), also show strong negative Nb and Ta anomalies, but should show the reverse. Amphibole, though an ideal mineral in many other respects to be involved in arc magma petrogenesis, does not yet appear to be able to account for Nb and Ta distributions.

A phase (or phases) which will sequester Nb, Ta and possible Ti without causing larger depletions in other HFS elements is required. Sphene, ilmenite and rutile almost certainly meet this criterion because of their high D values for Nb and Ta; garnet and pyroxene provides the means to retain other HFS elements such as Zr, P and Ti.

5. Potential sites for HFS element retention

Two fundamentally different series of models have been proposed to account for the low HFS element characteristic of arc magmas. In one model, the abundances are controlled by melting of mantle with residual titaniferous minerals ('titanites') which retain the HFS elements in the mantle source region (Saunders *et al.* 1980; Foley & Wheller 1990). In the other model, HFS elements are retained in the subducting oceanic crust (Saunders *et al.* 1980, 1988). The low abundance of HFS elements in the arc magmas is a consequence of this retention and of previous episodes of melt extraction from the wedge (Green 1972). In both models LIL-element abundances are significantly enhanced relative to that of the HFS elements because hydrous fluids selectively mobilize and transport LIL elements from the slab into the wedge and also within the wedge.

(a) *Titanite, zircon and apatite stability in the mantle wedge*

Titanium, being a stoichiometric component of rutile, sphene and ilmenite, is buffered at a specific activity in any liquid with which these minerals may equilibrate. In basalt or basaltic andesite, the natural melting products of lherzolite, the activity constant for Ti is low (Watson 1976; Ryerson & Watson 1986). This allows the TiO_2 content of the basic magma to increase to substantial concentrations before it is removed by fractionating phases such as titanomagnetite or ilmenite. Similarly, during melting, the first-formed melt in equilibrium with a Ti-bearing phase will be rich in TiO_2 and this concentration will not fall until all of the Ti-rich phase has been consumed in the melting process. To create a Nb or Ta anomaly, the modal abundance of the Ti-rich phase in the melting assemblage will have to be sufficient to survive melting to the point where the melt can separate from the residue and the mineral be left as a site for element sequestration. Moreover, before a TiO_2-rich phase can begin to equilibrate with a liquid, the other phases in the assemblage must be saturated with Ti. For instance both garnet and clinopyroxene can contain substantial Ti, especially at the high temperatures of the lherzolite solidus.

Green & Pearson (1986) showed that silica contents of liquids have a strong effect on titanite saturation, with higher silica liquids achieving saturation at lower TiO_2 contents. Indeed tholeiitic basalts require at least 1.5 % TiO_2, and possibly as much as 8 % TiO_2 in the source before saturation is achieved. Foley & Wheller (1990) have re-interpreted the experiments of Green & Pearson (1986) and Ryerson & Watson (1986) to suggest that at the predicted low solidus temperatures, high water content, high pressure and high fO_2 conditions accompanying arc magma formation, titanite saturation could be achieved in the mantle source regions, even at the low predicted TiO_2 concentrations. Residual apatite or zircon are unlikely to survive in the residue during normal melting of mantle lherzolite, although if melting occurs at sufficiently low temperatures (less than 1100 °C), and perhaps under hydrous conditions (as in the subarc environment), small amounts of apatite may survive melting (Watson 1980).

Critical to the stability of titanites and related phases in the mantle wedge is the nature of the melting process in the subarc region. For example, in the periphery of the melt régime, where the extent of melting may be very small (McKenzie 1984, 1985), minor phases, if initially present, may not be exhausted and this could lead to major LIL/HFS element fractionation in the extracted liquids. This process may be amplified if the melts are hydrous, as suggested by Foley & Wheller (1990). However, titanite phases must still exist in the mantle before melting. The minor phases could be inherited from earlier, presubduction lithospheric enrichment events (cf. McKenzie 1989), an apparent prerequisite for accounting for the LIL and HFS contents, and Sr- and Nd-isotope systematics, of continent-based arcs (Rogers & Hawkesworth 1989; Hawkesworth & Ellam 1989). Lithosphere-derived xenoliths, the MARID suite, which actually contain rutile, are found in kimberlite pipes, and were used by Saunders *et al.* (1980) as indirect support for their suggestion that titanite minerals might exist in the subarc wedge. Waters (1987) has subsequently argued that these xenoliths represent mantle material infiltrated by lamproitic magmas; alternatively that some may represent cumulates.

(b) *Titanite stability in the subducting slab*

The initial melting product of the wet, basaltic subducting slab is a silica-rich liquid similar to dacite (Green & Ringwood 1967). The activity coefficient for TiO_2 in such a liquid is higher than that in a basaltic liquid (Watson 1976); consequently TiO_2 concentrations required to saturate the dacitic liquid with respect to a Ti-rich phase will be lower than in the case of basalt produced from a lherzolite (Green & Pearson 1986). Moreover, the TiO_2 concentration in a basaltic source could be higher by a factor of 5–20 than that in a depleted lithosphere harzburgite. The lower melting temperatures also mean that less Ti is required to saturate the accompanying clinopyroxene and/or garnet, leading to higher modal proportions of the residual Ti-rich phase.

In most modern subduction zones, however, temperatures within the downgoing slab are too low to permit melting, although dehydration reactions will occur (Anderson *et al.* 1978; Peacock 1990). At high pressures, K, Rb, Ba and probably La and Ce can dissolve in water-rich fluids, whereas Ti, Nb, Ta and other HFS elements are likely to be highly insoluble (Mysen 1979; Tatsumi *et al.* 1986). This is the situation where the greatest decoupling of LIL from HFS elements will occur, providing the fluid can escape to transfer the LIL elements into the overlying mantle wedge, leaving Nb, Ta and Ti in the basaltic slab. Because most subduction-related basalts, basaltic andesites and boninites cannot easily be produced as direct melts of the slab, material must be transported into the mantle wedge, where it hybridizes with the mantle component (Wyllie & Sekine 1982): the MgO, Ni and Cr and other compatible elements being provided by the mantle wedge, with the slab providing the LIL elements and also the fingerprint of equilibration with a Ti-rich (and possibly other) minor phase.

Melting experiments carried out in basaltic systems confirm that ilmenite, rutile and sphene are stable above the solidus (Allan *et al.* 1975; Thompson 1975; Helz 1973; Hellman & Green 1979). Rutile replaces ilmenite with increasing pressure (Thompson 1975) and in wet melting experiments has been seen to replace sphene with both increasing pressure and temperature, sphene being stable to only 15 kbar (1.5 GPa) (Hellman & Green 1979). The phase ultimately responsible for the retention of Nb and Ta within the slab will then depend on the pressure and temperature experienced when the melt or fluid escapes from the slab. Zr, Hf and P, may be retained in garnet and clinopyroxene at high pressures, or of course in zircon and apatite at lower temperatures if their saturation points are reached.

HFS element retention in the slab cannot, alone, account for the low absolute abundances of these elements found in some primitive island arcs (Tonga: Ewart & Hawkesworth 1987; South Sandwich Islands: figure 3). Green (1972), Hole *et al.* (1984) and Ewart & Hawkesworth (1987) have proposed that the sources of these arc basalts had suffered previous melt extraction. Replenishment by LIL elements from the slab then produces the characteristic high LIL/HFS ratios in these magmas.

6. Fluxes in the subduction zone environment

Arc systems are a paradox in being voluminous zones of magmatism yet being regions where cool lithosphere is being returned to the mantle and isotherms are depressed. Hydrous fluids are clearly vital in lowering the solidus temperatures of slab and/or mantle wedge to enable melting to occur, but it is widely recognized that degrees of melting must be low and that the processes of segregation of these small melt fractions must be efficient. Fluids serve another important function in stabilizing hydrous mineral phases such as hornblende and phlogopite, which may exert significant control on the pathways followed by trace elements passing through the system.

Much fluid is lost from the oceanic lithosphere at relatively high levels in the subduction zone, as porosity is reduced and low-grade minerals dehydrate. However, fluids incorporated in hydrous minerals such as hornblende in the mantle wedge may be transferred down-dip to the site of arc-magma genesis by induced flow (drag) of metasomatized mantle adjacent to the slab. Water trapped in the cooler interior of the slab in amphiboles, or as high pressure forms of serpentine, may survive to be released at greater depths (perhaps to *ca.* 300 km; Ringwood 1990), thus extending the zone through which fluid fluxing and separation of LIL from HFS elements may occur.

There is conclusive proof from the presence of ^{10}Be that subducted young sedimentary material is being incorporated in the source of some arc magmas (Tera *et al.* 1986); Pb-isotopic ratios and the presence of negative Ce anomalies in some arc lavas strongly support a subducted sediment input into the magma source (Hole *et al.* 1984; Barreiro 1983; Woodhead & Fraser 1988); and the high volatile content in undegassed back-arc magmas indicates subduction of hydrous material (Muenow *et al.* 1980). However, most of these authors stress that the proportion of such subducted material involved in the formation of arc magmas is actually very small. For example, Hawkesworth & Ellam (1989) estimate that only about 3% of subducted Sr is required to satisfy the requirement of newly formed crust. Much of the subducted material must therefore bypass the magma-production zone to be carried down into the deep mantle, or it is efficiently removed beneath the forearc region.

The problem of quantifying the chemical flux from the slab is especially difficult in continent-based arcs, where there may be a significant chemical component from the mantle wedge, and where crustal contamination of magmas may be important.

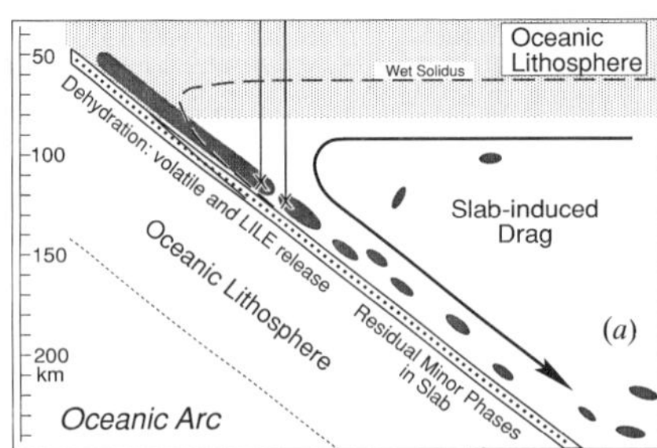

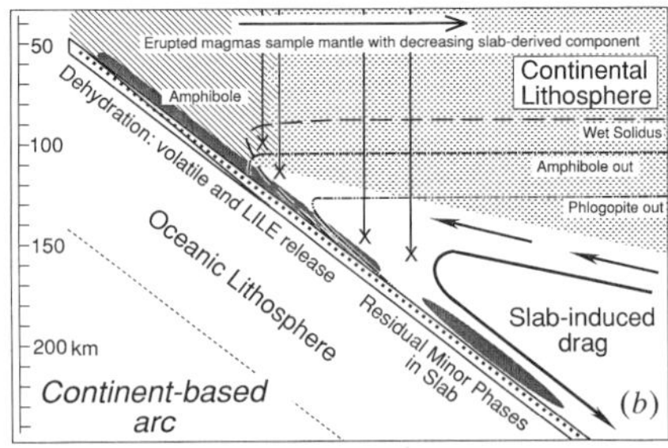

Figure 4. Schematic diagrams illustrating the alternative processes in (a) youthful, oceanic subduction systems, where the arc is underlain by asthenospheric and depleted lithospheric mantle, and where the trace element and isotopic compositions of the arc magmas are dominated by the slab-derived chemical flux; and (b) a mature continent-based arc system, where a wedge-derived flux is dominant over the slab-derived flux. The ultimate origin of the wedge-derived component is unclear; it may be inherited in the form of small-degree melts percolating from the asthenosphere, far removed from any subduction zone; or it may be inherited from ancient subduction zone systems. Light stipple: oceanic lithosphere; intermediate stipple: subcontinental lithosphere; dark stipple: immediate zone of metasomatism of mantle by slab-derived fluids; diagonal ornament: zone of amphibole formation; large crosses; zones of initial magma generation. In either case (a) or (b), fluids must play an important role in fractionating the LIL and HFS elements, either at the slab-mantle interface, or in the mantle wedge.

With primitive oceanic arcs, Ewart & Hawkesworth (1987) estimated that 84% of the Sr in basalts from Tafahi, in the Tonga Island, is from the slab, and Hole *et al.* (1984) estimated that 50% of the Sr in Mariana arc magmas may be slab (oceanic crust plus sediments) derived, the remainder coming from the mantle wedge.

A possible scenario for young oceanic arcs is summarized in figure 4 a. HFS elements are retained in the subducting slab: for Ti, Nb and Ta initially in sphene and, at greater depths, in rutile; Zr, Hf, and P enter zircon and apatite or clinopyroxene and garnet. Dehydration alone will permit mobile LIL elements, but not insoluble HFS elements, to be transported from the slab to the mantle wedge, thus producing the required decoupling from LIL elements with or without minor phases. Slab melting will not decouple the LIL and HFS elements unless residual minor phases are present. The fact that voluminous Archaean crustal compositions have marked negative Nb and Ta anomalies (Weaver & Tarney 1984), and were generated through hydrous melting of a basaltic source, suggests that these minor phases are commonly stable during wet melting.

Fluids from the slab interact with and metasomatize the mantle wedge, which initially may have been depleted by previous melt extraction, and this may then melt under the influence of fluids to form parental arc basalts. The fluids and their associated chemical flux enable growth of hornblende (and at a deeper level, phlogopite: Sekine & Wyllie 1982), but if there is slab-induced convection dragging down the metasomatized mantle, there may be continual pressure-induced breakdown of hornblende, releasing fluids for melting or for further fluxing the wedge, or promoting ascent of low-density diapirs. In intraoceanic arcs the asthenospheric nature of the wedge facilitates slab-induced convection and replenishment of the arc mantle source, although being oceanic it may not have a high content of LIL elements, and continual fluxing at the wedge-slab interface may be required to achieve the progressive LIL element enrichments observed with time in such arcs (cf. Tarney *et al.* 1981; Saunders & Tarney 1984).

Phil. Trans. R. Soc. Lond. A (1991)

Continent-based arcs, on the other hand, are associated with a foundation of thick, variably old, lithospheric mantle (figure 4*b*). Neither the thickness nor the thermal profile are well constrained. Melting of this lithosphere is much more difficult than with the asthenospheric mantle wedge of intraoceanic arcs, and indeed is most likely where extensional decompression permits uprise of asthenosphere into the lithosphere. It may be for this reason that emplacement of calc-alkaline granitoids is normally associated with extensional activity at Cordilleran margins, and with the extensional collapse of collisional orogenic belts. Although crustal contamination does occur, there is now good evidence that the magmas have inherited a significant proportion of their trace element budget from the subcontinental lithosphere. For instance, the systematic correlation between increasing $^{87}Sr/^{86}Sr$ ratio, Sr abundances and decreasing ϵ_{Nd} in arc magmas from northern Chile cannot readily be explained by adding a high Sr/Nd component from the slab (Hawkesworth 1979) as this would produce a convex-upward mixing trend on the $^{87}Sr/^{86}Sr$–ϵ_{Nd} isotope plot, which is not observed. Secondly, arc magmas from northern Chile show a progressive increase in HFS/REE ratio (e.g. Nb/Sm) with time which correlates positively with $^{87}Sr/^{86}Sr$ and negatively with ϵ_{Nd} (Rogers & Hawkesworth 1989). Again, these data are difficult to reconcile with crustal contamination, or input into the source solely from the subducted slab. Note, that this apparent temporal variation in northern Chile is also spatial; the oldest magmas in the west have the lowest $^{87}Sr/^{86}Sr$ ratio. Moreover, Hickey-Vargas *et al.* (1989) and Stern *et al.* (1990) describe similar spatial variations in Quaternary to Recent volcanic rocks from S. Chile. The primary control on these variations may therefore be spatial rather than temporal and reflect mineralogical control and/or increasing involvement of the subcontinental lithosphere.

The consistently lower LIL/Nb ratios of continental arc magmas relative to oceanic ones (figure 3) implies that the magmas have tapped mantle source material with an ocean island basalt (OIB) component. This trace element signature is most unlikely inherited from the slab since, as pointed out by Hickey-Vargas *et al.* (1989), the OIB-like signature becomes pronounced in those parts of the system furthest from the trench. Magmas generated closest to the trench will inherit a greater proportion of slab-derived flux; this is consistent with the lower $^{87}Sr/^{86}Sr$, lower Nb/REE ratios and higher ϵ_{Nd} values in these magmas. Those generated more distally will have a progressively diminished slab-derived component, until they become fully OIB-like, as in the Patagonian basalts of southern S. America (Stern *et al.* 1990).

An important constraint on the lithosphere contribution is that continental margin arcs like the Andes have been producing voluminous magmas continually for several hundred million years, so the supply of trace element and major element components must continually be replenished, and on a massive scale. This is most easily accomplished if subduction does induce convection in the mantle wedge so that subcontinental lithosphere is continually drawn into the melting zone. Because mantle-enrichment processes, some carbonatitic, are continually adding new OIB components to the base of the lithosphere through hot-spot plume activity (Saunders *et al.* 1988; Hawkesworth *et al.* 1990), the replenishment problem is readily satisfied. The trace element and isotopic differences observed between oceanic and continental arcs (Hawkesworth 1979) then reflects the fact that enrichment processes have affected subcontinental lithosphere over a considerably longer period than oceanic lithosphere. Interestingly, many island arc and continental margin volcanics have the isotopic and trace element characteristics of PREMA (prevalent mantle in the terminology of Hart & Zindler (1989)) rather than MORB, which either reflects the mixture of OIB components (EM1, EM2, HIMU) which have been added to the base of the lithosphere, or that PREMA is indeed a prevalent component that is being continually released from the lower mantle (Hart & Zindler 1989) and rises to underplate the lithosphere. The major element differences between the dominantly mafic compositions of primitive arc magmas and the silicic compositions of continental margin arc magmas may be a function of the higher degrees of melting provided by the hotter asthenospheric mantle in oceanic regions.

Finally, because mantle enrichment processes generally occur through veining of the refractory harzburgitic lithosphere by much more fertile carbonatitic fluids or OIB material, this may enable a titanite phase to remain stable in the residue during hydrous melting-out of calc-alkaline magmas. A consequence would be that it is not only the slab itself, but also the lithospheric mantle dragged down by the subducting slab, which carries the positive Nb anomaly that compensates for the negative Nb anomalies in subduction-related magmas (cf. Saunders *et al.* 1988).

7. Summary

Critical points from the above arguments are as follows.

1. Although sediments are subducted in significant volumes at subduction zones, and their presence in the magma source regions can be detected in arc chemistry, the more surprising fact is that the sediment contribution is usually relatively minor. This argues against a major contribution to modern arc volcanism through melting of the slab, though slab melting may occur at the initiation of subduction, and it may have been more common in the early Precambrian.

2. The lithosphere is the dominant material source of subduction magmas, induced convection continually dragging in new lithospheric material into the melting zone. This most easily accounts for the compositional differences between oceanic and continental arcs. OIB components added to the lithosphere play an important role in replenishing the source regions for arc magmatism.

3. Hornblende and phlogopite are important minor phases which control some aspects of the chemistry of subduction zone magmas.

4. Titanite phases, stabilized by the hydrous environment in the slab and probably in the wedge, confer a characteristic chemical signature on all subduction-related magmas. These residual minerals are carried deep into the mantle.

References

Allen, J. C., Boettcher, A. L. & Marland, G. 1975 Amphiboles in andesite and basalt. I. Stability as a function of P-T-fO_2. *Am. Mineral.* **60**, 1069–1085.

Anderson, R. N., Delong, S. E. & Schwartz, W. M. 1978 Thermal model for subduction with dehydration in the downgoing slab. *J. Geol.* **86**, 731–739.

Barreiro, B. A. 1983 Lead isotopic compositions of South Sandwich Island volcanic rocks and their bearing on magmagenesis in intra-oceanic island arcs. *Geochim. cosmochim. Acta* **47**, 817–822.

Boettcher, A. L. & Wyllie, P. J. 1968 Jadeiite stability measured in the presence of silicate liquids in the system $NaAlSiO_4$-SiO_2-H_2O. *Geochim. cosmochim. Acta* **32**, 999–1012.

Carmichael, I. S. E., Nicholls, J. & Smith, A. L. 1970 Silica activity in igneous rocks. *Am. Mineral.* **55**, 246–263.

Defant, M. J. & Drummond, M. S. 1990 Derivation of some modern arc magmas by melting of young subducted lithosphere. *Nature* **347**, 662–665.

Ewart, A. 1982 The mineralogy and petrology of Tertiary–Recent orogenic volcanic rocks: with special reference to the andesitic–basaltic compositional range. In *Andesites* (ed. R. S. Thorpe), pp. 25–95. Chichester: Wiley.

Ewart, A. W. & Hawkesworth, C. J. 1987 The Pleistocene to Recent Tonga-Kermadec arc lavas: interpretation of new isotope and rare earth element data in terms of a depleted source model. *J. Petrol.* **28**, 495–530.

Falloon, T. J. & Green, D. H. 1990 Solidus of carbonated fertile peridotite under fluid-saturated conditions. *Geology* **18**, 195–199.

Foley, S. F. & Wheller, G. E. 1990 Parallels in the origin of the geochemical signature of island arc volcanics and continental potassic igneous rocks: the role of residual titanites. *Chem. Geol.* **85**, 1–18.

Gill, J. B. 1981 *Orogenic andesites and plate tectonics.* Springer-Verlag: Berlin.

Green, D. H. 1972 Magmatic activity as the major process in the chemical evolution of the Earth's crust and mantle. *Tectonophys.* **13**, 47–71.

Green, D. H. 1973 Contrasted melting relationships in a pyrolite upper mantle under mid-ocean ridge, stable crust and island arc environments. *Tectonophys.* **17**, 285–297.

Green, T. H. 1982 Anatexis of mafic crust and high pressure crystallization of andesite. In *Andesites* (ed. R. S. Thorpe), pp. 465–487. Chichester: Wiley.

Green, D. H. & Ringwood, A. E. 1967 An experimental investigation of the gabbro to eclogite transition and its petrological implications. *Geochim. cosmochim Acta* **31**, 767–833.

Green, T. H. & Pearson, N. J. 1986 Ti-rich accessory phase saturation in hydrous mafic–felsic compositions at high P,T. *Chem. Geol.* **54**, 185–201.

Hart, S. R. & Zindler, A. 1989 Constraints on the nature and development of chemical heterogeneities in the mantle. In *Mantle convection* (ed. W. R. Peltier), pp. 261–387. New York: Gordon & Breach.

Hawkesworth, C. J. & Ellam, R. M. 1989 Chemical fluxes and wedge replenishment rates along recent destructive plate margins. *Geology* **17**, 46–49.

Hawkesworth, C. J. 1979 $^{143}Nd/^{144}Nd$, $^{87}Sr/^{86}Sr$ and trace element characteristics of magmas along destructive plate margins. In *Origin of granite batholiths: geochemical evidence* (ed. M. P. Atherton & J. Tarney), pp. 76–89. Nantwich: Shiva.

Hawkesworth, C. J. 1982 Isotopic characteristics of magmas erupted along destructive plate margins. In *Andesites* (ed. R. S. Thorpe), pp. 549–571. Chichester: Wiley.

Hawkesworth, C. J., Kempton, P. D., Rogers, N. W., Ellam, R. M. & van Calsteren, P. W. 1990 Continental mantle lithosphere, and shallow level enrichment processes in the Earth's mantle. *Earth planet. Sci. Lett.* **26**, 256–268.

Hellman, P. L. & Green, T. H. 1979 The role of sphene as an accessory phase in the high-pressure partial melting of hydrous mafic compositions. *Earth planet. Sci. Lett.* **42**, 191–201.

Helz, R. T. 1973 Phase relations of basalts in their melting range at $P_{H_2O} = 5$ kb as a function of oxygen fugacity. *J. Petrol.* **14**, 249–302.

Hickey, R. L., Frey, F. A., Gerlach, D. C. & Lopez-Escobar, L. 1986 Multiple sources for basaltic arc rocks from the southern volcanic zone of the Andes ($34°$–$41°$ S): trace element and isotopic evidence for contributions from subducted ocean crust, mantle and continental crust. *J. geophys. Res.* **91**, 5963–5983.

Hickey-Vargas, R., Moreno-Roa, H., Lopez-Escobar, L. & Frey, F. A. 1989 Geochemical variations in Andean basaltic and silicic lavas from the Villarrica-Lanin volcanic chain ($39.5°$ S): An evaluation of source heterogeneity, fractional crystallization and crustal assimilation. *Contr. Mineral. Petrol.* **103**, 361–386.

Hole, M. J., Saunders, A. D., Marriner, G. F. & Tarney, J. 1984 Subduction of pelagic sediment: implications for the origin of Ce-anomalous basalts from the Mariana Islands. *J. geol. Soc. Lond.* **141**, 453–472.

Irving, A. J. & Frey, F. A. 1984 Trace element abundances in megacrysts and their host basalts: Constraints on partition coefficients and megacryst genesis. *Geochim. cosmochim. Acta* **48**, 1201–1221.

Lambert, I. B. & Wyllie, P. J. 1972 Melting of gabbro (quartz eclogite) with excess water to 35 kilobars with geological applications. *J. Geol.* **80**, 693–720.

Marriner, G. F. & Millward, D. 1984 The petrology and geochemistry of Cretaceous to Recent volcanism in Colombia: the magmatic history of an accretionary plate margin. *J. geol. Soc. Lond.* **141**, 473–486.

Martin, H. 1987 Effect of steeper Archean geothermal gradient on geochemistry of subduction-zone magmas. *Geology* **14**, 753–756.

McKenzie, D. 1984 The generation and compaction of partially molten rock. *J. Petrol.* **25**, 713–765.

McKenzie, D. 1985 The extraction of magma from the crust and mantle. *Earth planet. Sci. Lett.* **74**, 81–91.

McKenzie, D. 1989 Some remarks on the movement of small melt fractions in the mantle. *Earth planet. Sci. Lett.* **95**, 53–72.

Muenow, D. W., Liu, N. W. K., Garcia, M. O. & Saunders, A. D. 1980 Volatiles in submarine volcanic rocks from the spreading axis of the East Scotia Sea back-arc basin. *Earth planet. Sci. Lett.* **47**, 272–278.

Mysen, B. O. 1979 Trace element partitioning between garnet peridotite minerals and water-rich vapour: experimental data from 5 to 30 kbar. *Am. Mineral.* **64**, 274–288.

Mysen, B. O. & Boettcher, A. L. 1975 Melting of a hydrous upper mantle: II. Geochemistry of crystals and liquids formed by anatexis of mantle peridotite at high pressures and high temperatures as a function of controlled activities of water, hydrogen and carbon dioxide. *J. Petrol.* **16**, 549–593.

Nakamura, Y. & Kushiro, I. 1974 Composition of the gas phase in Mg_2SiO_4-SiO_2-H_2O at 15 kbar. *Carnegie Instn Wash., Yearbk* **73**, 25–258.

Peacock, S. M. 1990 Fluid processes in subduction zones. *Science, Wash.* **248**, 329–337.

Pearce, J. A. 1983 Role of sub-continental lithosphere in magma genesis at active continental margins. In *Continental basalts and mantle xenoliths* (ed. C. J. Hawkesworth & M. J. Norry), pp. 230–249. Nantwich: Shiva.

Pearce, J. A. & Norry, M. J. 1979 Petrogenetic implications of Ti, Zr, Y and Nb variations in volcanic rocks. *Contr. Mineral. Petrol.* **69**, 33–47.

Ringwood, A. E. 1990 Slab–mantle interactions. 3. Petrogenesis of intraplate magmas and structure of the upper mantle. *Chem. Geol.* **82**, 187–207.

Rogers, G. & Hawkesworth, C. J. 1989 A geochemical traverse across the north Chilean Andes: evidence for crust generation from the mantle wedge. *Earth planet. Sci. Lett.* **91**, 271–275.

Rogers, G. & Saunders, A. D. 1989 Magnesian andesites from Mexico, Chile and the Aleutian Islands: implications for magmatism associated with ridge-trench collision. In *Boninites and related rocks* (ed. A. J. Crawford), pp. 416–445. London: Unwin Hyman.

Ryerson, F. J. & Watson, E. B. 1986 Rutile saturation in magmas: implications for Ti-Nb-Ta depletion in island arc basalts. *Earth planet. Sci. Lett.* **86**, 225–239.

Saunders, A. D., Tarney, J. & Weaver, S. D. 1980 Transverse geochemical variations across the Antarctic Peninsula: implications for the genesis of calc-alkaline magmas. *Earth planet. Sci. Lett.* **46**, 344–360.

Saunders, A. D., Rogers, G., Marriner, G. F., Terrell, D. J. & Verma, S. P. 1987 Geochemistry of Cenozoic volcanic rocks, Baja California, Mexico: implications for the petrogenesis of post-subduction magmas. *J. Volc. geoth. Res.* **32**, 223–243.

Saunders, A. D., Norry, M. J. & Tarney, J. 1988 Origin of MORB and chemically-depleted mantle reservoirs: trace element constraints. *J. Petrol., Special Issue* 415–455.

Sekine, T. & Wyllie, P. J. 1982 Phase relationships in the system $KAlSiO_4$-Mg_2SiO_4-SiO_2-H_2O as a model for hybridization between hydrous silicate melts and peridotite. *Contr. Mineral. Petrol.* **79**, 368–374.

Stern, C. R., Frey, F. A., Futa, K., Zartman, R. E., Peng, Z. & Kyser, T. K. 1990 Trace-element and Sr, Nd, Pb and O isotopic compositions of Pliocene and Quaternary alkali basalts of the Patagonian Plateau lavas of southernmost South America. *Contr. Mineral. Petrol.* **104**, 294–308.

Sudo, A. & Tatsumi, Y. 1990 Phlogopite and K-amphibole in the upper mantle: implications for magma genesis in subduction zones. *Geophys. Res. Lett.* **17**, 29–32.

Sun, S.-S. 1980 Lead isotopic study of young volcanic rocks from mid-ocean ridges, ocean islands and island arcs. *Phil. Trans. R. Soc. Lond.* A **297**, 409–445.

Sun, S.-S. & McDonough, W. F. 1989 Chemical and isotopic systematics of oceanic basalts: implications for mantle composition and processes. *Geol. Soc. Lond., Spec. Publ.* **42**, 313–345.

Takahashi, E. & Kushiro, I. 1983 Melting of a dry peridotite at high pressures and basaltic magma genesis. *Am. Mineral.* **68**, 859–879.

Tarney, J., Saunders, A. D., Weaver, S. D., Donnellan, N. C. B. & Hendry, G. L. 1979 Minor element geochemistry of basalts from Leg 49, North Atlantic Ocean. *Init. Repts DSDP* **49**, 657–691. Washington, D.C.: U.S. Government Printing Office.

Tarney, J. & Weaver, B. L. 1987 Geochemistry of the Scourian Complex: petrogenesis and tectonic models. In *Evolution of the Lewisian and comparable Precambrian high grade terrains* (ed. R. G. Park & J. Tarney). *Geol. Soc. Lond. Spec. Publ.* **27**, 45–56.

Tarney, J., Wood, D. A., Saunders, A. D., Varet, J. & Cann, J. R. 1979 Nature of mantle heterogeneity in the North Atlantic: evidence from Leg 49. In *Results of Deep Sea Drilling in the Atlantic* (ed. M. Talwani), *Maurice Ewing Series* **2**, 285–301. American Geophysical Union.

Tarney, J., Wood, D. A., Saunders, A. D., Cann, J. R. & Varet, J. 1980 Nature of mantle heterogeneity in the North Atlantic: evidence from deep sea drilling. *Phil. Trans. R. Soc. Lond.* A **297**, 179–202.

Tarney, J., Saunders, A. D., Mattey, D. P., Wood, D. A. & Marsh, N. G. 1981 Geochemical aspects of back-arc spreading in the Scotia Sea and Western Pacific. *Phil. Trans. R. Soc. Lond.* A **300**, 263–285.

Tatsumi, Y., Hamilton, D. L. & Nesbitt, R. W. 1986 Chemical characteristics of fluid phase released from a subducted lithosphere and origin of arc magmas: evidence from high-pressure experiments and natural rocks. *J. Volc. geoth. Res.* **29**, 293–309.

Taylor, S. R. & McLennan, S. M. 1985 *The continental crust: its composition and evolution.* Oxford: Blackwell.

Tera, F., Brown, L., Morris, J., Sacks, I. S., Klein, J. & Middleton, R. 1986 Sediment incorporation in island-arc magmas: inferences from [10]Be. *Geochim. cosmochim. Acta* **50**, 535–550.

Thompson, R. N. 1975 Primary basalts and magma genesis. II. Snake River Plain, Idaho, USA. *Contr. Mineral. Petrol.* **52**, 213–232.

Thompson, R. N., Morrison, M. A., Dickin, A. P. & Hendry, G. L. 1983 Continental flood basalts...arachnids rule OK? In *Continental basalts and mantle xenoliths* (ed. C. J. Hawkesworth & M. J. Norry), pp. 158–185. Nantwich: Shiva.

Thompson, R. N., Morrison, M. A., Hendry, G. L. & Parry, S. J. 1984 An assessment of the relative roles of crust and mantle in magma genesis: an elemental approach. *Phil. Trans. R. Soc. Lond.* A **310**, 549–590.

Thorpe, R. S., Francis, P. W. & O'Callaghan, L. 1984 Relative roles of source composition, fractional crystallisation and crustal contamination in the petrogenesis of Andean volcanic rocks. *Phil. Trans. R. Soc. Lond.* A **310**, 675–692.

Waters, F. G. 1987 A suggested origin of MARID xenoliths in kimberlites by high-pressure crystallization of an ultrapotassic rock such as lamproite. *Contr. Mineral. Petrol.* **95**, 523–533.

Watson, E. B. 1976 Two-liquid partition coefficients: experimental data and geochemical implications. *Contr. Mineral. Petrol.* **56**, 119–134.

Watson, E. B. 1980 Apatite and phorphorus in mantle source regions: and experimental study of apatite/melt equilibria at pressures to 25 kbar. *Earth planet. Sci. Lett.* **51**, 322–335.

Weaver, B. L. & Tarney, J. 1984 Estimating the composition of the continental crust: an empirical approach. *Nature* **310**, 575–577.

Wedepohl, K. H. 1970 *Handbook of geochemistry.* Springer-Verlag: Berlin.

Wendlandt, R. F. & Eggler, D. H. 1980 The origins of potassic magmas. 2. Stability of phlogopite in natural spinel lherzolite and in the system $KAlSiO_4$-MgO-SiO_2-H_2O-CO_2 at high pressures and high temperatures. *Am. J. Sci.* **280**, 421–458.

Wood, D. A., Joron, J.-L. & Treuil, M. 1979*a* A re-appraisal of the use of trace elements to classify and discriminate between magma series erupted in different tectonic settings. *Earth planet. Sci. Lett.* **45**, 326–336.

Wood, D. A., Joron, J.-L., Treuil, M., Norry, M. J. & Tarney, J. 1979*b* Elemental and Sr isotope variations in basic lavas from Iceland and the surrounding ocean floor: the nature of mantle source inhomogeneities. *Contr. Mineral. Petrol.* **70**, 319–339.

Woodhead, J. D. & Fraser, D. G. 1985 Pb, Sr and ^{10}Be isotopic studies of volcanic rocks from the northern Mariana islands. Implications for magma genesis and crustal recycling in the western Pacific. *Geochim. cosmochim. Acta* **49**, 1925–1930.

Wyllie, P. J. & Sekine, T. 1982 The formation of mantle phlogopite in subduction zone hybridization. *Contr. Mineral. Petrol.* **79**, 375–380.

Element fluxes associated with subduction related magmatism

By C. J. Hawkesworth[1], J. M. Hergt[1], R. M. Ellam[2]
and F. McDermott[1]

[1] *Department of Earth Sciences, The Open University, Milton Keynes MK7 6AA, U.K.*
[2] *Scottish Universities Research and Reactor Centre, East Kilbride,
Glasgow G75 0QU, U.K.*

Destructive plate margin magmas may be subdivided into two groups on the basis of their rare earth element (REE) ratios. Most island arc suites have low Ce/Yb, and remarkably restricted isotope ratios of $^{87}Sr/^{86}Sr = 0.7033$, $^{143}Nd/^{144}Nd = 0.51302$, $^{206}Pb/^{204}Pb = 18.76$, $^{207}Pb/^{204}Pb = 15.57$, and $^{208}Pb/^{204}Pb = 38.4$. However, they also have Rb/Sr (0.03), Th/U (2.2) and Ce/Yb (8.5) ratios which are significantly less than accepted estimates for the bulk continental crust. The high Ce/Yb suites have higher incompatible element contents, more restricted heavy REE, and much more variable isotope ratios. Such rocks are found in the Aeolian Islands, Grenada, Indonesia and Philippines, and their isotope and trace element features have been attributed both to contributions from subducted sediment, and/or old trace element enriched material in the mantle wedge. It is argued that for isotope and trace element models the slab component can usefully be taken to consist of subducted sediment and altered mid-ocean ridge basalts, since these may contain *ca.* 80% of the water in the subducted slab, and the distinctive trace element features of arc magmas are generally attributed to the movement of material in hydrous fluids. The isotope data indicate that not more than 15% of the Sr and Th in an average arc magma were derived from subducted material, and that the rest were derived from the mantle wedge. The fluxes of elements which cannot be characterized isotopically are more difficult to constrain, but for most minor and trace elements the slab derived contribution in arc magmas is too small to have a noticeable effect on the residual slab.

1. Introduction

Destructive plate margins are major sites of terrestrial magmatism which have long had a key role in most models for the generation of continental crust and the development of chemical heterogeneities in the upper mantle. However, several studies have recently emphasized that the net flux of material from the mantle to the crust above recent subduction zones is both more basaltic, and it has lower Rb/Sr and Th/U, than most estimates of the bulk composition of the continental crust (Kay & Kay 1986; Kushiro 1987; Ellam & Hawkesworth 1988; Ellam *et al.* 1990). Others have argued that the inferred fractionation in Sm/Nd between the mantle and the crust cannot be achieved in this tectonic setting, and that such fractionation primarily reflects the addition of small degree melts, generated in within plate environments, to the continental crust (O'Nions & McKenzie 1989). A high priority is therefore to establish the size and nature of the element fluxes between the mantle and the crust in different tectonic settings, and destructive plate margins retain a unique position because that is where new continental crust is generated, and where crustal material is recycled into the upper mantle. This contribution reviews selected data on subduction related magmas, and what are early tentative steps towards evaluating the fluxes of different elements in magmas generated along destructive plate margins (after Kay 1980). Such fluxes in turn constrain the composition of material returned to the deep mantle, and hence the likely contribution of subducted crust in the generation of oceanic basalts.

Radiogenic isotopes are sensitive tracers for the different materials which may contribute in the generation of magmatic rocks, and in the last few years there have been a number of detailed geochemical and isotope studies on subduction related rocks. Several minor and trace element features of subduction related rocks remain consistently different from those of mid-ocean ridge basalts (MORB) and ocean-island basalts (OIB), and in particular they are characterized by high (large ion lithophile)/(high field strength elements) (LIL/HFS) element ratios, and relatively low Nb, Ta, and perhaps Ti abundances (Pearce 1982). In contrast, the majority of destructive plate margin rocks have Nd, Sr and Pb isotope ratios which are broadly similar to those of OIB (Morris & Hart 1983), and to those MORB which have more enriched isotope signatures (higher $^{87}Sr/^{86}Sr$ and lower $^{143}Nd/^{144}Nd$). None the less, some subduction related rocks exhibit steep arrays on Pb isotope diagrams (Kay *et al.* 1978; Woodhead & Fraser 1985; White & Dupré 1986), consistent with the introduction of radiogenic Pb from subducted sediments, and more recently it has been demonstrated that young arc rocks often exhibit high ^{10}Be which requires a contribution from young (less than 5 Ma) sedimentary material, presumably in the subducted slab (Tera *et al.* 1986; Morris *et al.* 1990). U/Th ratios are highly fractionated in both altered MORB and in sedimentary carbonates, and so contributions from each material in young arc rocks should also be readily detected by Th isotope studies (Gill & Williams 1990; McDermott & Hawkesworth 1991). In general, however, the isotope data on subduction related rocks indicate that the contributions from sediments and altered oceanic crust in the subducted slab are typically much less than estimates of the 'subduction component' calculated on the basis of minor and trace element variations (Pearce 1983). It is this discrepancy which lies at the heart of recent debates on the processes responsible for the isotope and trace element signatures of subduction related magmas, and which is critical in any attempt to estimate element fluxes in this tectonic environment.

2. Minor and trace element variations

(a) *Rare earth elements*

The rare earth elements (REE) are widely used in models of petrogenesis and earth evolution, because they are a geochemically coherent group and they include the radioactive decay scheme of ^{147}Sm to ^{143}Nd. Figure 1 summarizes Ce and Yb variations in destructive plate margin rocks, and the data broadly fall into two groups: one in which Ce and Yb vary together, and a second group characterized by much higher Ce contents at similar Yb. Such differences require major differences in either the REE profiles of the source rocks for the two groups, and/or in the bulk distribution coefficients and the degree of melting. Rocks in the high Ce/Yb group have consistently higher $^{87}Sr/^{86}Sr$ ratios (figure 1*b*), and so the simplest interpretation

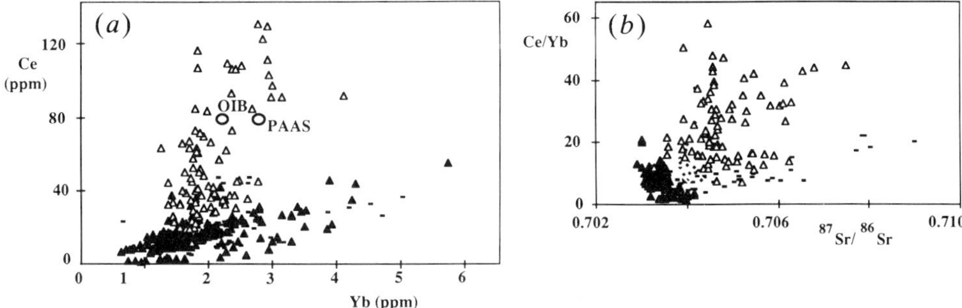

Figure 1. (*a*) Ce and Yb contents of destructive margin magmas with not more than 62 % SiO_2, compared with those for average ocean island basalt (OIB) from Sun & McDonough (1989), and post-Archaean average shale (PAAS, Taylor & McLennan 1989). The symbols illustrate that the arc rocks tend to fall into low and high Ce/Yb groups: △, Aeolian Islands, Grenada, Indonesia, and the Philippines; ▲, Aleutians, Manam Island, Marianas, New Britain, S. Andes, S. Sandwich Islands, Tonga–Kermadecs; –, Lesser Antilles, apart from Grenada. (*b*) Ce/Yb vs. $^{87}Sr/^{86}Sr$ for arc magmas, illustrating the restricted range in $^{87}Sr/^{86}Sr$ in the low Ce/Yb suites, and the much greater range in $^{87}Sr/^{86}Sr$ in the high Ce/Yb rocks. The symbols are as in (*a*).

is that they contain at least a contribution from material which was both old (i.e. old enough to have developed different isotope ratios), and light REE enriched. Such features have been attributed both to subducted sediments and to trace element enriched source regions in the mantle wedge.

The paradox of subduction related magmas is that the classic island arc suites, which appear to represent widespread and relatively simple products of destructive plate margin magmatism, do not have some of the key trace element features of the bulk continental crust. In the data base used here these suites are represented by rocks from Tonka-Kermadec, the Marianas, Manam Island, New Britain, the Aleutians, the South Sandwich Islands, and the Northern Lesser Antilles; they tend to have relatively restricted radiogenic isotope ratios (figures 1*b* and 3), and both Ce and Yb vary together (figure 1*a*). However, they have average Rb/Sr (0.03), Th/U (2.15), and Ce/Yb (8.5) which are all less than most estimates of the average continental crust (0.12, 3.8 and 15, respectively, Taylor & McLennan 1985). Note also that estimates of such element ratios in the bulk crust are consistent with the observed isotope variations in both crustal and mantle rocks widely used in Earth evolution models (Allègre 1982).

In general, the rocks with higher Ce/Yb in figure 1*a* also have higher Rb/Sr and Th/U ratios. Thus the destructive margin rocks which have trace element ratios more typical of the bulk crust are those which are slightly unusual in that they have initial isotope ratios indicating that they already contain trace element enriched material. If such material is subducted sediment, there is clearly some danger of concluding that subduction related magmas only have the trace element ratios of continental crust when they contain a significant contribution from pre-existing crust, in the form of subducted sediment. In some areas the high Ce/Yb ratios may reflect old trace element enriched source regions in the mantle wedge, and in that model the element ratios which are similar to those in the continental crust were set up by the introduction of small degree melts to the mantle wedge before the onset of subduction.

In summary, the low Ce/Yb group appears to be typical of the simpler island arc systems, and that is the one we focus on in seeking to evaluate the element fluxes in

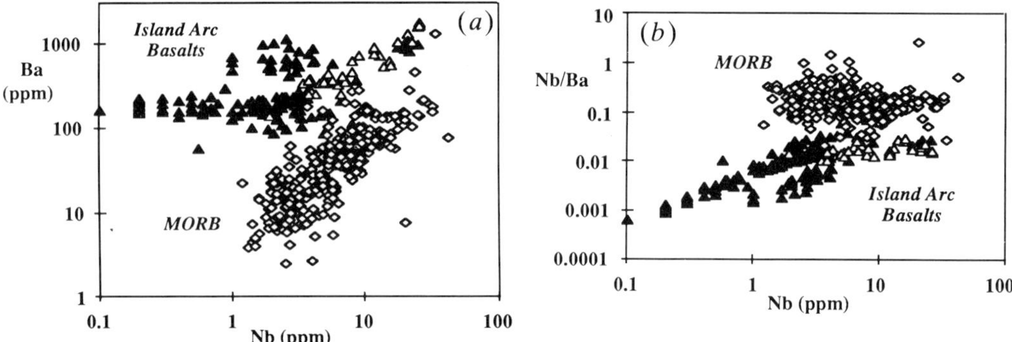

Figure 2. Ba, Nb, and Nb/Ba variations in destructive plate margin and MOR basalts. All samples have not more than 53 % SiO_2: ▲, △, island arc rocks with Ce/Yb $\leqslant$ 15 and $\geqslant$ 15, respectively (see figure 1); ◇, samples of MORB.

subduction zone magmas. The higher Ce/Yb rocks are often developed in more complex tectonic areas, and their trace element signatures are considered elsewhere (Hawkesworth *et al.* 1991).

(b) LIL and HFS elements

Figure 2 contrasts the variation of a typical LIL (Ba) and HFS (Nb) element in MOR and subduction related basalts. Any discussion of LIL/HFS element fractionation processes in the upper mantle requires rocks which have been relatively little affected by low pressure fractionation of Ti oxides, and yet Ti oxides start to crystallize at significantly different SiO_2 contents in tholeiitic and calc alkaline suites. All the samples selected to evaluate LIL/HFSE ratios have less than 53 % SiO_2, and consideration of the individual suites indicates that while some of them show the effects of Ti oxide fractionation the effects are small compared with the overall trends.

On the plot of Ba against Nb the MOR basalts define a positive array, consistent with suggestions that Ba and Nb have similar bulk distribution coefficients during partial melting in the upper mantle. The subduction related basalts are displaced to high Ba/Nb ratios, and their field has a much shallower slope than that for MORB (figure 2a). It is noticeable that Ba is relatively constant at low Nb contents, and this could only be ascribed to partial melting if the bulk D for Ba is *ca.* 1 during melting above subduction zones. This is regarded as unlikely, and so the preferred interpretation is that the almost constant Ba abundances reflect the addition of Ba-rich material associated with subduction. Moreover, the amount of Ba which has been added would appear to have been relatively constant at these different destructive plate margins.

A consequence of the relatively constant Ba contents in low Nb arc rocks is that their distinctive high LIL/HFS element ratios (in this case high Ba/Nb) are best developed in rocks with low HFS element abundances (figure 2b). Secondly, the proportions of an element from subducted material and the mantle wedge depend on how much its abundance varies in the wedge, as well on the size of the slab derived flux. Ba can be much more depleted in the wedge (see the MORB array in figure 2) than, for example, Sr, at least while clinopyroxene is present. Thus the ratio of subduction related Sr, to mantle wedge derived Sr, in arc rocks will tend to be less than the same ratio for Ba.

In summary, it is useful to identify three ways in which elements and isotopes appear to behave in subduction related rocks.

(i) Some elements are thought to be derived entirely from the mantle wedge, and therefore to reflect partial melting processes and the pre-subduction trace element contents of the mantle wedge (e.g. Nb, Ta, Ti).

(ii) There is then a group of predominantly LIL elements which exhibit relatively high abundances in subduction related rocks, but the resulting high LIL/HFSE ratios are best developed in rocks with low HFS element contents (i.e. Ba, K, Th, Sr).

(iii) There are also a number of tracers which appear to be so sensitive to the contribution from the subducted slab, that their abundances in arc rocks may be largely independent of the contribution from the mantle wedge, for example, [10]Be, [207]Pb, and perhaps highly incompatible trace elements, such as B (Morris *et al.* 1990; White & Dupré 1986).

A key point to emerge from such considerations of minor and trace elements in subduction related rocks is that the relative contributions of subducted material and the mantle wedge are very different for different elements. Thus in most models Nb and Ti, for example, are derived wholly from the mantle wedge, whereas Ba and B may be largely derived from the introduction of material which is related to subduction. In between there are a number of elements in which the inferred contribution from subducted material and the mantle wedge may vary significantly.

3. Radiogenic isotopes

Subduction related magmas may contain contributions from fresh and hydrothermally altered oceanic crust, subducted sediments, and variably enriched or depleted material in the mantle wedge. These tend to exhibit at least some distinctive parent/daughter trace element ratios, and so with time they will be characterized by distinctive radiogenic isotope signatures (table 1). Moreover, radiogenic isotopes are generally regarded as highly sensitive tracers in the identification of different source components in magmatic rocks. None the less, the general observation is that for Sr, Nd, Pb, and Th isotopes, the isotope ratios measured in low Ce/Yb subduction related rocks are surprisingly similar to those in ocean island basalts (Morris & Hart 1983).

(a) Sr, Nd *and* Pb *isotopes*

Figure 3 summarizes much of the available Nd and Sr isotope data on subduction related rocks, with different symbols for the high and low Ce/Yb suites identified in figure 1. The low Ce/Yb suites have remarkably restricted ^{87}Sr/^{86}Sr and ^{143}Nd/^{144}Nd with average values of 0.7033 ($\pm$0.0002) and 0.51302 ($\pm$0.00004) (one s.d.), and most of the observed isotope variations are in the high Ce/Yb rock suites of Grenada, the Aeolian Islands, Philippines and Indonesia. Subduction related rocks are characterized by high Sr/Nd ratios which suggests that material derived from the subducted oceanic crust has high Sr/Nd (Pearce 1983; Ellam & Hawkesworth 1988). Mixing between such high Sr/Nd material and a lower Sr/Nd end-member from, for example, the mantle wedge should result in curved mixing lines on figure 3, and in many suites these have not been observed. One exception is in the rocks from Tonga where ^{87}Sr/^{86}Sr does vary with LIL/HFS element ratios, and which suggests ^{87}Sr/^{86}Sr in the high Sr/Nd component in that area is 0.7042 (Ewart & Hawkesworth 1987).

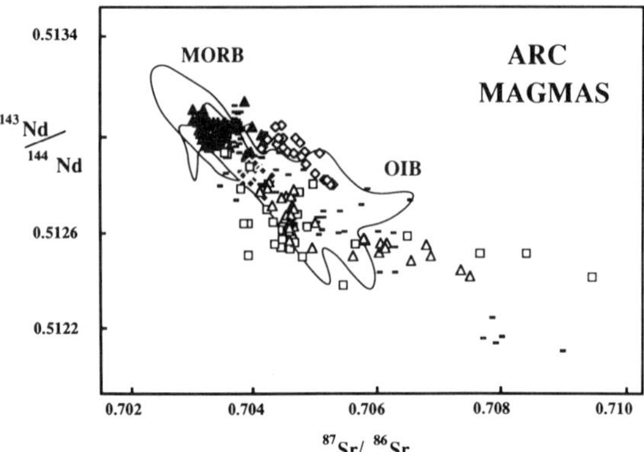

Figure 3. Nd and Sr isotope ratios in destructive margin rocks, compared with the fields for MORB and OIB, after Zindler & Hart (1986). The samples denoted by filled triangles, horizontal bars, and the filled diamonds are as for figure 1; and the open symbols are for the high Ce/Yb arc rocks: □, the Philippines; △, Aeolian Islands; ◇, Grenada.

Table 1.

	average arc basalt	melt from depleted mantle	PAWMS	PATS	altered MORB	altered MORB and sediment[b]	slab contribution[d]
Rb	9.7	0.34	3.6	160	9.04	10.4	13
Ba	183	3.79	1338	650	16.85	76	56
Th	0.78	0.07	0.23	14.6	0.072	0.22	0.08
U	0.36	0.03	0.05	3.1	0.321	0.338	0.09
Nb	1.4	1.4	1.25	19	2.8	2.9	0
La	5.1	1.5	25.8	38	3.3	4.5	1.7
Ce	11.9	4.5	9.6	80	12.7	13.3	5.0
Pb	2.0	0.18	10	20	0.3	0.89	0.62
Sr	407	54	500[a]	200	118	134	63
Nd	8.7	4.4	19.3	32	6.8	7.5	2.8
Zr	46	45	21.6	210	100	98	14
Y	18	17	46.3	27	42	42	12
Yb	1.5	1.8	5.6	2.8	3.8	3.8	0.9
H$_2$O		0.17	(6.3)	(6.3)	2.7		
^{87}Sr/^{86}Sr	0.7034	0.7028	0.710[a]	0.717	0.7046	0.7056	
^{143}Nd/^{144}Nd	0.51302	0.51300	0.5123	0.5123	0.51308	0.51303	
^{206}Pb/^{204}Pb	18.76	18.46	18.7	18.9	18.5	18.68	
^{207}Pb/^{204}Pb	15.57	15.49	15.7	15.7	15.5	15.64	
^{208}Pb/^{204}Pb	38.37	38.0	38.9	38.9		38.59	
(^{230}Th/^{232}Th)	1.3	1.3	0.66	0.6	13	4.4[c]	

[a] Values modified in the light of other sediment data (Ben Othman *et al.* 1989).

[b] Calculated as 1 part PATS, 4 parts PAWMS, and 95 parts altered MORB.

[c] Assuming U/Th fractionation does not occur until the depth of magma generation.

[d] Calculated contribution from subducted sediment and altered MORB, assuming that the relative mobilities depend on ionic radius/charge (after Tatsumi *et al.* 1986).

The Pb contents of oceanic sediments are much higher than those in basic magmas, and they tend to have relatively high ^{207}Pb/^{204}Pb (Ben Othman *et al.* 1989). Thus Pb isotopes have been widely used to evaluate contributions from subducted sediments in arc magmas (Kay *et al.* 1978; Woodhead & Fraser 1985; White & Dupré 1986).

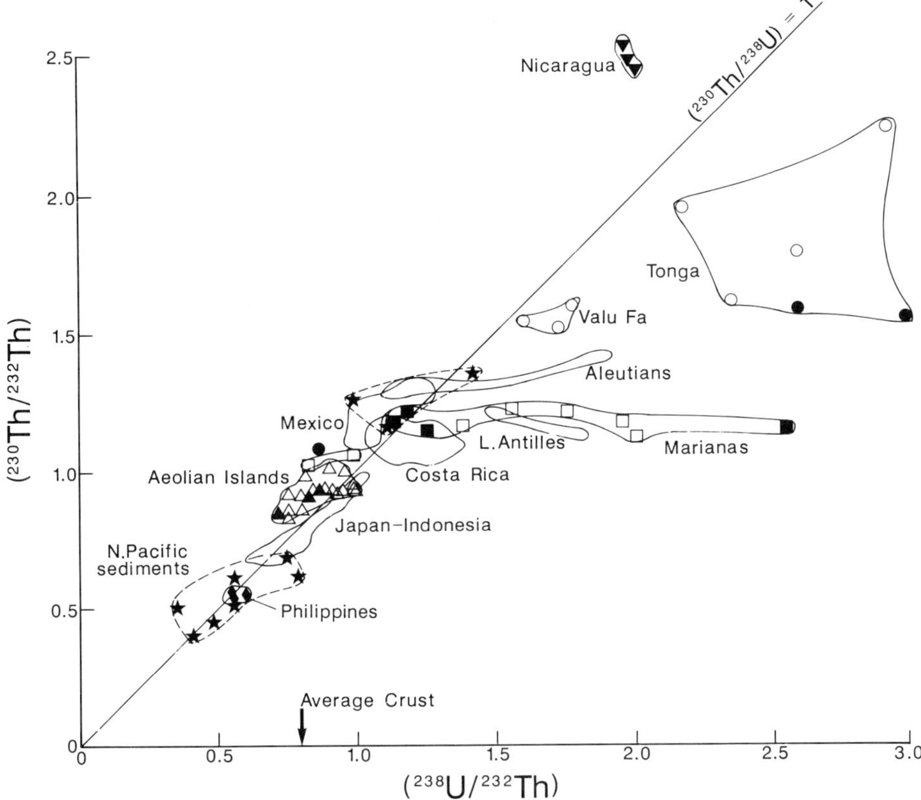

Figure 4. A (^{230}Th/^{232}Th)–(^{238}U/^{232}Th) diagram summarizing the available results for subduction related magmas and N. Pacific sediments (after Gill & Williams 1990; McDermott & Hawkesworth 1991). The overall range in (^{230}Th/^{232}Th) is similar to that in MORB and OIB, but destructive margin rocks are unusual in that some of them are displaced to relatively high (^{238}U/^{230}Th). Note that the majority of arc rocks have (^{238}U/^{232}Th) values which are significantly higher than most estimates of the bulk continental crust (Taylor & McLennan 1985). The dashed fields enclosing the 'star' symbols are N. Pacific sediments.

However, the marked Pb isotope variations are again largely confined to the high Ce/Yb suites, and the lower Ce/Yb rocks have a restricted range and an average Pb isotope composition of ^{206}Pb/^{204}Pb = 18.76 (± 0.13), ^{207}Pb/^{204}Pb = 15.57 (± 0.02), and ^{208}Pb/^{204}Pb = 38.4 (± 0.18). Zindler & Hart (1986) noted the high frequency of Nd and Sr isotope ratios in intraplate oceanic basalts at about 0.5130 and 0.7033, and suggested that they reflected the existence of a mantle component which they termed 'prevalent mantle' (PREMA). Given the diversity of materials which may contribute to arc magmatism, it is surprising to note that the average of the low Ce/Yb arc rocks is indistinguishable from PREMA for Nd and Sr isotopes, and similar for Pb (PREMA has Pb isotope ratios of *ca.* 18.3, 15.46, and 37.9).

(b) Th isotopes

^{230}Th is generated within the natural decay chain from ^{238}U to ^{206}Pb, and it has a half-life of 75200 years. In secular equilibrium (^{230}Th) = (^{238}U) (the parentheses denote activities), and equilibrium is restored within *ca.* 300000 years of any change in Th/U. The combination of Th and Pb isotopes therefore offers a powerful new

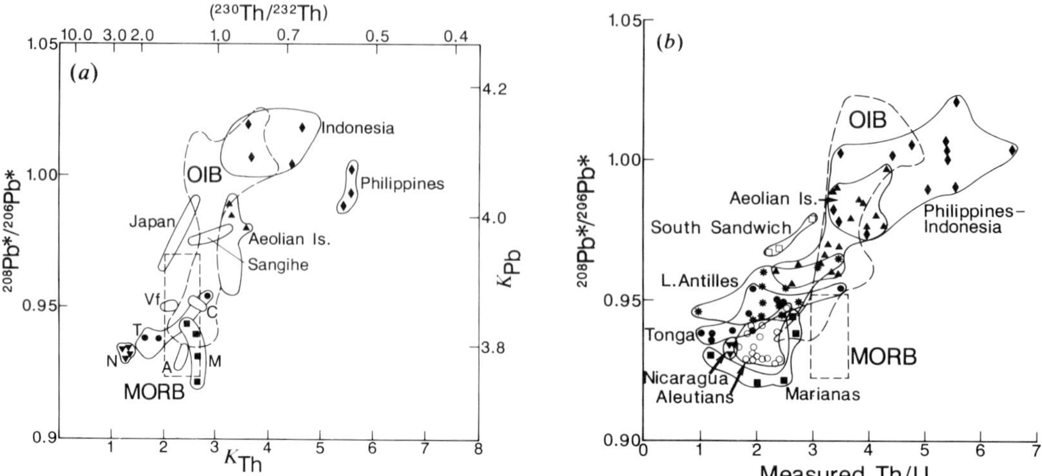

Figure 5. (a) ^{208}Pb*/^{206}Pb* against Th/U calculated from (^{230}Th/^{232}Th), which is taken to represent Th/U in the source of the magmas before any recent fractionation of Th/U. ^{208}Pb*/^{206}Pb* is the ratio of ^{208}Pb/^{204}Pb to ^{206}Pb/^{204}Pb relative to Canyon Diablo troilite (after Allègre *et al.* 1986), and so it reflects Th/U which have persisted for a minimum of several 100 Ma. Data symbols as in figure 4, N, Nicaragua; T, Tonga; A, Aleutians; M, Marianas; Vf, Valu fa; C, Costa Rica. (b) ^{208}Pb*/^{206}Pb* against measured Th/U ratios in the basalts. In both diagrams there is a striking positive correlation between ^{208}Pb*/^{206}Pb* and Th/U, and particularly in (a) the arc rocks plot in a similar array to MORB and OIB (McDermott & Hawkesworth 1991).

approach to investigate changes in Th/U over time scales ranging from tens of thousands to billions of years (Condomines *et al.* 1988; Gill & Williams 1990; McDermott & Hawkesworth 1991). Moreover, they are particularly suited to the study of arc magmas because Th/U ratios are highly fractionated in both altered MORB (Hart & Staudigel 1989) and marine carbonates (Chen *et al.* 1986) which are presumably present in the subducted oceanic crust.

On a (^{230}Th/^{232}Th)–(^{238}U/^{232}Th) diagram samples in secular equilibrium plot on the equiline (figure 4). Many young MORB and OIB are displaced to the left of the equiline, indicating that U/Th in the liquid is less than that in the source during partial melting in the upper mantle (Condomines *et al.* 1988). Some island arc rocks are unusual in that they plot to the right of the equiline, and this has been regarded as a feature of subduction related magmatism (Allègre & Condomines 1982). In more detail it is clear that the high (^{230}Th/^{232}Th) values, and any significant displacement to high U/Th, occur in the low Ce/Yb arc suites (figure 1). The high Ce/Yb rocks have lower U/Th, similar to that of the bulk crust, and more of them are in secular equilibrium. The range in Th isotopes in arc rocks is similar to that in MORB and OIB (Gill & Williams 1990; McDermott & Hawkesworth 1991), and this contrasts with the very high values of (^{230}Th/^{232}Th) $\geqslant$ 10 which may be inferred for altered MORB and marine carbonates from their measured U/Th ratios (Hart & Staudigel 1989).

Pb isotopes, expressed as ^{208}Pb*/^{206}Pb* because that reflects variations in Th/U on the time scale of 100 s of Ma (Allègre *et al.* 1986), correlate with both measured Th/U and κ_{Th} calculated from (^{230}Th/^{232}Th) (figure 5). Thus the major variation in Th/U in subduction related rocks is between different arc systems, it broadly correlates with Ce/Yb, and it has been present for long enough to affect the Pb isotope ratios. As indicated above, such relatively long-lived chemical differences may either be present in the mantle wedge, and/or reflect contributions from sediments with old source ages, but the key point is that these old variations in Th/U are both larger, and they appear to reflect different processes, than the displacement to low Th/U (high $^{238}U/^{232}Th$) observed in some arc rocks (figure 4).

Finally in this section, we note that, as with Nd and Sr isotopes (figure 3), the Th and Pb isotope ratios of many arc rocks plot within the mantle array defined by MORB and OIB (figure 5). This similarity between many of the isotope ratios of subduction related rocks, and those of MORB and OIB, appears to be a general feature, and it suggests either (i) the various components which might contribute to the generation of destructive margin magmas mix together in such proportions that the resulting isotope ratios are fortuitously similar to those in OIB; (ii) ocean island basalts contain a significant contribution from subducted oceanic crust recycled at subduction zones, and so it is to be expected that subduction related and ocean island basalts have similar isotope signatures; or (iii) the contribution of subducted materials in destructive plate margin magmas is much less than that implied by models in which the 'excess' LILE contents are derived from the subducted slab, and instead the ocean island-like isotope signature of many destructive margin magmas is a feature of the mantle wedge. The latter in turn implies that some of the isotope characteristics of ocean island basalts are also derived from relatively shallow levels in the Earth's mantle.

4. Discussion

Several studies have sought to estimate the relative contributions of the mantle wedge and subducted slab in the generation of new crust, and it is now accepted that those relative contributions are different for different elements. In his pioneering work Kay (1980) argued that the excess K responsible for the high K/REE ratios in arc magmas was derived from subducted material, and that in the case of K as much as 90% was probably of continental origin via the oceans. Pearce (1983) extended that approach and proposed that the excess LIL elements responsible for the high LIL/Nb ratios in arc rocks were from the subduction zone, and he estimated that 70–90% of the Sr, Th, Ba, Rb and K in a typical Chilean basalt could have been derived from subducted material. Such calculations are discussed further below (see figure 7), but the problem is very simply how these high values for the slab contribution can be reconciled with the radiogenic isotope data which for the most part is similar to those from many OIB (e.g. figures 3 and 5). One solution was to argue that the slab derived flux beneath different arcs was relatively constant, and that it was best estimated in areas where the mantle wedge was highly depleted (Hawkesworth & Ellam 1989). In such areas all the measured LILE contents of the arc magmas might be inferred to have been derived from the slab, and the relative contribution of such a flux in an estimate of average new crust was sufficiently small (not more than 25%) that it could be reconciled with the available isotope data.

The isotope ratios of different materials that are likely to be in the mantle wedge and the subducted slab are now sufficiently well known that in principle isotope-trace-element diagrams can be used to evaluate the sources of different elements in arc magmas. Subducted related magmas are characterized by high Sr/Nd and Ba/Nb ratios and, as illustrated for Ba–Nb (figure 2), the high ratios are best developed in low Nd and Nb rocks. A graph of $^{87}Sr/^{86}Sr$ against Nd/Sr (figure 6a) demonstrates that the distinctive high Sr/Nd component in arc magmas tends to have low $^{87}Sr/^{86}Sr$ of *ca.* 0.704, and that the high $^{87}Sr/^{86}Sr$ ratios of some arc rocks are associated with low Sr/Nd. Similarly, Pb isotope ratios expressed as $\Delta7/4$ (Hart

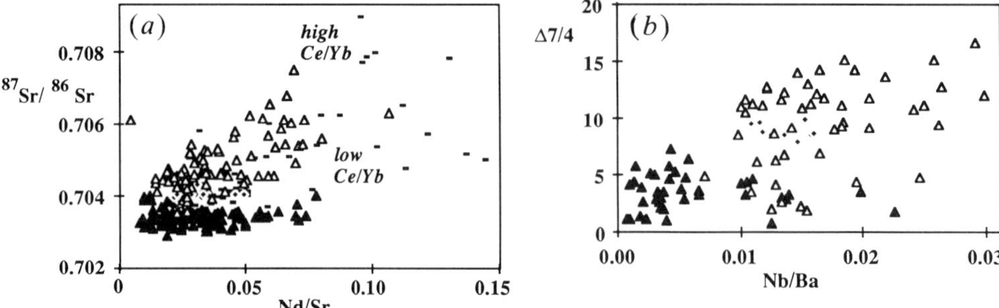

Figure 6. (a) ^{87}Sr/^{86}Sr against Nd/Sr, and (b) Δ7/4 vs. Nb/Ba, for arc rocks with not more than 62 % SiO$_2$. The symbols are as for figure 1, and Δ7/4 is the size of the displacement in ^{207}Pb/^{204}Pb above the Northern Hemisphere Reference Line for oceanic basalts, after Hart (1984). The rocks with the higher LIL/HFS element ratios (i.e. low Nd/Sr and Nb/Ba) have the less radiogenic Sr and Pb isotope ratios.

1984) have been widely regarded as sensitive indicators of contributions from subducted sediments, but the rocks with high Δ7/4 values which might suggest a larger sediment contribution tend to have low rather than high Ba/Nb (figure 6b). Thus in both these examples the more extreme isotope signatures are *not* associated with the distinctive high LIL/HFSE signature of the arc rocks, but instead the latter appears to have relatively unradiogenic Sr and Pb isotope ratios. For Pb isotopes this conclusion is particularly surprising, because the low Δ7/4 and Nb/Ba occur in the rocks with the lowest incompatible element contents, and they should therefore be most sensitive to any contribution from subducted sediment.

The proportion of different rock types in the subducted slab is difficult to constrain, but a reasonable estimate may be 20 % altered MORB, 1 % sediment, with the rest being unaltered MORB and gabbro. For isotope and trace element ratios the distinctive material is the altered MORB and sediments, and it is most unlikely that contributions from unaltered MORB in the subducted slab and MORB-type mantle in the overlying wedge can be readily distinguished. Unaltered MORB contains relatively little H$_2$O, and since the distinctive LIL/HFS element ratios of arc magmas appear to be linked to the release of hydrous fluids (Tatsumi *et al.* 1986), the slab contribution in arc magmas will be dominated by those from altered MORB and sediment, which contain *ca.* 80 % of the H$_2$O in the subducted oceanic crust (table 1). Thus, for the purposes of this discussion we shall simply investigate the contribution of altered MORB and subducted sediment in arc magmas.

Table 1 summarizes the isotope ratios and trace element abundances in altered MORB, subducted sediment, and an average arc basalt calculated for those rocks with $\leqslant 3$ p.p.m. Nb, and/or $\leqslant 65$ p.p.m. Zr. The estimated isotope ratios of the subduction component were used to calculate the maximum contribution from such material in the average arc magma, assuming that the isotope ratios of the arc magma reflect mixing between the subduction component and a depleted MORB-type magma. Some isotope ratios are more sensitive to the size of the slab contribution than others, because that depends on the differences between the isotope ratios and the element abundances in the slab component, the depleted mantle and the arc magma. Thus the slab contributions calculated from Nd and even Pb isotopes, are much less well constrained than those for Sr and Th isotopes, which are 15 and 5 % respectively.

Phil. Trans. R. Soc. Lond. A (1991)

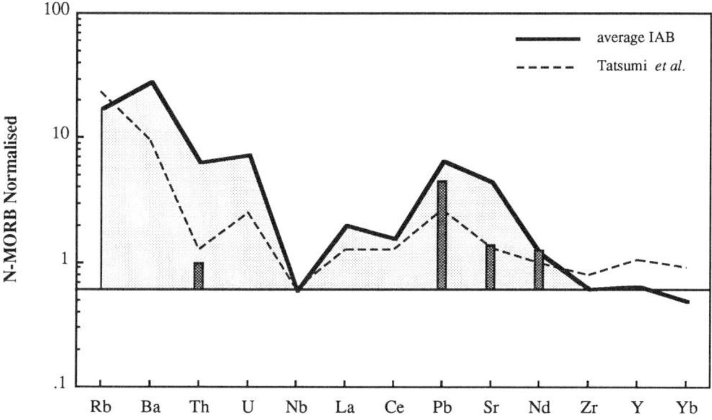

Figure 7. A MORB-normalized trace element diagram illustrating the composition of the average low Nb, low Ce/Yb arc basalt in table 1, and the maximum contribution of Th, Pb, Sr and Nd from altered MORB and sediment in the subducted slab, calculated from the isotope data. The shaded area is the slab contribution calculated by the method of Pearce (1983). The dashed line is the estimated slab contribution, from altered MORB and subducted sediment (table 1), which is consistent with both the isotope data and the relative mobilities of different trace elements during dehydration (Tatsumi *et al.* 1986).

Figure 7 presents the MORB-normalized trace element pattern of the average low Nb, low Ce/Yb arc magma in table 1. The trace element models suggests that the MORB-normalized element contents which are greater than that for Nb, Ta, and probably Zr and Ti, i.e. the shaded area in figure 7, are derived from subducted material (Pearce 1983). This indicates that 98% of the Ba, 90% of the Th, and Pb, 85% of the Sr, and 50% of the Nd in the average arc magma are derived from the slab, and yet for some elements these figures are much higher than the maximum slab contribution calculated from the isotope data (see also figure 7). One way to resolve the discrepancy might be to argue that the rest of the slab derived contribution inferred from the trace elements was from unaltered MORB, since this wasn't considered in the isotope calculations. However, for Th that would require unaltered MORB, which neither contains much water, nor presumably has any slab derived hydrous fluid fluxing through it, to be the source of *ca.* 90% of the Th in the arc magma.

Tatsumi *et al.* (1986) published some experimental results which indicate that the mobility of minor and trace elements during dehydration of serpentine depends largely on their ionic radios, although the conclusions drawn from our use of their data are not changed if we use the ratio of ionic radius to charge. Significantly, the differences in the ionic radii predict that Th will be less readily mobilized from the subducted slab during dehydration than Nd, and that Sr and Pb are more readily mobilized, in a way that is entirely consistent with the maximum slab contributions estimated from the isotope data (figure 7). Thus, the relationship between ionic radius/charge and mobility described by Tatsumi *et al.* (1986) can be used to calculate the possible contribution from subducted sediment and altered MORB (as in table 1) to the trace element budget of the arc magma, and that is illustrated in figure 7. The results are in reasonable agreement with the maximum slab contributions estimated from the isotope data, but the relative concentrations of Rb, Ba and K, for example, are much higher than those observed in the more depleted arcs, such as Tonga.

5. Summary

The radiogenic isotope ratios of the typical low Ce/Yb arc suites are very restricted relative to the range observed in oceanic basalts, and sediment and altered MORB in the subducted oceanic crust. It is argued that contributions from altered MORB and sediment will dominate the slab derived flux in arc magmas, and that the isotope data suggests that this represents not more than 15% of the Sr and the Th in an average arc magma. Such values are very much less than the estimates from trace element data alone, typically 85–90% respectively, but they are broadly consistent with earlier suggestions that the typical slab derived flux is best estimated in areas where the mantle wedge was highly depleted in incompatible trace elements before subduction (Hawkesworth & Ellam 1989). It is much more difficult to constrain the slab contribution for trace elements which cannot be characterized isotopically, and for the LIL elements significantly different results are obtained using the experimental data of Tatsumi *et al.* (1986), or observing the LIL element abundances in depleted arcs (Hawkesworth & Ellam 1989). None the less, both estimates require rather small proportions of the trace element inventory of the subducted lithosphere to be implicated in the production of the IAB trace element signature. Consequently, residual slab material subducted beyond the arc environment, and often invoked as OIB-source material, is relatively trace element rich and only marginally affected by removal of the slab flux, preserving more or less MORB-like incompatible trace element ratios.

We thank Julie Morris, Nick Rogers and Anthony Ware for their comments on earlier versions of the type, and many stimulating discussions on the problems of arc magmatism. The diagrams were prepared by Andrew Lloyd and the typescript by Janet Dryden.

References

Allègre, C. J. & Condomines, M. 1982 Basalt genesis and mantle structure studied through Th isotope geochemistry. *Nature* **299**, 21–24.

Allègre, C. J. 1982 Chemical geodynamics. *Tectonophysics* **81**, 109–132.

Allègre, C. J., Dupré, B. & Lewin, E. 1986 Thorium/uranium ratio of the Earth. *Chem. Geol.* **56**, 219–227.

Ben Othman, D., White, W. M. & Patchett, J. 1989 The geochemistry of marine sediments, island arc magma genesis, and crust-mantle recycling. *Earth planet. Sci. Lett.* **94**, 1–21.

Chen, J. N., Edwards, R. L. & Wasserburg, G. J. 1986 ^{238}U, ^{234}U and ^{232}Th in seawater. *Earth planet. Sci. Lett.* **80**, 241–251.

Condomines, M., Hemond, Ch. & Allègre, C. J. 1988 U–Th–Ra radioactive disequilibria and magmatic processes. *Earth planet. Sci. Lett.* **90**, 243–262.

Ellam, R. M. & Hawkesworth, C. J. 1988 Is average continental crust generated at subduction zones? *Geology* **16**, 314–317.

Ellam, R. M. & Hawkesworth, C. J. 1988 Elemental and isotope variations in subduction related basalts: evidence for a three component model. *Contrib. Mineral. Petrol.* **98**, 72–80.

Ellam, R. M., Hawkesworth, C. J. & McDermott, F. 1990 Pb isotope data from the Proterozoic subduction-related rocks: implications for crust-mantle evolution. *Chem. Geol.* **83**, 165–181.

Ewart, A. W. & Hawkesworth, C. J. 1987 The Pleistocene to Recent Tonga: Kermadec arc lavas: interpretation of new isotope and rare earth element data in terms of a depleted mantle source model. *J. Petrology* **28**, 495–530.

Gill, J. B. & Williams, R. W. 1990 Th isotopes and U-series studies of subduction-related volcanic rocks. *Geochim. cosmochim. Acta* **54**, 1427–1442.

Hart, S. R. & Staudigel, H. 1989 Isotope characterisation and identification of recycled components. In *Crust/mantle recycling at convergence zones* (ed. S. R. Hart & L. Gülen), pp. 15–28. Nato Workshop Volume, Kluwer Academic Publishers.

Hart, S. R. 1984 A large-scale isotope anomaly in the Southern Hemisphere mantle. *Nature* **309**, 753–757.

Hawkesworth, C. J. & Ellam, R. M. 1989 Chemical fluxes and wedge replacement rates along Recent destructive plate margins. *Geology* **17**, 46–49.

Hawkesworth, C. J., Hergt, J. M., McDermott, F. & Ellam, R. M. 1991 Destructive margin magmatism and contributions from the mantle wedge and subducted crust. *Spec. Publ. Geol. Soc. Austr.* (In the press.)

Hole, M. J., Saunders, A. D., Marriner, G. F. & Tarney, J. 1984 Subduction of pelagic sediments: implications for the origin of Ce-anomalous basalts from the Mariana Islands. *J. geol. Soc. Lond.* **141**, 453–472.

Kay, R. W. 1980 Volcanic arc magmas: implications of a melting–mixing model for element recycling in the crust–upper mantle system. *J. Geol.* **88**, 497–522.

Kay, R. W., Sun, S. S. & Lee-Hu, C. N. 1978 Pb and Sr isotopes in volcanic rocks from Aleutian Islands and Pribilof Islands, Alaska. *Geochim. cosmochim. Acta* **42**, 263–273.

Kay, R. W. & Kay, S. M. 1986 Petrology and geochemistry of the lower continental crust: an overview. In *The nature of the lower continental crust* (ed. J. B. Dawson, D. A. Carswell, J. Hall & K. H. Wedepohl), pp. 147–159. *Geol. Soc. Lond. Spec. Publ.* **24**.

Kushiro, I. 1987 A petrological model of the mantle wedge and lower crust in the Japanese island arcs. *Geochem. Soc. Spec. Pub.* **1**, 165–181.

McDermott, F. & Hawkesworth, C. J. 1991 Th, Pb and Sr isotope variations in young arc volcanics and oceanic sediments. *Earth planet. Sci. Lett.* (In the press.)

Morris, J. D. & Hart, S. R. 1983 Isotopic and incompatible trace element constraints on the genesis of island arc volcanics from Cold Bay and Amak Island, Aleutians, and implications for mantle structure. *Geochim. cosmochim. Acta* **47**, 2015–2030.

Morris, J. D., Leeman, W. P. & Tera, F. 1990 The subducted component in island arc lavas: constraints from Be isotopes and B–Be systematics. *Nature* **344**, 31–36.

O'Nions, R. K. & McKenzie, D. P. 1988 Melting and continent generation. *Earth planet. Sci. Lett.* **90**, 449–456.

Pearce, J. A. 1982 Trace element characteristics of lavas from destructive plate boundaries. In *Andesites: orogenic andesites and related rocks* (ed. R. S. Thorpe), pp. 525–548. New York: Wiley.

Pearce, J. A. 1983 Role of the sub-continental lithosphere in magma genesis at active continental margins. In *Continental basalts and mantle xenoliths* (ed. C. J. Hawkesworth & M. J. Norry), pp. 230–249. Nantwich, Cheshire: Shiva.

Sun, S. S. & McDonough, W. F. 1989 Chemical and isotopic systematics of oceanic basalts: implications for mantle composition and processes. In *Magmatism in ocean basins* (ed. A. D. Saunders & M. J. Norry), pp. 313–345. *Geol. Soc. Spec. Publ.* **42**.

Tatsumi, Y., Hamilton, D. L. & Nesbitt, R. W. 1986 Chemical characteristics of fluid phase released from a subducted lithosphere and origin of arc magmas: evidence from high-pressure experiments and natural rocks. *J. Volcanol. geotherm. Res.* **29**, 293–309.

Taylor, S. R. & McLennan, S. M. 1985 *The continental crust: its composition and evolution.* (312 pages.) Oxford: Blackwell.

Tera, F., Brown, L., Morris, J., Sacks, I. S., Klein, J. & Middleton, R. 1986 Sediment incorporation in island arc magmas: inferences from [10]Be. *Geochim. cosmochim. Acta* **50**, 535–550.

White, W. M. & Dupré, B. 1986 Sediments subduction and magma genesis in the Lesser Antilles: isotopic and trace element constraints. *J. geophys. Res.* **91**, B6, 5927–5941.

Woodhead, J. D. & Fraser, D. G. 1985 Pb, Sr and [10]Be isotopic studies of volcanic rocks from the northern Mariana Island. Implications for magma genesis and crustal recycling in the Western Pacific. *Geochim. cosmochim. Acta* **49**, 1925–1930.

Zindler, A. & Hart, S. 1986 Chemical geodynamics. *A. Rev. Earth planet. Sci.* **14**, 493–571.

Partial melting of subducted oceanic crust and isolation of its residual eclogitic lithology

By W. F. McDonough

*Research School of Earth Sciences, Australian National University, GPO Box 4,
Canberra ACT 2601, Australia*

Oceanic lithosphere is produced at mid-ocean ridges and reinjected into the mantle
at convergent plate boundaries. During subduction, this lithosphere goes through a
series of progressive dehydration and melting events. Initial dehydration of the slab
occurs during low pressure metamorphism of the oceanic crust and involves
significant dewatering and loss of labile elements. At depths of 80–120 km water
release by the slab is believed to lead to partial melting of the oceanic crust. These
melts, enriched in incompatible elements (excepting Nb, Ta and Ti), fertilize the
overlying mantle wedge and produce the enriched peridotitic sources of island arc
basalts. Retention of Nb, Ta and Ti by a residual mineral (e.g. in a rutile phase) in
a refractory eclogitic lithology within the sinking slab are considered to cause their
characteristic depletions in island arc basalts. These refractory eclogitic lithologies,
enriched in Nb, Ta and Ti, accumulate at depth in the mantle. The continued
isolation of this eclogitic residuum in the deep mantle over Earth's history produces
a reservoir which contains a significant proportion of the Earth's Ti, Nb and Ta
budget.

Both the continental crust and depleted mantle have subchondritic Nb/La and
Ti/Zr ratios and thus they cannot be viewed strictly as complementary geochemical
reservoirs. This lack of complementarity between the continental crust and depleted
mantle can be balanced by a refractory eclogitic reservoir deep in the mantle, which
is enriched in Nb, Ta and Ti. A refractory eclogitic reservoir amounting to *ca.* 2% of
the mass of the silicate Earth would also contain significant amounts of Ca and Al
and may explain the superchondritic Ca/Al value of the depleted mantle.

1. Introduction

The production of oceanic lithosphere at mid-ocean ridges and its subsequent
reinjection into the mantle at convergent plate boundaries are the two most
important crust–mantle differentiation process operating in the Earth today.
Hofmann & White (1980, 1982), Chase (1981) and Ringwood (1982) suggested that
these two processes are also ultimately responsible for the formation of intraplate
basalt source regions. The aim of this paper is to examine some of the consequences
of slab melting processes. Attention is focused on the fate of the subducting slab at
depths below which island arc basalts (IAB) are generated (i.e. over 100 km). There
is a need to understand the compositions and subsequent history of refractory
eclogitic materials produced during the partial melting of subducted oceanic crust in
the island arc environment.

Phil. Trans. R. Soc. Lond. A (1991) **335**, 407–418

The subduction of oceanic lithosphere into the mantle involves a continuous process of increasing metamorphism and dehydration within the slab. It is likely that components added to the uppermost portions of the oceanic crust during its migration from ridge to subduction trench undergo rapid mineralogical changes and substantial dehydration during the early phases of metamorphism. A significant amount of this hydrous component is transferred into the overlying lithosphere at relatively shallow levels (less than 80 km). When the slab reaches depths of 80–120 km increasing temperature, combined with the release of volatiles may lead to partial melting. At this stage, oceanic crust becomes transformed to quartz eclogite. Any partial melts derived from these lithologies migrate into the mantle wedge, enriching this region in incompatible elements.

It is widely considered that IAB form by the partial melting of refertilized peridotitic lithologies in the wedge overlying a subducting slab (Nicholls & Ringwood 1973). In most instances hydrous, silica-saturated basaltic melts of the island arc magma series are observed to display characteristic depletions† in Ti, Nb and Ta‡. It is widely believed that the depletions are caused by the retention of these elements in a titanite phase (such as rutile) during partial melting (Saunders *et al.* 1980). Thus, if slab melting occurs, the residual eclogite would be relatively enriched in Ti and Nb. (There are a few examples where such depletions are not found in these magmas and this is attributed to the subduction of young oceanic lithosphere at relatively high temperatures (Leeman *et al.* 1990).)

Refractory eclogitic lithologies thereby produced are transported deep into the mantle by the descending slab. This material is not recognized in mantle samples carried up as xenoliths, nor is its geochemical signature of Nb and Ti-enrichments observed in basalts derived from the mantle. Tarney *et al.* (1980) and Ringwood (1990) suggested that this refractory eclogitic material may become trapped deep in the mantle and accumulate to form a substantial geochemical reservoir. This material, which has been produced over geologic time (more than 2 Ga), would constitute an isolated reservoir in the mantle and contain enhanced levels of Nb and Ti relative to other incompatible (lithophile) elements.

In this model, the residual materials enriched in Nb and Ti occur not in the overlying lithospheric mantle, but in the subducted slab. It is difficult to completely assess the compositional variation in the continental lithospheric mantle; however, some constraints can be provided by examining the composition of peridotite xenoliths brought up in basalts. Peridotite xenoliths generally do not show any clear evidence for the storage of this residual, Nb, Ti-enriched, component (Jochum *et al.* 1989; McDonough 1990). The continental lithospheric mantle is believed to have an overall enrichment of incompatible elements, with a pattern which is similar to that seen intraplate basalts (McDonough 1990). Although this mantle reservoir has a superchondritic Nb/La ratio, it also possesses a subchondritic Ti/Zr ratio and a chondritic to subchondritic Ti/Eu ratio indicating that it is depleted in Ti and only

---

† The terms *enriched/enrichments* and *depleted/depletions* describe the abundance of an element relative to other elements or the absolute value of a ratio relative to the Primitive Mantle ratio. In this paper enrichments or depletions of elements are relative to elements with similar incompatibility (e.g. Nb/La and Ti/Zr) during melting of oceanic basalt sources (Sun & McDonough 1989).

‡ Throughout this discussion it is assumed that Nb and Ta are inseparable during most mantle processes, although it is recognized that these elements may be fractionated by certain phases. A Nb/Ta ratio of 17, the chondritic value, is assumed to apply to lavas ranging from mafic to ultramafic in composition and to peridotites (Jochum *et al.* 1989). Therefore, further discussions will be restricted to Nb and it is assumed that the behaviour of Ta is identical to that of Nb.

slightly enriched in Nb. Thus, this evidence does not imply that the continental lithospheric mantle represents an important Nb, Ti-enriched reservoir within the Earth.

It has recently been suggested that peridotite xenoliths commonly have depletions in high field strength elements (Ti, Zr, Nb and Ta) and that this material is representative of a world-wide shallow reservoir that is the source of IABs (Salters & Shimizu 1988). The basis for this model was that the relative abundances of Ti, Zr and the rare earth elements in clinopyroxenes from anhydrous spinel peridotite xenoliths reflected the bulk rock character. This assumption, however, has been shown to be incorrect for a number of spinel peridotite xenoliths (McDonough & Frey 1989). Additionally, it has been demonstrated that substantial quantities of Ti are found in orthopyroxenes coexisting with clinopyroxenes in these xenoliths, while more than 95% of the Eu, which has a similar incompatible melting behaviour to Ti, is hosted in the clinopyroxenes (McDonough *et al.* 1991). This indicates that, although clinopyroxenes from anhydrous spinel peridotite xenoliths contain nearly the total budget of the rare earth elements, this phase does not contain the same relative amount of Ti and therefore the incompatible element pattern of the clinopyroxene will not necessarily reflect that of the bulk rock. Given these observations, this model (Salters & Shimizu 1988) is not considered valid.

2. Evidence for a refractory eclogite reservoir

The continued storage and isolation of refractory eclogitic lithologies transported to the deep mantle by the subducting slab would have a significant influence on the distribution of some key elements (including Nb and Ti) in the mantle. The production and recycling of oceanic lithosphere is a process considered to have occurred throughout the Phanerozoic and probably much of the Precambrian. Accordingly, it is useful to examine the evidence for sequestering significant quantities of refractory eclogite, enriched in Nb and Ti in a deep mantle reservoir and to explore some geochemical consequences of such a reservoir.

There are several pieces of evidence which support the existence of a deep mantle reservoir composed of refractory eclogite enriched in Nb and Ti. The light rare earth (LREE) enriched continental crust is often believed to have a complementary composition to that of the depleted upper mantle, the source region of LREE-depleted mid-ocean ridge basalt (MORB). The average continental crust is believed to possess a subchondritic Nb/La ratio (Taylor & McLennan 1985) (table 1). Interestingly, N-type MORBs generally have chondritic to subchondritic Nb/La ratios, mostly in the range of 0.8–1.0 (Sun *et al.* 1979; Hofmann 1988) and their source regions would have at least a subchondritic ratio, given that Nb is widely regarded as being more incompatible than La during MORB genesis. This indicates that the Nb/La values of the depleted mantle and continental crust are not complementary relative to the primitive mantle (table 1). As noted earlier, the low La and Nb abundance and the relatively low, but superchondritic, Nb/La value of the continental lithospheric mantle (table 1) are not sufficient to balance the continental crust and depleted mantle with the primitive mantle. Thus, a fourth reservoir, having a superchondritic Nb/La composition and high Nb abundance, is required to balance the continental crust and depleted mantle (Saunders *et al.* 1988; Sun & McDonough 1989).

It has been suggested by Weaver *et al.* (1987) and Saunders *et al.* (1988) that refractory eclogitic lithologies produced by the processing of subducted oceanic crust

Table 1. *A mass balance calculation for elements in the continents and mantle reservoirs*

(Mass is expressed in 10^{25} g. The concentrations of Ti, Zr, Nb and La are expressed in p.p.m., and those of Ca and Al are given in wt %.)

relative contribution	mass (%)	Ti (%)	Zr (%)	Nb (%)	La (%)	Ca (%)	Al (%)
continental crust	0.5	2	5	8	14	1	2
continental l. mantle	2.5	1	2	3	4	1	1
depleted mantle	40	25	29	21	22	37	36
refractory eclogite[a]	2.0	17	9	13	5	5	6
undepleted mantle	55	55	55	55	55	55	55
silicate Earth	100						

	mass	Ti	Zr	Nb	La	Ca (%)	Al (%)	Ti/Zr	Nb/La	Ca/Al
silicate Earth	403	1275	11	0.71	0.69	2.67	2.45	116	1.03	1.09
continental crust	2.2	5400	100	11	16	5.3	8.4	54	0.69	0.63
continental l. mantle	10	550	8	1.0	0.90	1.4	1.1	69	1.11	1.27
depleted mantle	160	800	8	0.375	0.38	2.5	2.2	100	0.99	1.13
refractory eclogite[a]	8	10500	50	4.5	1.7	6.4	7.4	210	2.65	0.87
undepleted mantle	223	1275	11	0.71	0.69	2.67	2.45	116	1.03	1.09

[a] The proposed reservoir of refractory eclogite, enriched in Nb and Ti and carried in the down going slab, was produced during island arc magmatism; its composition and mass has been modelled in this calculation (see text for further explanation). The composition and mass of the continental crust are from Taylor & McLennan (1985). The composition and mass of the continental lithospheric mantle are from McDonough (1990). The composition of the depleted mantle is derived from a melting model based on a relatively primitive MORB composition (Sun & McDonough 1989); its mass is assumed to be about 40 % of the mass of the primitive mantle. The composition of the undepleted mantle is set equal to that of the primitive mantle (Sun & McDonough 1989) and its mass makes up the remainder of the silicate Earth system.

in the island arc environment are transported deep into the mantle and reincorporated in the source regions of oceanic basalts (MORB and oceanic intraplate basalts (OIB)). This hypothesis, however, is inconsistent with data for N-type MORBs which show that these basalts are not enriched in Nb (Sun *et al.* 1979; Hofmann 1988; Sun & McDonough 1989). It has been demonstrated above that N-type MORBs are, in fact, depleted in Nb with respect to La. Additionally, it can also be argued that N-type MORBs are not enriched in Ti. These basalts generally have chondritic to subchondritic Ti/Zr (figure 1a), Ti/Eu and Ti/Gd ratios. Significant fractionation of these element ratios is not expected during MORB genesis, because Ti, Zr, Eu and Gd have relatively similar incompatibility at these high degrees of melting (Sun & McDonough 1989). Thus, ratios of these elements in MORBs tend to reflect their source values. By comparison, the fractionation of Th and U, two elements with differing incompatibility and both of which are much more incompatible than Ti and related elements, is generally on the order of 10 % or less during MORB genesis (Condomines *et al.* 1988; Ben Othman & Allègre 1990). Ratios of Ti/Zr, Ti/Eu or Ti/Gd, which are *ca.* 10 % higher in the N-type MORB source are still in the range of chondritic to subchondritic compositions. These observations indicate that the source region of N-type MORBs is either depleted in Ti relative to Zr, Eu or Gd, or possesses relatively unfractionated Ti/Zr, Ti/Eu or Ti/Gd ratios. The average continental crust has a subchondritic Ti/Zr value (figure 1b) and is thus not complementary to the depleted mantle (table 1).

It is unlikely that refractory eclogitic lithologies produced by the processing of

Phil. Trans. R. Soc. Lond. A (1991)

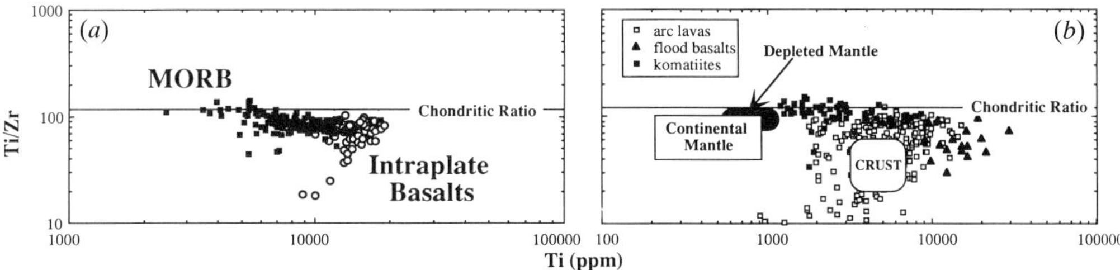

Figure 1. Ti/Zr against Ti (p.p.m.) plot for (*a*) mid-ocean ridge basalts (MORB) and intraplate basalts; (*b*) island arc basalts, flood basalts, and komatiites. The estimated fields for the bulk continental crust (crust), the continental lithospheric mantle and depleted mantle (MORB source region) are also shown in (*b*). Data are from the literature. The bulk silicate Earth has a chondritic Ti/Zr value of 115. Komatiites tend to have relatively chondritic Ti/Zr values, while all other rocks have subchondritic Ti/Zr values. To produce a balanced bulk silicate Earth there must be another reservoir with a significant complement of Ti and having a superchondritic Ti/Zr. A refractory eclogitic reservoir enriched in Nb and Ti and having a superchondritic Ti/Zr would be capable of providing this necessary complement.

subducted oceanic crust in the island arc environment are reincorporated in the source regions of intraplate basalts. There is little disagreement that the sources of intraplate basalts can be related to the recycling of oceanic crust into the mantle; however, models of this type differ greatly in explaining the timing and processes whereby their sources are refertilized (Hofmann & White 1980, 1982; Chase 1981; Ringwood 1982). It is widely held that the strong enrichments of incompatible elements in intraplate basalts coupled with their often depleted Nd and Sr isotopic ratios relative to bulk Earth ratios require that their source regions have been refertilized. Hofmann *et al.* (1986) made the important observation that intraplate basalts and MORBs possess similar Nb/U and similar Ce/Pb ratios, which are distinct from the primitive mantle ratio and those of average continental crust. The constancy of these ratios over a range of melting conditions can be explained by similar incompatibility of Nb and U (and likewise for Ce and Pb) during the genesis of these oceanic basalts (Hofmann *et al.* 1986). Both the MORB source, which is depleted in incompatible elements, and the intraplate basalt source, which is enriched in incompatible elements, share non-primitive mantle ratios of Ce/Pb and Nb/U that are for the most part complementary to the continental crust. These observations led Hofmann *et al.* (1986) to conclude that the source regions of oceanic basalts were rehomogenized after a major differentiation process which lead to extraction of much of the continental crust from the mantle.

A residual Nb–Ti-bearing phase, believed to be stable during the production of the refractory eclogitic lithology, is suggested to be responsible for causing the depletions of Nb and Ti in IAB (Saunders *et al.* 1980). This phase would be responsible for fractionating the Nb/U ratio between the melt and residue. A relatively constant Nb/U ratio in the source of MORB and intraplate basalts (Hofmann *et al.* 1986) is unlikely to be maintained if the sources of intraplate basalts were generated from these refractory eclogite lithologies.

Zindler & Hart (1986) defined a diversity of intraplate basalt compositional types including: HIMU (high μ (or high $^{206}U/^{204}Pb$)), EM (enriched mantle) and PREMA (prevalent mantle). These basalt types display systematic variations in their chemical and isotopic compositions (figures 2 and 3). Although differences are small, HIMU basalts have higher Nb/U, Nb/Th, U/Pb and Ce/Pb and lower Nb/La, Rb/Sr

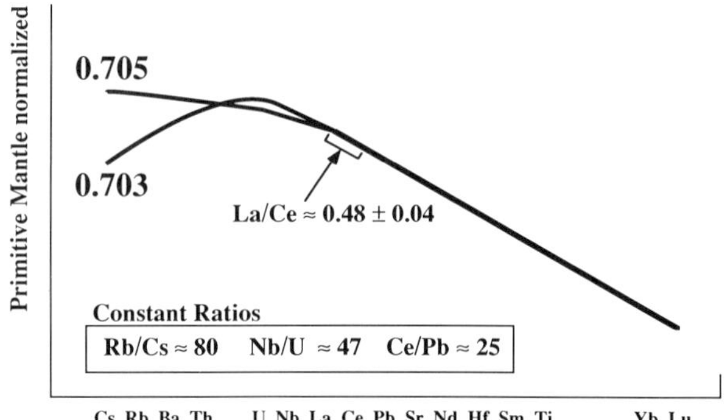

Figure 2. Schematic mantle normalized diagram illustrating the abundance pattern of incompatible elements in HIMU and EM-type intraplate basalts. The elements are ordered according to their relative incompatibility, with Cs, Rb and Ba being the most incompatible during melting (Sun & McDonough 1989). The idealized HIMU basalt has an incompatible element enriched pattern with elements more incompatible than Nb having a more depleted trend. This pattern is characteristically associated with high ^{206}Pb/^{204}Pb and low ^{87}Sr/^{86}Sr values. The idealized EM basalt is shown as also having an incompatible element enriched pattern, with elements more incompatible than Nb showing greater enrichments and higher ^{87}Sr/^{86}Sr values than the HIMU basalts. Constant ratios of Rb/Cs, Nb/U and Ce/Pb are found in most unaltered intraplate basalts and MORB. Additionally, most alkalic intraplate basalts have a relatively constant La/Ce value (Sun & McDonough 1989).

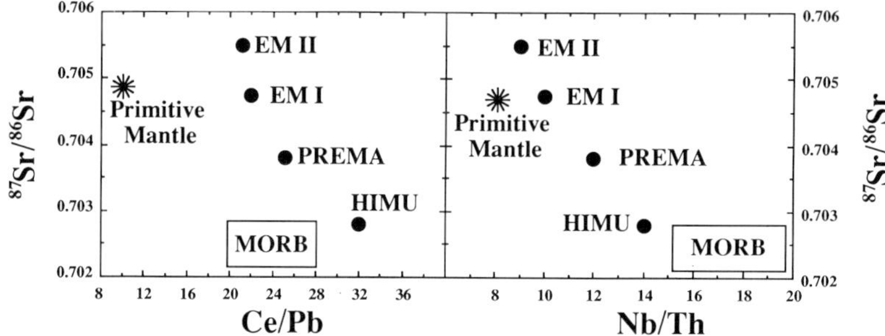

Figure 3. A schematic Ce/Pb and Nb/Th versus ^{87}Sr/^{86}Sr diagram showing the range of values found in intraplate basalts, MORB and the primitive mantle. The terms HIMU, PREMA, EM I and EM II are from Zindler & Hart (1986) and represent the range of intraplate mantle reservoirs yet identified. These reservoirs are generally distinguished by their Sr, Nd and Pb isotopic compositions. These diagrams, however, illustrate that these intraplate basalt types can also be distinguished by certain element ratios. The values of Ce/Pb, Nb/Th and ^{87}Sr/^{86}Sr for these mantle reservoirs and the primitive mantle are taken from table 3 in Sun & McDonough (1989). The continental crust has much lower Ce/Pb and Nb/Th values and a considerably higher ^{87}Sr/^{86}Sr value.

and K/Nb values than EM basalts (figure 2) which are coupled with lower ^{87}Sr/^{86}Sr (figure 3) and higher ^{206}Pb/^{204}Pb ratios in HIMU than in EM basalts (see also table 3 of Sun & McDonough 1989). This systematic co-variation in chemical and isotopic compositions is consistent with models which suggest that the peridotitic source of intraplate basalts has been refertilized early in its evolution by intramantle melting and fractionation processes (Ringwood 1982, 1990; Sun & McDonough 1989). The coherent regularities in chemical and isotopic composition within and between intraplate basalt types is unlikely to be produced by enrichment processes such as hydrothermal alteration of oceanic crust or contamination by continental-derived sediments.

Phil. Trans. R. Soc. Lond. A (1991)

Intraplate basalts show a range of Nb/La ratios from about 0.9 to 1.7, with many having a value of between 1.1 and 1.4 (as compared to MORB with ratios of 0.8–1.0). Weaver *et al.* (1987) and Saunders *et al.* (1988) suggest that the sources of typical oceanic intraplate basalts, particularly the HIMU basalts having high Nb/La values (more than 1.4), are derived by the addition of a component of recycled oceanic crust enriched in Nb and Ti. This interpretation is not unique; the high Nb/La ratios in these lavas do not necessarily reflect the source ratios (Sun & McDonough 1989). The relative abundance of Nb and La in intraplate basalts is very sensitive to the degree of partial melting involved in their genesis as well as the processes which have contributed to the refertilization of their source. Unlike MORBs, which are believed to be generated by large degrees of partial melting, many intraplate basalts are generated by lower degrees of melting and thus the source and lava Nb/La ratios will differ. Sun & McDonough (1989) have addressed this issue also. They observed that several Hawaiian basalt suites revealed markedly different Nb/La ratios and that a range of Nb/La ratios was produced by variable degrees of partial melting and, in some cases, by the (postulated) presence of a residual titanate mineral. Additionally, Sun & McDonough (1989) highlighted the fact that a Ua Pou (French Polynesia) tholeiite having HIMU chemical and isotopic characteristics possesses a subchondritic Nb/La value: i.e. it shows a Nb depletion relative to La. Moreover, intraplate basalts, including HIMU, EM and PREMA, do not show any evidence for excess Ti in their sources. They typically display subchondritic ratios of Ti/Zr (figure 1), Ti/Eu and Ti/Gd. Thus, these observations, in addition to the fact that the chemical and isotopic covariation found in intraplate basalts cannot be explained easily by such models, seem to imply that refractory eclogitic lithologies produced by the processing of subducted oceanic crust in the island arc environment are not reincorporated into the source regions of intraplate basalts.

It has recently been suggested that flood basalt volcanism may be the earliest phase of intraplate magmatism associated with the initiation of an ascending mantle plume (White & McKenzie 1989; Campbell & Griffiths 1990). If this is so these lavas may yield significant geochemical and isotopic information about the nature of the plume (intraplate) source region. Campbell & Griffiths (1990) suggested that the early phase picrites of the flood basalt provinces faithfully represent the geochemical composition of the plume source. In figure 4, examples are shown from the North Atlantic, Reunion-Deccan and Siberian provinces that have slight LREE-enrichments and do not display anomalous enrichments or depletions of Nb (i.e. they lack any observable 'Nb-kick'). These lavas have geochemical characteristics similar to EM-type intraplate basalts and their sources. Campbell & Griffiths (1990) cite other plume head picrites as possessing strong incompatible element-enrichments. However, they are not considered to be representative of these source regions and may have been contaminated by continental crust (Sun & McDonough 1989).

Archaean komatiites that are not contaminated by continental crust and N-type MORB are believed to be derived by large degrees of partial melting. Large volumes of mantle may have been sampled during the generation of these lavas. Interestingly, the mantle-normalized patterns of incompatible elements of both the komatiite and MORB samples shown in figure 4 are rather uniform and show a smoothly increasing depletion of elements with increasing incompatibility and no 'Nb-kick'. Such patterns are not expected for a mantle source region which has been modified by the addition of recycled oceanic crust enriched in Nb and Ti. A peridotite sample has a similar pattern to those shown by these lavas (figure 4), but at lower absolute

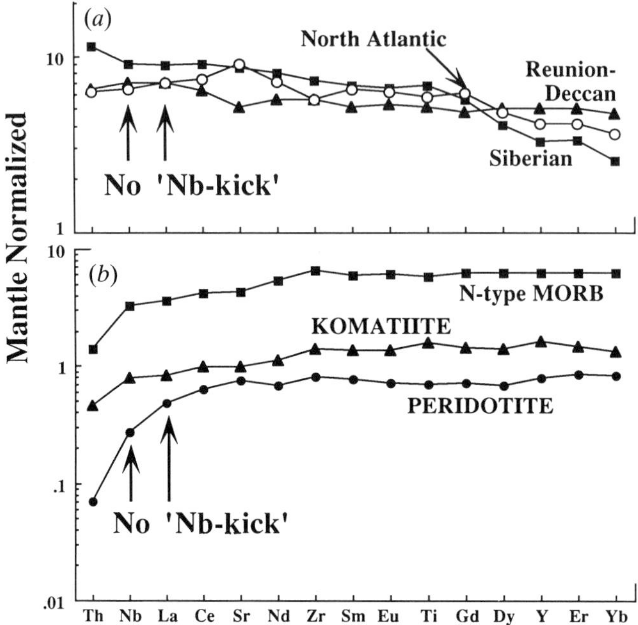

Figure 4. Mantle normalized diagram for mafic to ultramafic lavas and for a peridotite xenolith. The three plume head picrites (*a*), the komatiite and the N-type MORB represent magmas generated by large degrees of partial melting. The plume head picrites have relatively uniform incompatible element-enriched patterns, while the N-type MORB and komatiite have depleted patterns. Both the enriched and depleted patterns show no obvious 'Nb-kick' (i.e. no obvious depletion or enrichment of Nb relative to adjoining incompatible elements).

concentrations. Jochum *et al.* (1989) suggested that this peridotite pattern is chemically analogous to MORB sources. These data provide additional evidence that the depleted mantle does not possess obvious enrichments in Ti or Nb.

The overall variation of Ti/Zr ratios in a wide variety of lavas, as well as the average ratios for the continental crust, continental lithospheric mantle and depleted mantle (figure 1) further reveals that the continental crust and depleted mantle do not have complementary compositions. The lack of complementarity in both Nb/La and Ti/Zr ratios between the continental crust and depleted mantle can be balanced by invoking a refractory eclogitic reservoir enriched in Nb and Ti which is stored deep in the mantle. A refractory eclogitic reservoir, enriched in Nb relative to La and enriched in Ti relative to Zr, would be capable of providing the necessary complement to balance the continental crust and depleted mantle relative to the bulk silicate Earth (table 1). Table 1 presents the parameters for a simple mass balance calculation. The composition and masses of the continental crust and continental lithospheric mantle are taken from the literature (Taylor & McLennan 1985; McDonough 1990). The mass of the depleted mantle is assumed to be about 40% of the primitive mantle; however, estimates of the size of this reservoir vary greatly (between 20 and 80%). The composition of the depleted mantle is derived from a simple melting model, assuming that MORBs are generated by about 10% partial melting. The mass of the undepleted mantle is set at 55% and its composition is assumed to be equal to the primitive mantle.

From this model the composition and mass of the refractory eclogite reservoir can be estimated. The composition of this reservoir is derived by partially melting an N-

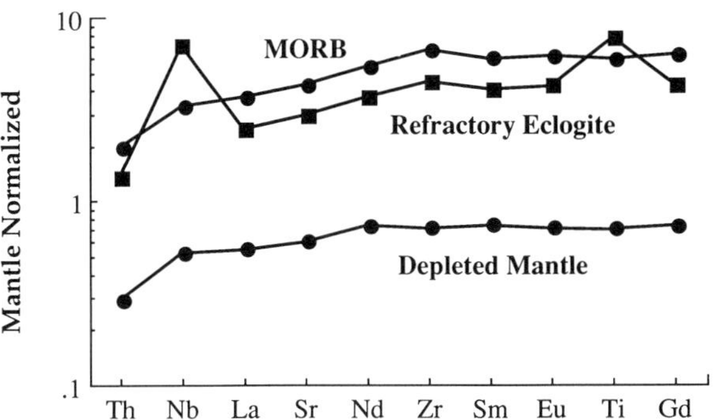

Figure 5. A mantle normalized diagram showing the relative incompatible element patterns of MORB, the depleted mantle (MORB source region) and an estimate of the refractory eclogite reservoir. Mantle normalizing values are from Sun & McDonough (1989).

type MORB which is representative of the upper 1–2 km of oceanic crust. Before this crust is melted it is estimated to have about 1.4 wt % TiO_2, 85 p.p.m. Zr, 2.9 p.p.m. La and 2.8 p.p.m. Nb (Hofmann 1988; Staudigel *et al.* 1989; Sun & McDonough 1989). This lithology is subjected to about 10% partial melting, leading to a depletion of 40% or more of its inventory of incompatible elements (except for Ti and Nb which are enriched in the residue by 25 and 50%, respectively). Figure 5 compares the incompatible element patterns of N-type MORB, its depleted mantle source and the above refractory eclogite. The total mass of oceanic crust produced and recycled back into the mantle throughout time is estimated to be between 5 and 10% of the mass of the silicate Earth, and is equivalent to about $2–4 \times 10^{26}$ g of basalt. This value was derived by assuming that the present thickness of oceanic crust, plate production and consumption rate and area of seafloor (Parson 1981) has remained constant over the past 3.5 Ga. The proportion of subducted oceanic crust involved in island arc magmatism is estimated to be about 25% of the total amount of subducted oceanic crust, which is equivalent to the top *ca.* 1.7 km of a 7 km thick crust. The total mass of refractory eclogitic material is therefore estimated to be *ca.* 2% of the mass of the silicate Earth (table 1 shows that this is equivalent to about four times the mass of the continental crust).

These parameters provide a plausible numerical solution to balance the different silicate reservoirs in the Earth. This solution is non-unique; however, there is only a limited family of solutions which can satisfy the empirical constraints. The composition of the primitive mantle, continental crust, continental lithospheric mantle and undepleted mantle are reasonably well constrained in both their absolute and relative values. Likewise, the relative concentrations of elements in the depleted mantle and refractory eclogite are reasonably well constrained. The only parameter which can be changed is the inventory of elements in the depleted mantle and refractory eclogitic reservoir and these vary as a function of the assumed ratio of mass of depleted mantle to mass of undepleted mantle. The above calculation illustrates that a reservoir composed of refractory eclogitic material, enriched in Nb and Ti and comparable in size to the continental lithospheric mantle, satisfactorily accounts for the inventory of these elements in the entire silicate Earth.

3. Discussion

One of the consequences of the model proposed here is that continued storage of refractory eclogitic lithologies, characterized by enrichments in Ti and Nb, would produce a secular change in the composition of the depleted mantle. This process may also affect the abundance of other elements. Using primitive mantle xenoliths and komatiitic lavas, Palme & Nickel (1985) argue that the present day depleted mantle has a non-chondritic Ca/Al ratio. It is generally assumed, however, that the whole mantle started out with a chondritic Ca/Al ratio. Basalts and eclogites both have relatively high Ca and Al contents compared with the mantle and thus progressive accumulation of eclogitic material into an isolation reservoir could change the Ca/Al ratio of the depleted mantle with time. The refractory eclogitic reservoir was assumed to have average CaO and Al_2O_3 contents of about 9 and 14 wt %, respectively (table 1) comparable with the compositions of MORB and eclogitic xenoliths. If the refractory eclogitic reservoir comprises *ca.* 2 % of the mass of the silicate Earth, then it would contain a significant amount of Ca and Al relative to the bulk silicate Earth. This reservoir together with the continental crust would complement the depleted mantle and could produce a balance for the Ca/Al ratio of the bulk silicate Earth.

There may also be additional evidence for a secular change in the composition of the depleted mantle from Ti/Zr systematics. The sources of komatiites and MORBS are considered to have different Ti/Zr compositions. Moreover, it can be shown that Archaean komatiites are derived from depleted mantle sources. A comparison between MORB and komatiites (figure 1) suggests that there has been a secular decrease in the Ti/Zr ratio of the depleted mantle. This change can be attributed to the progressive removal of Ti relative to Zr from the depleted mantle and its storage in a refractory eclogite reservoir.

It is not possible to confidently establish evidence for a secular change of La/Nb ratios in the depleted mantle. More abundant and generally more precise data for Nb in komatiites, MORB and peridotites are needed before this idea can be properly tested. This remains an important goal in mantle geochemistry today.

The geochemical evidence for a refractory eclogitic reservoir isolated in the deep mantle is believed to be strong; however, the processes by which this reservoir have been isolated from the down going slab and kept at depth are poorly understood. Ringwood (1982) and Ringwood & Irifune (1988) suggest that the transportation and trapping of this material in a deep mantle reservoir was facilitated by density segregation. Studies of the viscosities and density contrasts of eclogitic and peridotitic materials and segregation mechanisms involved are likely to provide important insights into this regard.

The author is grateful to Professor A. E. Ringwood for the helpful discussions which have contributed greatly to this paper. Helpful suggestions given to me by Roberta Rudnick and Shen-su Sun have also improved this paper. I thank them all for their constructive criticism.

References

Ben Othman, D. & Allègre, C. J. 1990 U–Th isotopic systematics at 13° N east Pacific ridge segment. *Earth planet. Sci. Lett.* **98**, 129–137.

Campbell, I. H. & Griffiths, R. W. 1990 Implications of mantle plume structure for the evolution of flood basalts. *Earth planet. Sci. Lett.* **99**, 79–93.

Chase, C. G. 1981 Oceanic island Pb: two-stage histories and mantle evolution. *Earth planet. Sci. Lett.* **52**, 277–284.

Condomines, M., Hemond, C. & Allègre, C. J. 1988 U–Th–Ra radioactive disequilibria and magmatic processes. *Earth planet. Sci. Lett.* **90**, 243–262.

Hofmann, A. W. 1988 Chemical differentiation of the Earth: the relationship between mantle, continental crust and oceanic crust. *Earth planet. Sci. Lett.* **90**, 297–314.

Hofmann, A. W., Jochum, K. P., Seufert, M. & White, W. M. 1986 Nb and Pb in oceanic basalts: new constraints on mantle evolution. *Earth planet. Sci. Lett.* **79**, 33–45.

Hofmann, A. W. & White, W. M. 1980 The role of subducted oceanic crust in mantle evolution. *Carnegie Inst. Washington Yearbook* **79**, 477–483.

Hofmann, A. W. & White, W. M. 1982 Mantle plumes from ancient oceanic crust. *Earth planet. Sci. Lett.* **57**, 421–436.

Jochum, K. P., McDonough, W. F., Palme, H. & Spettel, B. 1989 Compositional constraints on the continental lithospheric mantle from trace elements in spinel peridotite xenoliths. *Nature* **340**, 548–550.

Leeman, W. P., Smith, D. R., Hildreth, W., Palacz, Z. & Rogers, N. 1990 Compositional diversity of late Cenozoic basalts in a transect across the southern Washington Cascades: implications for subduction zone magmatism. *J. geophys. Res.* **95**, 19561–19582.

McDonough, W. F. 1990 Constraints on the composition of the continental lithospheric mantle. *Earth planet. Sci. Lett.* **101**, 1–18.

McDonough, W. F. & Frey, F. A. 1989 REE in upper mantle rocks. In *Geochemistry and mineralogy of rare earth elements* (ed. B. Lipin & G. R. McKay), pp. 99–145. Chelsea, Michigan: Mineralogical Society of America.

McDonough, W. F., Stosch, H.-G. & Ware, N. 1991 Relative distribution of Ti and the REE in peridotite minerals. (In preparation.)

Nicholls, I. A. & Ringwood, A. E. 1973 Effect of water on olivine stability in the production of SiO_2-undersaturated magmas in the island arc environment. *Geochim. cosmochim. Acta* **81**, 285–300.

Palme, H. & Nickel, K. G. 1985 Ca/Al ratio and composition of the Earth's upper mantle. *Geochim. cosmochim. Acta* **49**, 2123–2132.

Parson, B. 1981 The rates of plate creation and consumption. *Geophys. Jl R. astr. Soc.* **67**, 437–448.

Ringwood, A. E. 1982 Phase transformations and differentiation in subducted lithosphere: implications for mantle dynamics, basalt petrogenesis, and crustal evolution. *J. Geol.* **90**, 611–643.

Ringwood, A. E. 1990 Slab-mantle interactions. 3. Petrogenesis of intraplate magmas and structure of the upper mantle. *Chem. Geol.* **82**, 187–207.

Ringwood, A. E. & Irifune, T. 1988 Nature of the 650 km seismic discontinuity: implications for mantle dynamics and differentiation. *Nature* **331**, 131–136.

Salters, V. J. M. & Shimizu, N. 1988 World-wide occurrence of HFSE-depleted mantle. *Geochim. cosmochim. Acta* **52**, 2177–2182.

Saunders, A. D., Norry, M. J. & Tarney, J. 1988 Origin of MORB and chemically-depleted reservoirs: trace element constraints. *J. Petrol.* Special Lithosphere Issue, pp. 415–445.

Saunders, A. D., Tarney, J. & Weaver, S. D. 1980 Transverse geochemical variations across the Antarctic Peninsula: implications for the genesis of calc-alkaline magmas. *Earth planet. Sci. Lett.* **46**, 344–360.

Staudigel, H., Hart, S. R., Schmincke, H.-U. & Smith, B. M. 1989 Cretaceous ocean crust at DSDP sites 417 and 418: carbon uptake from weathering versus loss by magmatic outgassing. *Geochim. cosmochim. Acta* **53**, 3091–3094.

Sun, S.-S. & McDonough, W. F. 1989 Chemical and isotopic systematics of oceanic basalts: implications for mantle composition and processes. In *Magmatism in the ocean basins* (ed. A. D. Saunders & M. J. Norry). *Geol. Soc. Lond. Spec. Publ.* **42**, 313–345.

Sun, S.-S., Nesbitt, R. W. & Sharaskin, A. Y. 1979 Geochemical characteristics of mid-ocean ridge basalts. *Earth planet. Sci. Lett.* **44**, 119–138.

Tarney, J., Wood, D. A., Saunders, A. D., Cann, J. R. & Varet, J. 1980 Nature of mantle heterogeneity in the North Atlantic: evidence from deep sea drilling. *Phil. Trans. R. Soc. Lond.* **A297**, 179–202.

Taylor, S. R. & McLennan, S. M. 1985 *The continental crust: its composition and evolution.* Oxford: Blackwell.

Weaver, B. L., Wood, D. A., Tarney, J. & Joron, J. L. 1987 Geochemistry of oceanic island basalts from the South Atlantic: Ascension, Bouvet, St Helena, Gough and Tristan da Cunha. In *Alkaline igneous rocks* (ed. J. G. Fitton & B. G. J. Upton). *Geol. Soc. Lond. Spec. Publ.* **29**, 253–267.

White, R. & McKenzie, D. 1989 Magmatism at rift zones: the generation of volcanic continental margins and flood basalts. *J. geophys. Res.* **94**, 7685–7729.

Zindler, A. & Hart, S. 1986 Chemical geodynamics. *A. Rev. Earth planet. Sci.* **14**, 493–571.